AF540289

COMPUTATIONAL CELL BIOLOGY

By

Dr. M. Prakash

Dept. of Zoology
M.M.H. Post Graduate College
Ghaziabad (U.P.)
(India)

DISCOVERY PUBLISHING HOUSE PVT. LTD.
NEW DELHI-110 002

First Published-2009

ISBN: 978-81-8356-472-4

Published by:
DISCOVERY PUBLISHING HOUSE PVT. LTD.
4831/24, Ansari Road, Prahlad Street
Darya Ganj, New Delhi-110002 (India)
Phone: 23279245 • Fax: 91-11-23253475
E-mail: dphbooks@rediffmail.com • dphtemp@indiatimes.com
web: www.discoverypublishinggroup.com

Printed at:
Sachin Printers
Delhi

Preface

The present title "*Computational Cell Biology*" is an exciting new field bringing together life scientists, mathematicians, computer scientists and engineers to explore a new and deeper understanding of biological systems. The present text has been compiled for computer science and mathematics graduate and level undergraduate students. It will also be of interest to molecular biologists interested in bioinformatics.

The rationable of the book is to present algorithmic ideas in computational biology and to show how they are connected to molecular biology and to biotechnology. To achieve this goal, the book has a substantial "computating biology without formulas" component that presents biological motivation and computational ideas in a simple way. This simplified presentation of biology and computing aims to make the book accessible to computer scientists entering this new areas and to biologists who do not have sufficient background for more involved computational techniques. Every chapter has an introductory section that describes both computational and biological ideas without any formulas. Thus book concentrate on computational ideas rather than details of the algorithms and makes special efforts to prevent there ideas in a simple way.

To make the work more comprehensive and informative, the author has consulted many authoritative books, research journals, abstracts, monographs etc. He is grateful to all those great scholars whose work are cited or substantially reproduced.

There can be no claim to originality except in the manner of treatment and much of the information has been obtained from the books and scientific journals available in the different libraries.

The author expresses his thanks to his friends and colleagues whose continue inspirations have initiated him to bring out this book.

The author expresses his gratitude to Mr. Wasan and staff of M/s Discovery Publishing House Pvt. Ltd. for their whole hearted cooperation in the publication of this book.

In the mean time, the author will remain sincerely responsible for any shortcomings of the book and be grateful to the readers for their suggestions and constructive criticism for the continuous betterment of the book. He takes this opportunity to appeal to the readers to send their suggestions straightaway to his Publisher.

Author

CONTENTS

1 INTRODUCTION

All the living organisms share many common attributes, such as the capability to extract energy from nutrients, the power to respond to changes in their environments, and the ability to grow, to differentiate, and to reproduce. Biochemistry is the study of life at the molecular level. It investigates the phenomena of life by using physical and chemical methods dealing with (*a*) the structures of biological compounds (biomolecules), (*b*) biomolecular transformations and functions, (*c*) changes accompanying these transformations, (*d*) their control mechanisms, and (*e*) impacts arising from these activities.

The distinct feature of biochemistry is that it uses the principles and language of one science, chemistry, to explain the other science, biology at the molecular level. It can be divided into three principal areas: (1) Structural biology focuses on the structure of the components of living matter and the relationship between chemical structure and biological function. (2) Dynamic biology deals with the totality of chemical reactions known as metabolic processes that occur in living systems and their regulations. (3) Information biology is concerned with the chemistry of processes and substances that store and transmit biological information. The third area is also the province of molecular genetics, a field that seeks to understand heredity and the expression of genetic information in molecular terms. Among biomolecules, water is the most common compound in living organisms, accounting for at least 70% of the weight of most cells, because water is both the major solvent of organisms and a reagent in many biochemical reactions.

Most complex biomolecules are composed of only a few chemical elements. In fact, over 97% of the weight of most organisms is due to six elements (% in human): oxygen (62.81%), carbon (19.37%), hydrogen (9.31%), nitrogen (5.14%), phosphorus (0.63%), and sulfur (0.64%). In addition to covalent bonds (3000 ± 150 kJ/mol for single bonds) that hold molecules together, a number of weaker chemical forces (ranging from 4 to 30 kJ/mol) acting between molecules are responsible for many of the important properties of biomolecules.

Among these noncovalent interactions are van der Waals forces, hydrogen bonds, ionic bonds/electrostatic interactions, and hydrophobic interactions. All biomolecules are ultimately derived from very simple, low-molecular-weight *precursors* (M.W. = 30 ± 15), such as CO_2, H_2O, and NH_3, obtained from the environment. These precursors are converted by living matter via series of metabolic *intermediates* (M.W. = 150 ± 100), such as acetate, α-keto acids, carbamyl phosphate etc., into the building-block biomolecules (M.W. = 300 ± 150) such as glucose, amino acids, fatty acids and mononucleotides. They are then linked to each other covalently in a specific manner to form *biomacromolecules* (M.W. = $10^7 \pm 10^4$) or biopolymers. The unique chemistry of living systems results in large part from the remarkable and diverse properties of biomacromolecules. Macromolecules from each of the four major classes (*e.g.*, polysaccharides, lipid bilayers, proteins, nucleic acids) may act individually in a specific cellular process, whereas others associate with one another to form supramolecular structures (particle weight > 10^6) such as proteosome, ribosomes, and chromosomes. All of these structures are involved in important cellular processes.

Table 1.1. Energy Contribution and Distance of Noncovalent Interactions in Biomolecules

Chemical Force	*Description*	*Energy (kJ / mol)*	*Distance (nm)*	*Remark*
Van der Waals Interactions	Induced electronic interactions between closely approaching atoms/molecules.	0.4–4.0	0.2	The limit of approach is determined by the sum of their *vdW* radii and related to the separation (r) of the two atoms by r^{-6}.
Hydrogen Bonds	Formed between a covalently bonded hydrogen atom and an electronegative atom that serves as the hydrogen bond acceptor.	12–38	0.15–0.30	Proportional to the polarity of the donor and acceptor, stable enough to provide significant binding energy, but sufficiently weak to allow rapid dissociation.
Ionic bonds	Attractive forces between oppositely charged groups in aqueous solutions.	~20	0.25	Depending on the polarity of the interacting charged species and related to q_iq_j/Dr_{ij}.
Hydrophobic interactions	Tendency of nonpolar groups or molecules to stick together in aqueous solutions.	~25	—	Proportional to buried surface area for the transfer of small molecules to hydrophobic solvents, the energy of transfer is 80–100 kJ/mol/Å^2 that becomes buried.

The supramolecular complexes/systems are further assembled into organelles of eukaryotic cells and other types of structures. These organelles and substructures are enveloped by cell membrane into intracellular structures to form cells that are the fundamental units of living organisms. Viruses are supramolecular complexes of nucleic acids (either DNA or RNA) encapsulated in a protein coat and, in some instances, surrounded by a membrane envelope. Viruses infecting bacteria are called bacteriophages. The cell is the basic unit of life and is the setting for most biochemical phenomena. The two classes of cell, eukaryotic and prokaryotic, differ in several respects but most fundamentally in that a eukaryotic cell has a nucleus and a prokaryotic cell has no nucleus. Two prokaryotic groups are the *eubacteria* and the *archaebacteria (archaea).* Archaea, which include thermoacidophiles (heat and acid-tolerant bacteria), halophiles (salt-tolerant bacteria), and methanogens (bacteria that generate methane), are found only in unusual environments where other cells cannot survive.

Prokaryotic cells have only a single membrane (plasma membrane or cell membrane), though they possess a distinct nuclear area where a single circular DNA is localized. Eukaryotic cells are generally larger than prokaryotic cells and more complex in their structures and functions. They possess a discrete, membrane-bounded nucleus (repository of the cell's genetic material) that is distributed among a few or many chromosomes. In addition, eukaryotic cells are rich in internal membranes that are differentiated into specialized structures such as the endoplasmic reticulum and the Golgi apparatus. Internal membranes also surround certain organelles such as mitochondria, chloroplasts (in plants), vacuoles, lysosomes, and peroxisomes. The common purpose of these membranous partitions is the creation of cellular compartments that have specific, organized metabolic functions. All complex multicellular organisms, including animals (Metazoa) and plants (Metaphyta), are eukaryotes. Most biochemical reactions are not as complex as they may at first appear when considered individually. Biochemical reactions are enzyme-catalyzed, and they fall into one of six general categories: (1) oxidation and reduction, (2) functional group transfer, (3) hydrolysis, (4) reaction that forms or breaks carbon-carbon bond, (5) reaction that rearranges the bond structure around one or more carbons, and (6) reaction in which two molecules condense with an elimination of water. These enzymatic reactions are organized into many interconnected sequences of consecutive reactions known as metabolic pathways, which together constitute the metabolism of cells. Metabolic pathways can be regarded as sequences of the reactions organized to accomplish specific chemical goals.

To maintain homeostatic conditions (a constant internal environment) of the cell, the enzyme-catalyzed reactions of metabolism are intricately regulated. The metabolic regulation is achieved through controls on enzyme quantity (synthesis and degradation), availability (solubility and compartmentation), and activity (modifications, association/dissociation, allosteric effectors, inhibitors, and activators) so that the rates of cellular reactions and metabolic fluxes are appropriate to cellular requirements. An inquiry into the continuity and evolution of living organisms has provided great impetus to the progress of information biochemistry.

Double-stranded DNA molecules are duplicated semiconservatively with high fidelity. The triplet-code words of genetic information encoded in DNA sequence are transcribed into codons of messenger RNA (mRNA) which in turn are translated into an amino acid sequence of polypeptide chains. The semantic switch from nucleotides to amino acids is aided by a 64-membered family of transfer RNA (tRNA). The ensuing folding process of polypeptide chains produces functional protein molecules. The processes of information transmission involve the coordinated actions of numerous enzymes, factors, and regulatory elements. One of the exciting areas of studies in information biochemistry is the development of recombinant DNA technology which makes possible the cloning of tailored made protein molecules. Its impact on our life and society has been most dramatic.

COMPUTATIONAL SCIENCES

A computer is a machine that has the ability to store internally sequenced instructions that will guide it automatically through a series of operations leading to a completion of the task. A microcomputer, then, is regarded as a small stand-alone desktop computer (strictly speaking, a microcomputer is a computer system built around a microprocessor) that consists of three basic units:

1. The central processor unit (CPU) including the control logic that coordinates the whole system and manipulates data.
2. The memory consisting of random access memory (RAM) and read-only memory (ROM).
3. The buses and input/output interfaces (I/O) that connect the CPU to the other parts of the microcomputer and to the external world.

Computer science is concerned with four elements of computer problem solving namely problem solver, algorithm, language and machine.

An algorithm is a list of instructions for carrying out some process step by step. An instruction manual for an assay kit is a good example of an algorithm. The procedure is broken down into multiple steps such as preparation of reagents, successive addition of reagents, and time duration for the reaction and measurement of an increase in the product or a decrease in the reactant. In the same way, an algorithm executed by a computer can combine a large number of elementary steps into a complicated mathematical calculation. Getting an algorithm into a form that a computer can execute involves several translations into different languages — for example.

English → Flowchart language → Procedural language → Machine language

A flowchart is a diagram representing an algorithm. It describes the task to be executed. A procedure language such as *Fortran* and C enables a programmer to communicate with many different machines in the same language, and it is easier to comprehend than machine language. The programmer prepares a procedure language program, and the computer compiles it into a sequence of machine language instructions. To solve a problem, a computer must be given a clear set of instructions and the data to be operated on. This set of instructions is called a *program.*

The program directs the computer to perform various tasks in a predetermined sequence. It is well known at a very basic level that computers are only capable of processing quantities expressed in binary form—that is, in machine code. In general, the computational scientist uses a high-level language to program the computer. This allows the scientist to expr ess his/her algorithms in a concise and understood form. FORTRAN and C++ are the most commonly used high-level programming languages in scientific computations. Recent years have seen considerable progress in computer technology, in computer science, and in the computational sciences. To a large extent, developments in these fields have been mutually dependent.

Progress in computer technology has led to (*a*) increasingly larger and faster computing machines, (*b*) the supercomputers, and (*c*) powerful microcomputers. At the same time, research in computer science has explored new methods for the optimal use of these resources, such as the formulation of new algorithms that allow for the maximum amount of parallel computations. Developments in computer technology and computer science have had a very significant effect on the computational sciences including computational biology computational chemistry and computational biochemistry. The main tasks of a computer scientist are to develop new programs and to improve efficiency of existing programs, whereas computational scientists strive to apply available software intelligently on real scientific problems.

APPLICATION OF COMPUTER TECHNOLOGY

There is a general trend in biology toward more quantitative and sophisticated interpretations of experimental data. As a result, demand for accurate, complex, and elaborate calculation increases. Recent progress in computer technology, along with the synergy of increased need for complex biochemical models coupled with an improvement in software programs capable of meeting this need, has led to the birth of computational biochemistry. Computational biology can be considered as a second-generation interdisciplinary subject derived from the interaction between biology and computer science. It is a discipline of computational sciences dealing with all of the three aspects of biochemistry, namely, structure, reaction, and information.

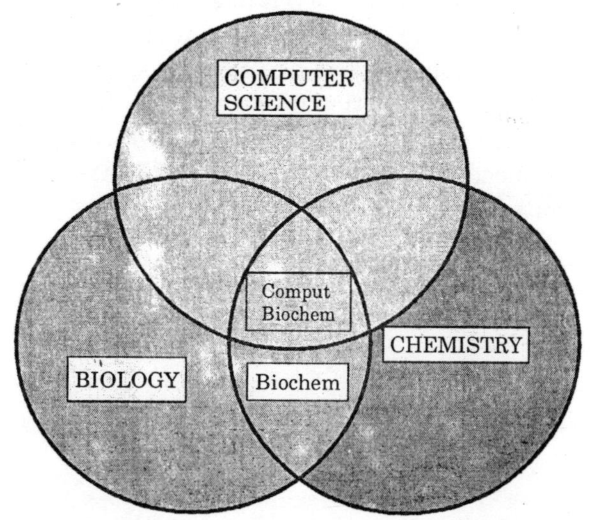

Fig. 1.1. Relationship showing computational cell biology as an interdisciplinary subject. Biochemistry is represented by the overlap (interaction) between biology and chemistry. A further overlap (interaction) between biochemistry and computer science represents computational biochemistry.

Computational cell biology is used when biochemical models are sufficiently well developed that they can be implemented to solve related problems with computers. It may encompass bioinformatics. Bioinformatics is information technology applied to the management and analysis of biological data with the aid of computers. Computational cell biology then applies computer technology to solve biochemical problems, including sequence data, brought about by the wealth of information now becoming available. The two subjects are highly intertwined and extensively overlapped. Computational cell biology is an emerging field.

The contribution of computational has contributed initially to its development; however, as the field broadens and grows in its importance, the involvement of "cell biology" increases prominently. In its early stage, computational biochemistry has been exclusively the domain of those who are knowledgeable in programming. This hindered the appreciation of computational cell biology in the early days. The wide availability of inexpensive microcomputers and application programs in biochemistry has helped to relieve these restrictions. It is now possible for biochemists to rely on existing software programs and Internet resources to appreciate computational cell biology in biochemical research and biochemical curriculum. Well-established techniques have been reformulated to make more efficient use of the new computer technology. New and powerful algorithms have been, successfully implemented. Furthermore, it is becoming increasingly important that biochemists are exposed to databases and database management systems due to exponential increase in information of biochemical relevance. Visual modeling of biochemical structures and phenomena can provide a more intuitive understanding of the process being evaluated. Simulation of biochemical systems gives the biochemist control over the behavior of the model. Molecular modeling of biomolecules enables biochemists not only to predict and refine three-dimensional structures but also to correlate structures with their properties and functions. The field has matured from the management and analysis of sequence data, *albeit* still the most important areas, into other areas of molecules. This text is an attempt to capture that spirit by introducing computational biomolecules from the biochemists' prospect. The material content deals primarily with the applications of computer technology to solve biochemical problems. The subject is relatively new and perhaps a brief description of the text may benefit the students.

CELLULAR DYNAMICS

Over the past several decades, progress in the measurement of rates and interactions of molecular and cellular processes has initiated a revolution in our understanding of dynamic phenomena in cells. Spikes or bursts of plasma membrane electrical activity or intracellular signaling via receptors, second messengers, or other networked biochemical pathways in single cells, or more complex processes that involve small clusters of cells, organelles, or groups of neurons, are examples of the complex behaviors that we know take place on the cellular scale. The vast amount of quantitative information uncovered in recent years, leveraged by the intricate mechanisms already shown to exist, results in an array of possibilities that makes it quite hard to evaluate new hypotheses on an intuitive basis.

Using mathematical analysis and computer simulation we can show that some seemingly reasonable hypotheses are not possible. Analysis and simulations that confirm that a given hypothesis is reasonable can often result in quantitative predictions for further experimental exploration. Rapid advances in computer hardware and software technology combined with pioneering work giving structure to the interface between mathematics and biology have put the ability to test hypotheses and evaluate mechanisms with simulations within the reach of all cell biologists and neuroscientists.

SCOPE OF CELLULAR DYNAMICS

Generally speaking, the phrase *dynamic phenomenon* refers to any process or observable that changes over time. Living cells are inherently dynamic. Indeed, to sustain the

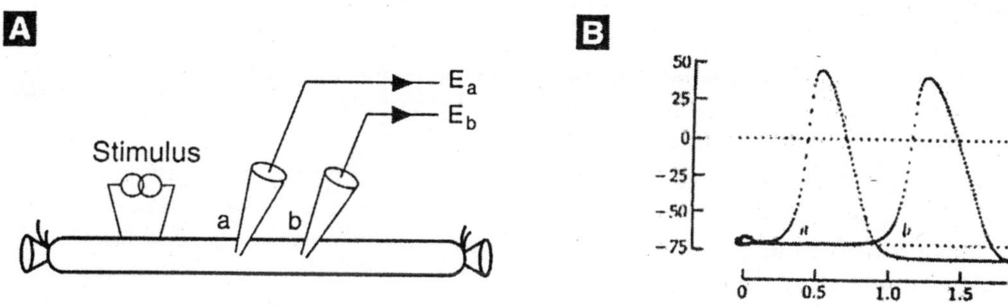

Fig. 2.1. (A) Schematic diagram of the recording electrodes (a and b) used to detect action potentials following a stimulus shock in an isolated giant axon from squid. (B) The membrane potential recorded at electrodes a and b in the upper panel following a depolarizing shock.

characteristic features of life such as growth, cell division, intercellular communication, movement, and responsiveness to their environment, cells must continually extract energy from their surroundings. This requires that cells function thermodynamically as open systems that are far from static thermal equilibrium. Much energy is utilized by cells in the maintenance of gradients of ions and metabolites necessary for proper function. These processes are inherently dynamic due to the continuous movement of ionic and molecular species across the cell membrane. Electrical activity of excitable cells is a widely studied example of cellular dynamics.

The classical behaviour of an action potential in the squid giant axon is shown in Fig. 2.1. This single spike of electrical activity, initiated by a small positive current applied by an external electrode, propagates as a traveling pulse along the axonal membrane. Hodgkin and Huxley were the first to propose a satisfactory explanation for action potentials that incorporated experimental measurements of the response of the squid axon to depolarizations of the membrane potential. Membrane transporters allow cells to take up glucose from the blood plasma.

Cells that use glycolytic enzymes to convert energy from carbon and oxygen bonds to phosphorylate adenosine diphosphate (ADP) and produce the triphosphate ATP. ATP, in turn, is utilized to pump Ca^{2+} and Na^{+} ions from the cell and K^{+} ions back into the cell, in order to maintain the osmotic balance that helps give red cells the characteristic shape. ATP is also used to maintain the concentration of 2,3-diphosphoglycerate, an intermediary metabolite that regulates the oxygen binding conformation of hemoglobin. The final products of glucose metabolism in red cells are pyruvate and lactate, which move passively out of the cell down a concentration gradient through specific transporters in the plasma membrane.

Because red cells possess neither a nucleus nor mitochondria, they are not capable of reproduction or energetically demanding processes. Nonetheless, by continually extracting energy from the transformation of glucose to lactate, red blood cells maintain the capacity to shuttle oxygen and carbon dioxide between the lungs and the capillaries. Remarkably, this efficient biochemical factory is only 5 μm across, with a volume of less than 10^{-14} L. Electrical activity and membrane transport are coupled cellular mechanisms. Experimental measurements of the membrane potential in pancreatic cells have revealed regular bursts of electrical activity corresponding to insulin secretion that is stimulated by increases in blood glucose levels. These oscillations occur at physiological levels of glucose. Recent work in vitro has shown that the rapid spikes of electrical activity, known as *action potentials,* are caused by rapid influx of Ca^{2+} from the exterior of the beta cell followed by a slower efflux of K^{+}. The periods of rapid spiking are referred to as *active phases* of the burst, which are separated by intervals referred to as *silent phases.*

A variety of mechanisms have been proposed to explain bursting behavior, and computer models of bursting were the first to predict that oscillations of Ca^{2+} within the cytoplasm should occur in phase with the electrical activity. Oscillations in Ca^{2+} were recorded for the first time in vitro eight years after they were predicted by a theoretical model. These oscillations are important physiologically, because cytoplasmic Ca^{2+} plays a major role in triggering insulin secretion. The control of cellular processes by interlocking molecular mechanisms can also produce spatiotemporal oscillatory Ca^{2+} signals that are independent of electrical activity.

The spiral pattern of cytoplasmic Ca^{2+} oscillations that occurs when an immature *Xenopus leavis* egg (an *oocyte)* is stimulated by a microinjection of inositol 1,4,5-trisphosphate (IP3).

IP3 is a phospholipid membrane metabolite that is widely involved in signaling by receptors in the plasma membrane and that triggers release of Ca^{2+} from the *endoplasmic reticulum* (ER). The ER is an intracellular compartment that functions, among other purposes, as a store for Ca^{2+}. The ER maintains an internal Ca^{2+} concentration ([Ca^{2+}] ER) that is comparable to that of the external medium (ca. 5 mM), whereas the cytoplasmic Ca^{2+} concentration ([Ca^{2+}]*i*) is typically 1000-fold smaller. The spiral waves of Ca^{2+} in *Xenopus* oocytes can be explained quantitatively by kinetic models of the feedback mechanisms responsible for uptake and release of Ca^{2+} from the ER. Simple models of regenerative Ca^{2+} release that are solved on a spatial domain provide insight into the processes of *self organization* that result in spiral waves. Circadian rhythms are regular changes in cellular processes that have a period of about 24 hours (from the Latin *circa,* about, and *dies,* day) and represent another dynamical phenomenon that is widely observed in cells.

A great deal about the mechanisms of circadian rhythms has been uncovered in recent years, and circadian biology offers a rich source of unsolved modeling problems. Internal clocks provide an organism with the ability to predict changes in the environment. For example, flower opening and insect-egg hatching occur in advance of the rising sun. Cell division in *Euglenids* may also synchronize to light-dark cycles. The dark /light bands correspond to periods of absence and presence of light that simulate the normal dark/light cycle during a day. The growth rate of *Euglena* is temperature dependent, and cell division synchronizes to a 24-hour dark/light cycle only when the temperature is in the range found in its natural environment. At this temperature the population doubling time is close to 24 hours.

Recent experiments with the fruit fly *Drosophila* and other organisms suggest that circadian rhythms like this are controlled by oscillations in gene transcription.

Fig. 2.2. A spiral wave of Ca^{2+} ions detected as the bright fluorescence from an indicator dye after microinjection of IP3 into an immature frog egg, the *Xenopus laevis* oocyte.

The cell division cycle is the process by which a cell grows and divides into two daughter cells that can repeat the process. The eukaryotic cell cycle consists of a regular sequence of events chromosome replication during a restricted period of the cycle (S phase), chromosome segregation during metaphase and anaphase (M phase, or *mitosis),* and finally *cytokinesis,* in which two daughter cells separate. This cycle involves a cascade of molecular events that center on the proteins Cdc2 and cyclin, which make up a complex known as M phase promoting factor, or MPF. This complex has been shown to oscillate in synchrony with cell division and to be regulated by a series of phosphorylation and dephosphorylation reactions.

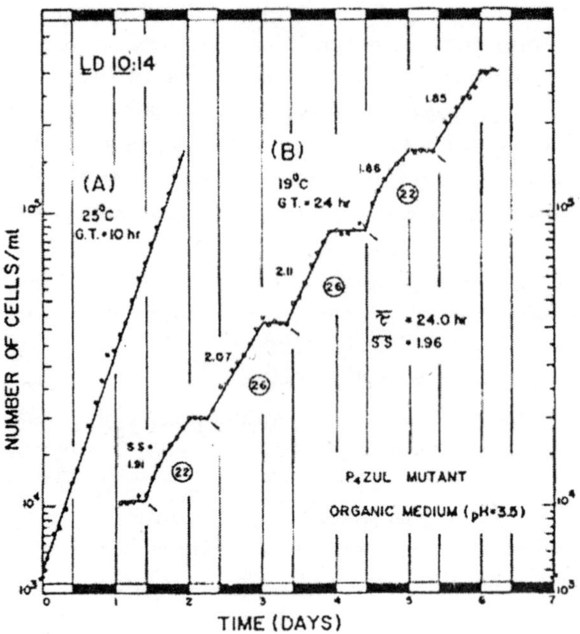

Fig. 2.3. Population growth of a mutant Euglena strain shows log growth at 25°C independent of the light/dark cycle, indicated by the alternating light/dark bands on the time axis. At 19°C a circadian (approximately 24 hour) growth rhythm develops that entrains with the light/dark cycle.

Related dynamical changes occur during *meiosis,* in which germ line cells produce eggs and sperm. After DNA replication is complete, each chromosome consists of two "sister chromatids," which must be separated during mitosis so that each daughter nucleus gets one and only one copy of each chromosome. Segregation of sister chromatids during mitosis is another complex dynamical process that involves self-organizing structures in the cell that pull sister chromatids apart. This wonderfully coordinated dynamical behavior is just one of many examples of motile cellular processes. Other important examples include muscle contraction, cell movement, and projections of cell membrane called *pseudopodia.*

COMPUTATIONAL MODELING IN BIOLOGY

Even the simplest of the dynamic phenomena described in the previous section are exceedingly complex, and computer models have proven to be an important tool in helping to dissect the molecular processes that control their evolution in time. In the physical sciences, theoretical methods in combination with experimental measurements have for many years provided rich insights into dynamical phenomena. The abundance of quantitative experimental data now available at the cellular level has opened the door to similar collaborations in neurobiology and cell physiology.

Cartoons, Mechanisms, and Models

The interplay of experiment, theory, and computation follows a conceptual framework similar to that which has proven successful in the physical sciences:

1. Taking clues from experimental work, the first step is to sort through possible molecular mechanisms and focus on the most plausible ones. In most cases, this step requires close consultation with experimentalists working on the problem.
2. The selection of mechanisms defines the basis for a schematic representation, or cartoon, that depicts the overall model. To be useful, the cartoon should be explicit enough to be translated into a series of elementary steps representing the individual mechanisms.
3. Next, the basic laws of physics and chemistry can be used to translate the elementary steps of the mechanism into mathematical expressions.

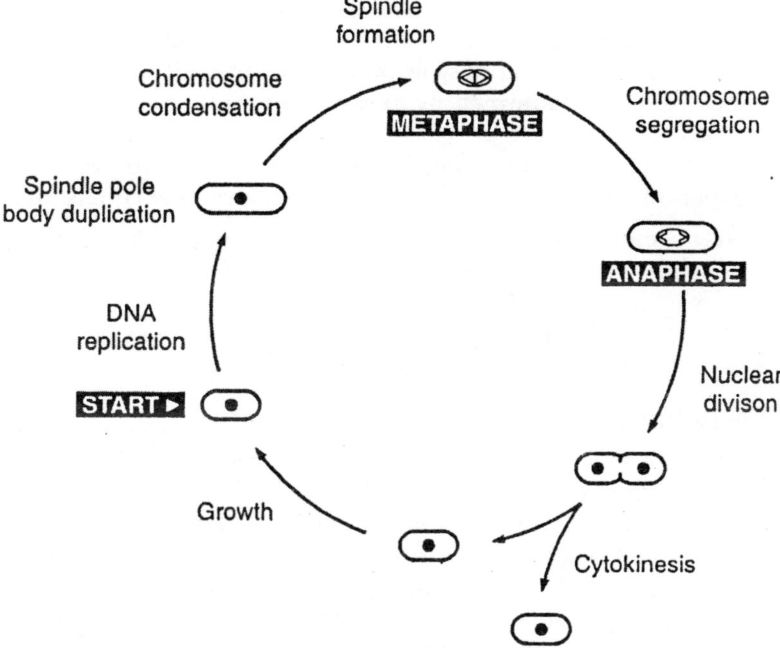

Fig. 2.4. The cell division cycle in fission yeast. Spindle formation initiates metaphase, where condensed chromosomes pair up for segregation into daughter cells. The cycle is completed when cytokinesis cleaves the dividing cell.

4. These expressions are then combined into time dependent differential equations that quantify the changes described by the whole model.
5. Finally, careful study of the differential equations reveals whether or not the cartoon is a successful model of the biological system.

The challenge of the theorist in biology then becomes similar to that in astrophysics or quantum mechanics: to analyze the equations, simplify them if possible, solve them, and, most importantly, make predictions that can be tested by further experiment. Further experiments may uncover inconsistencies in a model that will require changes. The process that we have outlined above and will revisit in later chapters is an iterative cycle of ever improving approximation where the mathematical or computer model plays the role of a quantitative hypothesis.

The Role of Computation

Advances in computer hardware and numerical analysis have made the solution of complicated systems of ordinary differential equations fast, accurate, and relatively easy. Indeed, the role of computation is critical because the differential equations describing biological processes nearly always involve control mechanisms that have nonlinear components. Simple linear differential equations often can be solved analytically, which means that we can obtain an exact solution using traditional mathematical methods. Nonlinearities often make it difficult or impossible to obtain an exact solution; however, we can/obtain quite good estimates using numerical methods implemented on computers.

Spatial variation is often an important feature in cellular mechanisms, so one is confronted with analyzing and solving spatially explicit *partial differential equations,* which can be still more complicated and less analytically tractable than ordinary differential equations. Computer models permit one to test conditions that may at present be difficult to attain in the laboratory or that simply have not yet been examined by experimentalists. Each numerical solution of the differential equations can therefore provide a *simulation* of a real or potential laboratory experiment. These simulations can be used to help assess parameters, such as diffusion constants or kinetic constants, that may be difficult to measure experimentally.

Numerical simulations can test how intervention by pharmacological agents might affect a process. With simulations one can test specific hypotheses about the role of individual mechanistic components or make predictions that can be tested in the laboratory. Often the most important result of a simulation is negative: A well-crafted model can rule out a particular mechanism as a possible explanation for experimental observations.

Mathematical at Techniques

The scope of mathematical techniques employed to investigate problems in mathematical biology spans almost all of applied mathematics. The modeling of processes is discussed in detail here, but only the basics of the mathematics and the elementary tools for the analysis of these models are introduced. Rigorous mathematics plays at least three important roles in the computational modeling of cell biology. One role is in the development of the techniques and algorithms that make up the tools of *numerical analysis.* In its essence, the computation of solutions to mathematical problems on computers is fundamentally a process of estimation, and the accuracy and efficiency of these methods of estimation are the subjects of much study. The process of developing model mechanisms that we described above is also fundamentally a process of approximation due to the simplifications that must be introduced to produce a useful model. Not only must these simplifications make sense in terms of the physical process being studied, but they must also be valid from a mathematical standpoint. It is one thing to solve the differential equations that result from the formulation of a model, but another thing to understand why a model behaves as it does.

Mathematicians have developed techniques and tools for the analysis of systems of differential equations that describe complex interrelated processes. These tools reveal the structure, properties, and dynamical behavior of the system much as anatomical, physiological, and molecular biological techniques reveal the physical basis of the model. In particular, such analysis reveals details about behaviors in a model such as the oscillations and other complex behaviors that often motivate study of such biological phenomena. Rigorous analysis of complicated differential equations requires specialized training, because there are many subtleties that are appreciated only with experience.

Similarly, choosing proper numerical methods and selecting valid simplifications requires caution. While the creation and manipulation of simple models is within the reach of all cell biologists, the careful scientist will seek collaborations with experienced mathematicians, particularly for the valid simplification of complicated models into more tractable ones. In the middle ground between established disciplines such as biology and mathematics, fruitful scientific work can be done, and all parties gain valuable insight from the interdisciplinary experience.

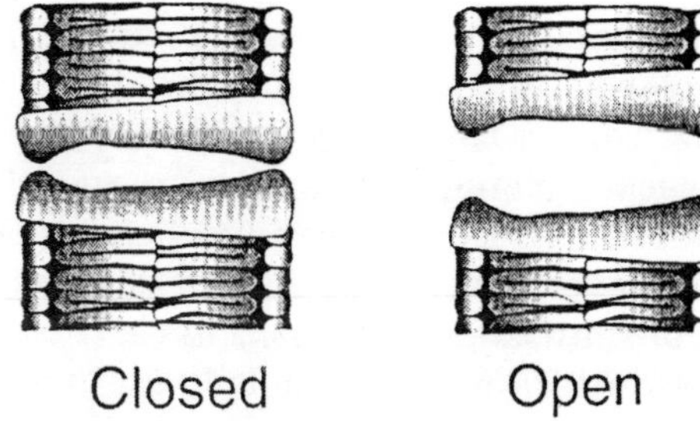

Fig. 2.5. Mechanistic cartoon of a gated ionic channel showing an aqueous pore that is selective to particular types of ions. The portion of the transmembrane protein that forms the "gate" is sensitive to membrane potential, allowing the pore to be in an open or closed state.

A MOLECULAR SWITCH

In this section we illustrate, with a simple model of channel conformation, the kinds of phenomena that are investigated in detail in subsequent chapters. We introduce some of the methods underlying the analysis of these models and also try to demonstrate the basic conceptual modeling framework utilized throughout the book. We begin with a simple channel because it is an intuitively clear example of transition between different molecular states corresponding to different conformations of a macromolecule. Let us be clear that we are modeling only proteins that are switching between an "open" state and a "closed" state and nothing more at this point. The simplest cartoon of gating is a channel with two states, one with the pore open and the other with it closed.

This kinetic "cartoon" is easily translated into a conventional kinetic model of the sort often employed in biochemistry. The model takes the form of the diagram in (2.1). Diagrams like this, which will be used extensively in this book, represent molecular states or entities by symbols and transitions between states by solid lines or arrows :

$$C \underset{k^-}{\overset{k^+}{\rightleftharpoons}} C \qquad \text{...(2.1)}$$

The C in (2.1) corresponds to the closed state, the O corresponds to the open state of the channel, and the arrows represent elementary molecular processes. These states represent a complex set of underlying molecular conformations. The transitions between C and O are *unimolecular* processes because they involve only the channel molecule. An important aspect of transitions between molecular states is that they are reversible, which is a consequence of microscopic reversibility of molecular processes. Using special techniques, transitions between closed and open states can be measured for single channel molecules. However, here the focus is on the average change for a collection of channels. Because it is not unusual to have several thousand channels of a given type in the plasma membrane of a cell, the average behaviour of the entire ensemble of channels is often what determines the cellular dynamics. The rate of an elementary process in a kinetic diagram is determined by the so-called *law of mass action.* (Despite the name, mass action is not technically a physical law, but rather is a constitutive relation that holds as a very good approximation for any well-mixed system.

The "law," which dates back to the early studies of chemical kinetics, states that the rate of a process is proportional to the product of the concentrations of the molecular species involved in the process. If we define k^+ as the proportionality constant or the *rate constant,* the rate of the transition from state C to state O is given by $J_+ = k^+[C]$, where the square brackets denote concentration, with [C] representing the concentration of channel molecules in state C. In this case k^+ is unimolecular, with practical units of s^{-1}. Similarly, the rate of the reverse reaction, $C \leftarrow O$, is given by $J = k^-[O]$ with the unimolecular rate constant k^-. To translate the mechanism in (2.1) into an equation, the law of mass action is applied to the concentration of channels in states C and O. For cellular mechanisms, a variety of measures of "concentration" can be used. For example, if the channels are in intact cells, concentration is often expressed in terms of total cell volume.

Another measure in common use involves total weight of protein in a sample. The total number of transporters is useful for single cells. Here we choose the latter to define concentration so that $[O] = fo = No/N$, where fo will refer to the fraction of open channels, and N and N_o the total number of channels and open channels, respectively (similarly, fc and N_c refer to the closed state. Because the kinetic model involves only interconversion of channel states, the total number of channels should be preserved. This introduces the idea of a conservation law, $N_c + N_o = N$, which states mathematically that channels are neither created nor destroyed. Using conservation relations, one of the dependent variables can be eliminated because $N_c = N - N_o$. The differential equation for N_c therefore becomes redundant, and the number of differential equations to be solved is reduced to only one along with the algebraic equation for N_c. The fraction of channels in the closed state is therefore $1 - fo$. Having established the correspondence of the diagram with rate expressions, it is easy to write down the differential equations that the diagram represents.

To do so one must keep track of the change that each elementary process in the diagram makes for each state, which we refer to as a *flux.* Thus the process connecting states O and C causes a loss of state O in the reverse direction and again in the forward direction. These small whole numbers that correspond to losses or gains of a state (*e.g.,* –1 for state C in the process $C \rightarrow O$) are called the *stoichiometric coefficients* for the mechanism. Using the coefficients in conjunction with the kinetic diagram, the ordinary differential equations follow for the rate of change in the states:

$$\text{flux } O \rightarrow C = j_- = k^- f_o, \quad ...(2.2)$$

$$\text{flux } C \rightarrow O = j_+ = k^+ (1 - f_o). \quad ...(2.3)$$

The difference between the two fluxes represents the change in f_o over time:

$$\begin{aligned} \frac{dfo}{dt} &= j_+ - j_- \\ &= -k^- f_o + k^+ (1 - fo) \\ &= -(k^- + k^+)\left(f_o - \frac{k^+}{(k^- + k^+)}\right) \end{aligned} \quad ...(2.4)$$

defining $\tau = 1/(k^- + k^+)$ and $f_\infty = k^+/(k^- + k^+)$, we have

$$\frac{dfo}{dt} = \frac{-(f_o - f_\infty)}{\tau} \quad ...(2.5)$$

The example that we have presented here shows the basic framework under which all of the models in this book are developed and understood. With the formulation of the model equations, the first steps in the modeling process are completed. What remains is the analysis of the equations.

ANALYZING DIFFERENTIAL EQUATIONS

Many students have worked with differential equations in their studies of the physical sciences or elementary mathematics, and they may have been introduced to solution techniques explicitly in an advanced calculus course. In general, the differential equations that arise for the rate of change in cellular properties will be complicated and difficult or impossible to solve exactly using analytical techniques. One example that can be solved is the simple linear equation.

$$\frac{dX}{dt} = \frac{-X}{\tau} \qquad ...(2.6)$$

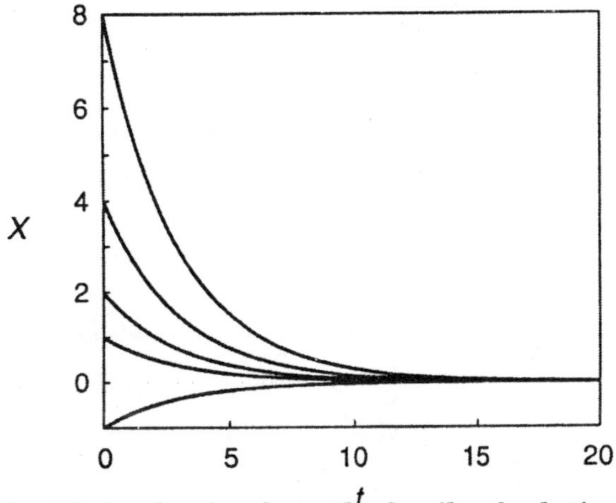

8
6
4
2
0
X
0 5 10 15 20
t

Fig. 2.6. A selection from the family of solution to (1.6) for τ = 2.

In order to solve (2.6), we seek a function $X(t)$ whose derivative is proportional to itself with proportionality constant $1/\tau$. Remembering that the derivative of the exponential function is still an exponential function, we guess that the exact solution to (1.6) is

$$X = Ce^{-t/\tau} \qquad ...(2.7)$$

It can be verified that (2.7) is indeed a solution of (2.6) by differentiating the solution with respect to *t*, thereby recovering our original equation

$$\frac{dX}{dt} = \frac{d}{dt}(Ce^{-t/\tau}) = -\frac{1}{\tau}(Ce^{-t/\tau}) = \frac{-X}{\tau} \qquad ...(2.8)$$

Because a differential equation describes how a variable changes for all time, the solution for a differential equation has to prescribe both how a variable evolves and at what value it begins. Thus there is a family of solutions to a differential equation, and the correct one is chosen by specifying an *initial condition.* This is an important concept that is particularly relevant to the numerical solution of differential equations. We choose a *particular solution* to (2.6) from the *general solution* given by (2.7) by adjusting the constant *C* such the initial condition $X = X(0)$ is satisfied at time $t = 0$, or

$$X(0) = Ce^{-0/t}, \qquad ...(2.9)$$

which means that $C = X(0)$. Together this gives the particular solution

$$X = X(0)e^{-t/\tau} \qquad ...(2.10)$$

for the initial condition $X(0)$. Representatives from the family of exact solutions for different values of $X(0)$ are shown in Fig. 2.8 for $\tau = 2$.

A deep understanding of the rules of differentiation and integration from calculus is not required for the level of modeling and mathematics that will be encountered in the first half of this book. The exponential function is encountered frequently, and so a review of its properties as given in elementary calculus textbooks is advisable. It is also helpful when analyzing experimental literature to understand the difference between the time constant τ and the half-time $\tau_{1/2}$ in the context of the exponential function. The half-time is the time at which the value of the function decays to $\frac{1}{2}$ of the initial value, or the solution of

$$X(0)e^{-t/\tau} = \frac{X(0)}{2} \quad ...(2.11)$$

Or

$$e^{-t/\tau} = \frac{1}{2} \quad ...(2.12)$$

which is $t = \tau \ln 2$. In contrast, after one time constant or at time $t = \tau$, the value of X is given by

$$X = X(0)e^{-\tau/\tau} = X(0)e^{-1} \approx X(0)\,.0.37 \quad ...(2.13)$$

Note that these principles apply to growing as well as decaying functions.

The solution *X(t)* always "relaxes" or decays to zero at long times. (Note that to mathematicians an equation *decays* to steady state whether it approaches a value greater or smaller than the initial value.) An equation related to the simple exponential decay equation given in (2.6) is the exponential approach to a steady state other than zero. The single channel model given in (2.5) describes exponential decay (or growth) to a steady state fraction of open channels, f_∞. We rewrite it here to aid the presentation:

$$\frac{df_o}{dt} = \frac{-(f_o - f_\infty)}{\tau} \quad ...(2.14)$$

It is easy to solve (2.14) if we make a *variable substitution.* If we create a new variable Z and define it as

$$Z = f_o - f_\infty \quad ...(2.15)$$

then because f_∞ is a constant, and therefore

$$\frac{dZ}{dt} = \frac{df_o}{dt} \quad ...(2.16)$$

we can rewrite (2.14) as

$$\frac{dZ}{dt} = \frac{Z}{\tau} \quad ...(2.17)$$

We saw in (2.10) that the solution to this equation is

$$Z = Z(0)e^{-t/\tau} \quad ...(2.18)$$

Recalling that $Z = f_o - f_\infty$ (and therefore that $Z(0) = f_o(0) - f_\infty$ we can once again exchange the variables and arrive at the analytic solution to (2.14) for a given initial condition $f_o(0)$:

$$fo(t) = f_\infty + (f_o(0) - f_\infty)e^{-t/\tau} \quad ...(2.19)$$

Numerical Integration

Even if (2.14) were more complicated and could not be solved exactly, a numerical approximation could still be calculated. The simplest and perhaps the oldest method of numerical solution goes back to the mathematician Euler and is easy to understand. The method is called the *forward Euler* method and it is a prototype for all other methods of solving ODEs numerically. Consider an approximation to the derivative in (2.6).

$$\frac{dX}{dt} \approx \frac{\Delta X}{\Delta t} = \frac{X(t+\Delta t)-X(t)}{\Delta t}, \quad ...(2.20)$$

where ΔX and Δt are small, but not infinitesimal like the differentials dX and dt. If this approximation to the derivative is substituted into the differential equation, the resulting equation can be solved for $X(t+\Delta t)$, giving

$$\frac{dX}{dt} = \frac{-X(t)}{\tau},$$

$$\frac{X(t+\Delta t)-X(t)}{\Delta t} = \frac{-X(t)}{\tau},$$

$$X(t+\Delta t) = X(t) - \frac{X(t)\Delta t}{\tau} \quad ...(2.21)$$

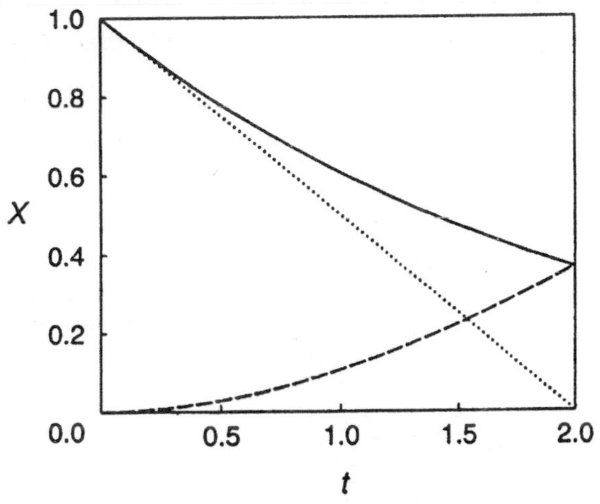

Fig. 2.7. The forward Euler method of numerical integration relies upon a series of short linear approximations using the derivative at the old time point. The solution to (1.6) with τ = 2 is shown in solid, the linear approximation using the derivative at τ = 0 (–0.5) is shown in dots, and the difference between these two curves is shown in dashes. Note that by t = 2, the magnitude of error between the actual and approximate functions is equal to the value of the actual function.

The smaller Δt is, the better the Euler approximation of the derivative will be. Also, because the Euler approximation gives a piecewise linear estimate of the solution, the further from linear the problem is, the smaller Δt must become to give an accurate solution. The essence of numerical integration is that we start at some value and crawl along the solution in increments of Δt by estimating the change over that interval. If Δt is very small, our estimate of the rate of change is good and our solution is accurate, but it may take a very long time to compute the solution.

This is termed *computationally expensive,* because it either requires a faster computer or a longer time to run. Two solutions to (2.19) obtained by integrating the equation using the Euler method. The time step was chosen to be dt = 0.03, and two different initial conditions were used, $f_o(0) = 1$ and $f_o(0) = 0$. Independent of the initial condition, $f_o(t)$ relaxes to its *steady state* value $f_\infty = 0.5$. It corresponds to the point where the rate vanishes. The rate at which the steady state is approached depends on the value of τ, which is 3 in these simulations. The solution to the equation in the Euler method depends on step size.

Only step sizes that are more than an order of magnitude smaller than the value of τ do a reasonable job of approximating the exact exponential solution, which for the parameter values used is $fo(t) = 0.5(1 + \exp(-t/3))$. Unreasonably large step sizes like $\Delta t = 6$ give approximations that are not even close to the exact solution. In fact, the numerical method has become *unstable,* and the computed solution oscillates around the true solution. There are many other methods of numerical integration that give better approximations to the derivative. These methods are generally more complicated, but have fewer restrictions on τ. These more complicated methods also address some other important problems such as *numerical stability*.

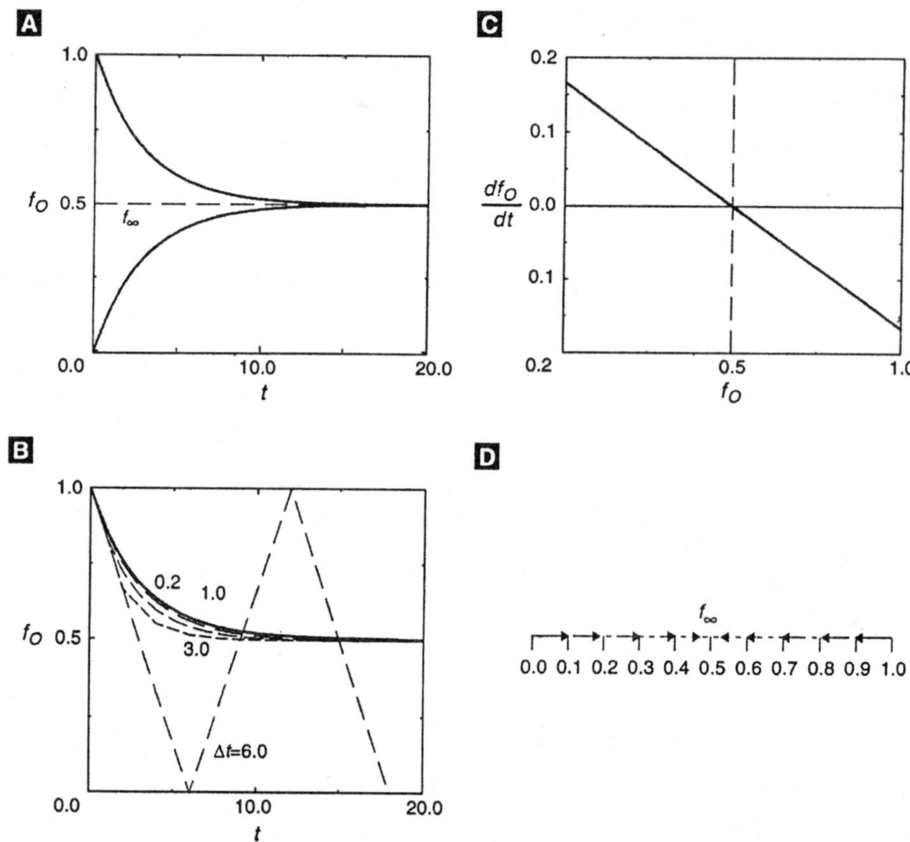

Fig. 2.8. (A) The exponential decay of the open fraction of channels. Initial conditions at either 1 or 0 both decay to the steady-state value of 0.5. (B) The effect of the step size in the Euler method for the simulation in panel A starting with the initial condition $f_o(0) = 1$. The exact solution, $f_o(t) = 0.5(1 + \exp(-1/3))$, is given by the full line. (C) Plot of the rate of change of n as a function of n. (D) Phase portrait with the arrows representing the direction and relative magnitude of the rate for each value of f_o. All the arrows point toward the steady state, $f_\infty = 0.5$.

There are many line texts on numerical analysis that discuss these issues and explain the various advantages and disadvantages of each method. As mentioned above, however, the best way to ensure an optimal and valid means of solution is to collaborate with a mathematician who has experience in scientific computation or numerical analysis. There is a different way to plot the results of solving the differential equation that frequently gives insight into the properties of the solution. This is demonstrated in Fig. 2.8C, where the function $df_o/dt = -(f_o - f_\infty)/\tau$ plotted versus the value of f_o for the two initial conditions in Fig. (2.8A). Since *fo* is restricted on physical grounds to be between 0 and 1, the plot shows that f_∞ is the unique steady state by making it clear that when $f_o > 0.5$, n decreases with time because $df_o/dt < 0$, and when $f_o < 0.5$, $f_o(t)$ increases with time. To further emphasize this concept, arrows in Fig. 2.8D show the direction in which f_o is changing. This type of plot is called a *phase portrait,* in one dimension. Phase portraits are particularly useful for analyzing ODEs with two variables, where they are typically called *phase plane* diagrams.

Numerical Packages

While it is important to understand the limitations of whichever numerical algorithm is used for the solution of a problem, fortunately it is not necessary to face the task of implementing these algorithms on a computer from scratch. Several excellent software programs have been developed that not only solve ODEs, but represent solutions graphically and allow their dynamical properties to be analyzed. These packages include very sophisticated commercial mathematical packages such as Matlab and Mathematica, which can often be obtained at reduced cost in the form of student versions. There are also myriad university produced packages such as Berkeley Madonna that are designed to solve ODEs.

A new direction for computational cell biology is the creation of several highly sophisticated packages such as The Virtual Cell, which is an integrated database and computational system expressly designed for cell biology modeling. Among the best for our purposes here is a public domain package, XPPAUT, that has been developed by Bard Ermentrout at the University of Pittsburgh. The name of the program evolved from a DOS version that was called *Phase-Plane,* refering to the program's ability to carry out phase plane analysis. A version that ran in X-windows under Unix or linux was then developed and was called X-PhasePlane (or XPP for short).

Finally, when the automatic bifurcation tool AUTO developed by E. Doedel, was added it became X-PhasePlane-Auto, or XPPAUT. XPPAUT is an excellent tool for solving and analyzing ordinary differential equations, one that provides very sophisticated tools without automating the process too much to be useful for our didactic objectives. Moreover, XPPAUT is free and available for both Unix/Linux, Windows, and MacOSX. We adopt it as the basic software program for use with this text.

GRAPHIC MOLECULAR STRUCTURE

Computer graphics has changed the way in which chemical structures are presented and perceived. The facile conversion of macromolecular sequences into three-dimensional structures that can be displayed and manipulated on the computer screen have greatly improved the biochemist's understanding of biomolecular structures. Tools for graphical visualization and manipulation of biomolecular structures are described. *Molecular graphics* refers to a technique for the visualization and manipulation of molecules on a graphical display device. The technique provides an exciting opportunity to augment the traditional description of chemical structures by allowing the manipulation and observation in real time and in three dimensions, of both molecular structures and many of their calculated properties.

Recent advances in this area allow visualization of even intimate mechanisms of chemical reactions by graphical representation of the distribution and redistribution of electron density in atoms and molecules along the reaction pathway. All graphics programs must be able to import commands defining representations and translate these into a picture according to the representations specified. The graphics programs offer various choices of renderings of the model, with color coding of atoms or groups and with selective labeling. These include the following.

Ball-and-Stick Models

Traditionally, drawings of small molecules have represented either (a) each atom by a sphere or (*b*) each bond by a line segment. Bond representations give a clearer picture of the topology or connectivity of a structure. In a simple picture of line drawings, there is a direct correspondence; that is, one line segment equals one bond. There are basically two ways to extract bonds from a given set of atomic coordinates:

1. *Screen the atoms by distance.* This is the most general approach. For every pair of atoms in the structure, the distance between them is calculated. If the distance is less than the sum of the van der Waals radii of the two atoms, a bond between them is assumed. For proteins or nucleic acids, this approach can be specialized by checking only atoms in the same residue/nucleotide, plus the atoms in the peptide/nucleotide bonds between successive residues/ nucleotides.
2. *Create an explicit list of bonded pairs.* For protein containing only standard amino acids and common ligands, this can be done once and for all. For each residue type, one can

make a list of pairs of atom names, each pair corresponding to a bond. Then in drawing a picture of a protein for each residue, one can search the coordinates of each pair of atoms in the list of bonds and add the appropriate line segment to the drawing. Similar considerations apply to nucleic acids. In pdb files, explicit connectivity lists are provided.

A ball-and-stick drawing is a simple skeletal model, in which the representation of the bond is generalized to a cylinder, and a disc is added at the position of each atom. Additional information may thereby be displayed in that different atom types may be distinguished by size and shading, and bonds of different appearance may be drawn. To create the line segments corresponding to a pure skeletal model, one needs only copy the coordinates of each pair of bonded atoms as the line segment end-points. To create ball-and-stick pictures, one must :

1. Draw a circle at the position of each atom with the facility to vary the atomic color and radius.
2. Determine the line segments that represent the bond and attach these segments at the edge of the circle.

Wire models and ball-and-stick models are extremely useful because of the great detail they contain. They are particularly useful in connection with blow-ups of selected protein/ nucleic acid molecules.

Shaded-Sphere Pictures. Each atom is shown to represent complete chemical detail of molecules. This category includes three basic types of picture:

1. *Line drawings :* Each atom is represented as a disc. The picture is a limit of the ball-and-stick drawings as the radius of each atomic ball is made in proportional to the van der Waals radius.
2. *Colour raster devices :* A raster device can map an array stored in memory on to the screen so that the value of each element of the array controls the appearance of the corresponding point on the screen. It is possible to draw each atom as a shaded sphere, or even to simulate the appearance of the Corey-Pauling-Koltun (CPK) physical models to maintain most of the familiar colour scheme (C = black, N = blue, O = red, P = Green, and S = yellow, etc.). In such representation, atoms are usually opaque, so that only the front layer of atoms is visible. However, clipping with an inner plane or rotation can show the packing in the molecular interior.
3. *Real-time rotation and clipping:* Facilities available on vector graphics devices are very useful in connection with another technique for representing atomic and molecular surfaces. The spatter-painting of the surface of a sphere by a distribution of several hundred dots produces a translucent representation of the surface (dot-surface). It is possible to combine dot-surface representations with skeletal models to show both the topology of the molecule and its space-filling properties. Dot-surface pictures used on an interactive graphics device with a color screen have been helpful in solving problems of docking ligands to proteins and exploring the goodness of fit in interfaces.

Basically, any colour can be matched by a suitable combination of three primary colors, RGB (red, green, and blue). These three primary colours on a binary status (on/off) provide eight colors. Video monitors generate a large number of colour combining various amount of the three primary components. The observed colour of an object depends on the spectrum of the light it emits, transmits, or reflects. Observable colors may be distinguished on the basis of three characteristics :

1. *Hue (colour):* Colour in the most common colloquial sense (*i.e.,* red, green and blue) describes different hues. For monochromic light, different wavelengths correspond to different hues, but different spectra can give the same perceived hue. A flat spectrum appears achromatic : white, gray, or black.
2. *Tone (value) :* Roughly speaking, the total amount of light per unit area (*i.e.,* multiplying a spectrum by a constant) changes the tone. Members of the series white → gray → black differ in tone. Empirical scales of tone are not linear in integrated intensity; moreover, changes in tone can alter perceived hue.
3. *Saturation (intensity or chroma) :* The difference between a colour and the gray with the same tone exemplifies saturation. A pure or saturated colour can be diminished in saturation by adding white light and normalizing the result to the same perceptual tone.

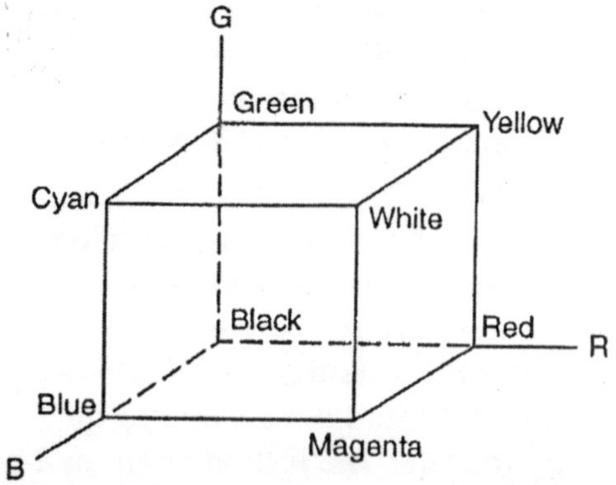

Fig. 3.1. Cube model for RGB system. The RGB cube model illustrates the definition of colours by the three primary components along the three axes R, G, and B. Each colour point is represented by a triple (r, g, b). The three primary colours are red (1, 0, 0), green (0, 1, 0), and blue (0, 0, 1). Other binary-status (0/1 for r, g, b) colours are cyan (0, 1,1), magenta (1, 0, 1), yellow (1,1, 0), white (1, 1, 1), and black at origin (0, 0, 0). Different colours are expressed by a combination of r, g, and b values varied between 0 and 1. For example, gray colours correspond to the main diagonal between black and white.

As a result, colours may be thought of as point in a three-dimensional space, the axes of which might be the primary colours (primaries): red, green, and blue. Each colour is a vector, the components of which are the intensities of the primaries required to match it. For displays generated by three-gun (red, green, and blue) monitors, these are the numbers specified. The literature of computer graphics reveals considerable efforts to achieve truly convincing representations of real objects.

MOLECULAR STRUCTURES

There are three levels of structural graphics: 1D formula or string/character format (*e.g.,* SMILES string), 2D chemical structures (*e.g.,* ISIS draw), and 3D molecular structures (*e.g.,* PDB atomic coordinate display). For proteins and nucleic acids, the primary structures represented in coding sequences are 1D formats. The sequences represented in the chemical linkages of amino acid or nucleotide structures are 2D drawings, and the full atomic representations of the 3D structures constitute 3D graphics. The common 1D presentations for nucleotide sequences are in GenBank, EMBL, and fasta formats. The common files for amino acid sequences are PIR, Swiss-Prot, GenPept and fasta formats.

```
BASE COUNT 132 a      169 c      172 g      111 t
  ORIGIN
      1  tcccgctgtg  tgtacgacac  taggcaacatg aggtctttgc  taatcttggt gctttgcttc
     61  ctgcccctgg  ctgctctggg  gaaagtcttt  ggacgatgtg  agctggcagc ggctatgaag
    121  cgtcacggac  ttgataacta  tcggggatac  agcctgggaa  actgggtgtg tgttgcaaaa
    181  ttcgagagta  acttcaacac  ccaggctaca  aaccgtaaca  ccgatgggag taccgactac
    241  ggaatcctac  agatcaacag  ccgctggtgg  tgcaacgatg  gcaggacccc aggctccagg
    301  aacctgtagca acatcccgtg  ctcagccctg  ctgagctcag  acataacagc gagcgtgaac
    361  tgcgcgaaga  agtcgtcag   cgatggaaac  ggcatgagcg  cgtaaatcgc ctggcgcaac
    421  cgctgcaagg  gtaccgacgt  ccaggcgtgg  atcagaggct  gccggctgtg aggagctgcc
    481  gcacccggcc  cgcccgctgc  acagccggcc  gctttgcgag  cgcgacgcta cccgcttggc
    541  agttttaaac  gcatccctca  ttaaaacgac  tatacgcaa   cgcc
```

Fig. 3.2. GenBank format for nucleotide sequence of chicken egg-white lysozyme.

```
SQ  Sequence 584 BP; 132 A; 169 C; 172 G; 111 T; 0 other;
    tcccgctgtg   tgtacgacac   tggcaacatg  aggtctttgc  taatcttggt  getttgcttc   60
    ctgccccctgg  ctgctctggg   gaaagtcttt  ggacgatgtg  agctggcagc  ggctatgaag  120
    cgtcacggac   ttgataacta   tcggggatac  agcctgggaa  actgggtgtg  tgttgcaaaa  180
    ttcgagagta   acttcaacac   ccaggctaca  aaccgtaaca  ccgatgggag  taccgactac  240
    ggaatcctac   agatcaacg    ccgctggtgg  tgcaacgatg  gcaggacccc  aggctccagg  300
    aacctgtgca   acatcccgtg   ctcagccctg  ctgagctcag  acataacagc  gagcgtgaac  360
    tgcgcgaaga   agatcgtcag   cgatggaaac  ggcatgagcg  cgtgggtcgc  ctggcgaaca  420
    cgctgcaagg   gtaccgacgty  ccaggcgtgg  atcgaggct   gccggctgtg  aggagctgcc  480
    gcacccggcc   cgcccgctgc   acagccggcc  gctttgcgag  cgcgacgcta  cccgcttggc  540
    agttttaaac   gcatccctca   ttaaaacgac  tatacgcaaa  cgcc                    584
```

Fig. 3.3. EMBL format for nucleotide sequence of chicken egg-white lysozyme.

```
>gi|63580|emb|v00428.1.|GGLYSI Gallus gallus mRNA coding for 1ysozyme
TCCCGCTGTGTGTACGACACTGGCAACATGAGGTCTTTGCTAATCTTGGTGCTTTGCTTCCTGCCCCTGG
CTGCTCTGGGGAAAGTCTTTGGACGATGTGAGCTGGCAGCGGCTATGAAGCGTCACGGACTTGATAACTA
TCGGGGATACAGCCTGGGAAACTGGGTGTGTGTTGCAAAATTCGAGAGTAACTTCAACACCCAGGCTACA
AACCGTAACACCGATGGGAGTACCGACTACGGAATCCTACAGATCAACAGCCGCTGGTGGTGCAACGATG
GCAGGACCCCAGGCTCCAGGAACCTGTGCAACATCCCGTGCTCAGCCCTGCTGAGCTCAGACATAACAGC
GAGCGTGAACTGCGCGAAGAAGATCGTCAGCGATGGAAACGGCATGAGCGCGTGGGTCGCCTGGCGCAAC
CGCTGCAAGGGTACCGACGTCCAGGCGTGGATCAGAGGCTGCCGGCTGTGAGGAGCTGCCGCACCCGGCC
CGCCCGCTGCACAGCCGGCCGCTTTGCGAGCGCGACGCTACCCGCTTGGCAGTTTAAACGCATCCCTCA
TTTAAAACGACTATACGCAAACGCC
```

Fig. 3.4. Fasta format for nucleotide sequence of chicken egg-white lysozyme.

```
> SUMMARY        # length 147      #molecular_weight        16238
SEQUENCE
             5         10         15         20         25         30
  1 M R S L L I L V L C F L P L A A L G K V F G R C E L A A A M
 31 K R H G L D N Y R G Y S L G N W V C A A K F E S N F N T Q A
 61 T N R N T D G S T D Y G I L Q I N S R W W C N D G R T P G S
 91 R N L C N I P C S A L L S S D I T A S V N C A K K I V S D G
121 N G M N A W V A W R N R C K G T D V Q A W I R G C R L
```

Fig. 3.5. Swiss-Prot format for amino acid sequence of chicken egg-white lysozyme.

```
Length: 147 AA Molecular weight: 16238 Da
        10          20          30          40          50          60
         |           |           |           |           |           |
MRSLLILVILC FLPLAALGKV  FGRGELAAAM  KRHGLDNYRG  YSLGNWVCAA  KFESNFNTQA
        70          80          90         100         110         120
         |           |           |           |           |           |
TNRNTDGSTD  YGILQINSRW  WCNDGRTPGS  RNLCNIPCSA  LLSSDITASV  NCAKKIVSDG
       130         140         147
         |           |           |
NGMNAWVAWR  NRCKGTDVQA  WIRGCRL
```

Fig. 3.6. Swiss-prot format for amino acid sequence of chicken egg-white lysozyme.

```
ORIGIN
      1 mrsllilvlc flplaalgkv fgrcelaaam krhgldnyrg yslgnwvcva kfesnfntga
     61 tnrntdgstd ygilqinsrw wcndgrtpgs rnlcnipcsa llssditasv ncakkivsdg
    121 ngmsawvawr nrckgtdvqa wirgcrl
```

Fig. 3.7. GenPept format for amino acid sequence of chicken egg-white lysozyme.

>gi|63581|emb|CAA23711.1|coding sequence lysozyme *[Gallus gallus]*

```
MRSLLILVLCFLPLAALGKVFGRCELAAAMKRHGLDNYRGYSLGNWVCVAKFESNFNTQATNRNTDGSTD
YGILQINSRWWCNDGRTPGSRNLCNIPCSALLSSDITASVNCAKKIVSDGNGMSAWVAWRNRCKGTDVQA
WIRGCRL
```

Fig. 3.8. Fasta format for amino acid sequence of chicken egg-white lysozyme.

Protein topology cartoons (TOPS) are two-dimensional schematic representations of protein structures as a sequence of secondary structure elements in space and direction. The TOPS of trypsin domains have the following symbolisms:

1. Circular symbols represent helices (α and 3_{10}).
2. Triangular symbols represent β strands.

3. The peptide chain is divided into a number of fragments, and each fragment lies in only one domain.
4. Each fragment is labeled with an integer (i), beginning at N_i and ending at C_{i+1} with the first fragment being $N_1 \rightarrow C_2$.
5. If the chain crosses between domains, it leaves the first at C_{i+1} to join the next N_{i+2}.
6. Each secondary structure element has a direction (N to C) that is either *up* (out of the plane of the diagram) or *down* (into the plane of the diagram).
7. The direction is up if the N-terminal connection is drawn to the edge of the symbol and the C-terminal connection is drawn to the center of the symbol. Otherwise, the direction is down if the N-terminal connection is drawn to the center of the symbol and the C-terminal connection is drawn to the edge.
8. For β strands, up strands are indicated by upward-pointing triangles whereas down strands are indicated by downward-pointing triangles.

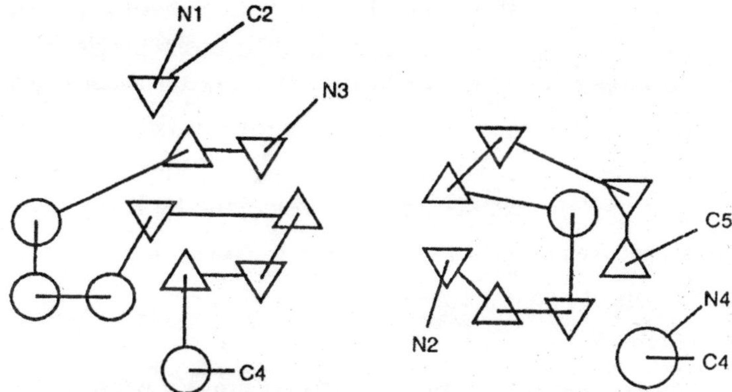

Fig. 3.9. TOPS diagrams for trypsin domains.

The most obvious data in a typical 3D structure record are the atomic coordinate data, the locations in space of the atoms of a molecule represented by *(x, y, z)* triples, and distances along each axis to some arbitrary origin in space. The coordinate data for each atom are attached to a list of labeling information in the structure record such as that derived from the protein or nucleic acid sequence. Three dimensional molecular structure database records employ two different "minimalist" approaches regarding the storage of bond data.

The chemistry rule applies observable physical principles of chemistry to record molecular structures without bond information. There is no residue dictionary required to interpret data encoded by this approach, just a table of bond lengths and bond types for every conceivable pair of bonded atoms. This approach is the basis for 3D biomolecular structure file format of Protine Data Bank. The other approach is used in the database records of the Molecular Modeling Database (MMDB) at NCBI, which uses a standard residue dictionary of all atoms and bonds in the biomacromolecules of amino acid and nucleotide residues plus end-terminal variants. The software that reads in MMDB data which are derived from the data of PDB can use the bonding information supplied in the dictionary to connect atoms together, without trying to enforce the rules of chemistry. Almost all known protein structures have been determined by X-ray crystallography. A few contain details derived from neutron diffraction, and a few have been determined from nuclear magnetic resonance (NMR).

Recently, the theoretical models derived from molecular modeling are added. The resolution of a structure is a measure of how much data were collected. The more data collected, the more detailed the features in the electron density map to be fitted, and, of course, the greater the ratio of observations to parameters to be determined (*i.e.,* the atomic coordinates and *R*-factors). Resolution is expressed in angstroms (Å), which is a measure of distance. The lower the number, the higher the resolution. Protein structures are generally determined to a resolution between 1.7 and 3.5 Å; those determined at 2.0 Å or better are considered high-resolution. The *R*-factor of a structure determination is a measure of how well the model reproduces the experimental intensity data.

Other things being equal, the lower the *R*-factor, the better the structure. The *R*-factor is a fraction expressed as a percentage; $R = 0\%$ would be an impossible ideal case (no disorder, no experimental error), and $R = 58\%$ for a collection of atoms placed randomly in unit cell of the crystal. The Protein Data Bank (PDB) is the collection of publicly available structures of proteins, nucleic acids, and other biological macromolecules initiated by Brookhaven National Laboratory and now maintained by the Research Collaboratory for Structural Bioinformatics (RCSB). The PDB coordinates of biomacromolecules can be classified into the following:

1. Protein structures determined by X-ray or neutron diffraction or NMR which may include co-factors, substrates, inhibitors, or other ligands.
2. Oligonucleotide or nucleic acid structures determined by X-ray crystallography.
3. Carbohydrate structures determined by X-ray diffraction.
4. Hypothetical models of protein structures.
5. Bibliographic entries.

Each set of coordinates deposited with the PDB becomes a separate entry. Each entry is associated with an accession PDB code with a unique set of four alphanumeric characters. PDB and its mirror sites offer a text search engine that uses an index of all the textual information in each PDB record (*e.g.,* PDB ID); an example of such an index is 1LYZ for hen's egg-white lysozyme. The first character is a version number. An identifier beginning with the number 0 signifies that the entry is purely bibliographic. The pdb file is a text file with an explanatory header followed by a set of atomic coordinates. The atomic coordinates are subjected to a set of standard stereochemical checks and are translated into a standard entry format; for example, Fig. 3.10 shows partial coordinate file for 1LYZ.pdb or pdb1 YZ.ent.

The PDB format includes information about the structure determination, bibliographic references describing the structure, types/locations of secondary structures, and the atomic coordinates. Most software programs created for molecular graphics or other computational analysis of protein structures can read files in PDB format, with file extension of either .pdb or .ent. The 3D graphical representations of these biomacromolecules can be displayed with RasMol (Sayle and Milner-White, 1995), Cn3D or KineMage (Richardson and Richardson, 1992; Richardson and Richardson, 1994) as shown in Fig. 4.11. For online visualization of 3D structures, the MIME types for PDB format is chemical/x-pdb, which enables the display of 3D structure on the Web with RasMol. Once the atomic coordinates are known, the reader can manipulate the image in the browser to rotate the molecule, view it from a different perspective, or change the manner in which the structure is presented. A comprehensive list of chemical MIME media type (Rzepa et al., 1998) is available from http://www.ch.ic.ac.uk/chemime/.

DISPLAY OF MOLECULAR STRUCTURES

String and Sequences

One of the common character formats or chemical nomenclature of a valence model which is recognizable by a number of 2D-structure drawing programs is SMILES. The general rules for biochemical compounds are as follows:

1. *Hydrogen atom:* Hydrogen atoms are not normally specified. An explicit hydrogen specification is required for charged hydrogen, isotopic hydrogen, or chiral assignment. It is written within brackets [ ] if an isotope, chiral, or charge are inferred—that is, [H^+] for a proton.
2. *Atom specification:* Elemental identity is represented by a standard atomic symbol without explicit hydrogen if the number of attached hydrogens conforms to the lowest normal valence consistent with explicit bonds—that is, C(4), N(3,5), 0(2), P(3,5), and S(2,4,6).

HEADER	HYDROLASE		(O-GLYCOSYL)		01-FEB-75				1LYZ	
COMPND	LYSOZYME		(E.C.3.2.1.17)						1LYZ	
SOURCE	HEN	(GALLUS SGALLUS)			EGG	WHITE			1LYZ	
...	...									
...	...									
ATOM	258	N	PHE	34	2.966	25.379	11.370	7.00	1.50	1LYZ
ATOM	259	CA	PHE	34	3.955	25.906	11.749	6.00	1.50	1LYZ
ATOM	260	C	PHE	34	4.614	25.247	12.887	6.00	1.50	1LYZ
ATOM	261	O	PHE	34	5.801	25.379	12.887	8.00	1.50	1LYZ
ATOM	262	CB	PHE	34	3.494	27.356	12.002	6.00	1.50	1LYZ
ATOM	263	CG	PHE	34	3.362	27.949	10.612	6.00	1.50	1LYZ
ATOM	264	CD1	PHE	34	2.241	29.004	10.297	6.00	1.50	1LYZ
ATOM	265	CD2	PHE	34	4.614	27.949	9.728	6.00	1.50	1LYZ
ATOM	266	CE1	PHE	34	1.978	29.663	9.160	6.00	1.50	1LYZ
ATOM	267	CE2	PHE	34	4.417	28.674	8.402	6.00	1.50	1LYZ
ATOM	268	CZ	PHE	34	6.164	29.400	8.212	6.00	1.50	1LYZ
ATOM	269	N	GLU	35	3.889	24.653	13.771	7.00	1.50	1LYZ
ATOM	270	CA	GLU	35	4.680	23.796	14.971	6.00	1.50	1LYZ
ATOM	271	C	GLU	35	5.273	22.412	14.529	6.00	1.50	1LYZ
ATOM	272	O	GLU	35	6.526	22.149	14.845	8.00	1.50	1LYZ
ATOM	273	CB	GLU	35	4.087	23.599	15.982	6.00	1.50	1LYZ
ATOM	274	CG	GLU	35	3.757	23.665	17.308	6.00	1.50	1LYZ
ATOM	275	CD	GLU	35	4.680	24.522	17.940	6.00	1.50	1LYZ
ATOM	276	OE1	GLU	35	5.603	24.588	17.687	8.00	1.50	1LYZ
ATOM	277	OE2	GLU	35	4.153	25.642	18.509	8.00	1.50	1LYZ

```
ATOM    278   N    SER  36    4.680  21.687  13.834  7.00  1.50  1LYZ
ATOM    279   CA   SER  36    4.878  20.435  13.455  6.00  1.50  1LYZ
ATOM    280   C    SER  36    5.142  20.039  11.876  6.00  1.50  1LYZ
ATOM    281   O    SER  36    5.669  18.919  11.370  8.00  1.50  1LYZ
ATOM    282   CB   SER  36    4.285  19.446  14.024  6.00  1.50  1LYZ
ATOM    283   OG   SER  36    2.966  19.116  13.266  8.00  1.50  1LYZ
... ...
TER     1002       LEU  129                                    1LYZ
HETATM  1003  O    HOH  1     1.437  16.676  19.902  7.36  1.59  11LYZ
HETATM  1004  O    HOH  2     -.616  11.133  19.523  8.12  1.80  21LYZ
... ...
CONECT  48    47   981                                         1LYZ
... ...
END                                                            1LYZ
```

Fig. 3.10. PDB file (partial) for 3D structure of hen's egg-white lysozyme (1LYZ.pdb). The abbreviated file shows partial atomic coordinates for residues 34-36. Informational lines such as AUTHOR (contributing authors of the 3D structure), REVDAT, JRNL (primary bibliographic citation), REMARK (other references, corrections, refinements, resolution and missing residues in the structure), SEQRES (amino acid sequence), FTNOTE (list of possible hydrogen bonds), HELIX (initial and final residues of α-helices), SHEET (initial and final residues of /β-sheets), TURN (initial and final residues of turns, types of turns), and SSBOND (disulfide linkages) are deleted here for brevity. Atomic coordinates for amino acid residues are listed sequentially on ATOM lines. The following HETATM lines list atomic coordinates of water and/ or ligand molecules.

3. *Bond specification:* Single, double, triple, and aromatic bonds are represented by the symbols —, =, #, and :, respectively, for example, CC=O for acetaldehyde. Generally, single and aromatic bond symbols are omitted.
4. *Branch specification:* Branches are enclosed in nested or stacked paren-theses—for example, C(C)CC(N)C(=O)O for valine.
5. *Ring specification :* Ring closure bonds are specified by appending matching digits to the specifications of the joined atoms, with the bond symbol preceding the digit—for example, N1CCCC1C(=O)O for proline.
6. *Aromaticity* (A ring having sp^2 hybridized carbons with $4N + 2$ p-electrons): Aromatic atoms are specified with lowercase atomic symbols with appen-ding matching digits following the joined atomic symbols—for example, OclccccclCC(N)C(=O)O or OC1=CC=CC==C1CC(N)C(==O)O for tyrosine.
7. *Disconnection:* The period or dot is used to represent disconnections—for example,-C=COP(=O)(=O)OC(=O)[O-].[Na+] for sodium phosphorenolpyruvate.
8. *Isotope :* Isotopic specification is indicated by prefixing the atomic symbol with a number equal to the integral isotopic mass—for example, [^{2}H] for deuterium and [^{13}C] for carbon-13.

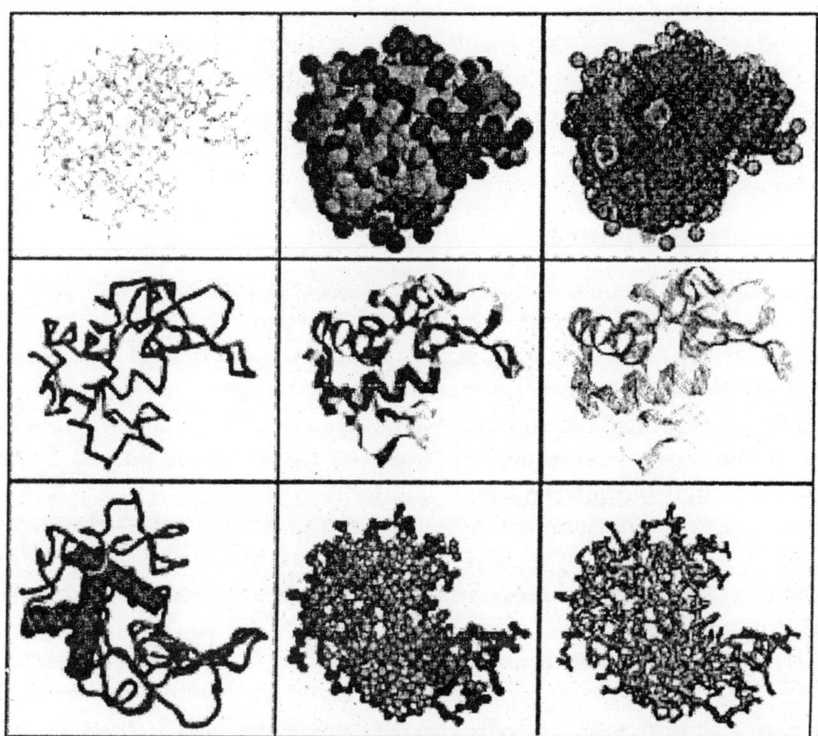

Fig. 3.11. Graphic representations of protein 3D structure. Three-dimensional graphics of hen egg white lysozyme as visualized with RasMol (first and second rows, 1LYZ.pdb) and Cn3D (third row, 1 LYZ.val) are shown from left to right (colour type) in wireframe (atom), spacefill (atom), dots (residue), backbone (residue), ribbons (secondary structure), strands (secondary structure), secondary structure (secondary structure), ball-and-stick (residue), and tubular (domain) representations.

9. *Isomerism (geometric) :* Configuration around double bonds is specified by the characters / and \ indicating relative directionality between the connected (by double bond) atoms — for example,

   CCCCCCCC/C=C\CCCCCCCC(=O)O

   for oleic acid with *cis* double bond.

10. *Isomerism (chiral):* The most common type of chirality in biochemistry is tetrahedral'. The tetrahedral chiral specification (@ or @ @) is written as an atomic property following the atomic symbol of the chiral atom. Looking at the chiral center from the direction of the "from" atom (preceding the chiral atom), @ (or @1) means "the other" three atoms (following the chiral atom) are listed anti-clockwise; @@ (or @2) means clockwise. If the chiral atom has a nonexplicit hydrogen, it will be listed inside the chiral atom's brackets as [C@H]—for example,

    [C@2H]O1 ([C@2H] (O) [C@2H] (O) [C@2H] (O) [C@2H] 1CO

    for β-D-glucopyranose that has all hydroxyl groups in equatorial configuration.

The default sequence format (nucleotide/amino acid) retrieved from three integrated database retrieval systems are GenBank/GenPept from Entrez as well as DBGet, and EMBL/Swiss-Prot from EBI. Though different formats can be specified at the time of retrieval. The formats supported by ReadSeq are IG/Stanford, GenBank/GB, NBRF, EMBL, GCG, DNASrider, Fitch, Pearson.Fasta, Philip3.2, Philip. PIR/CODATA, MSF, ASN.1, and PAUP/ NEXUS.

Drawing of Molecular Structures

The 1D nucleotide/amino acid sequences in character format (without index, *e.g.*, fasta format) can be converted into the 2D chemical structures with ISIS Draw, which can be downloaded from MDL Information System for academic use. Install the package by issuing Run command, C:\Isis\Draw23.exe. Launch IsisDraw to open the Draw window). Retrieve nucleotide/amino acid sequence file in fasta format (remove the heading, line) or prepare text file of sequence in one-letter characters. Rename the file as seqname.seq. Prepare to import the sequence by checking (√) Show sequence bond, Show leaving groups, Amino acid-WDNA-/ RNA-1 letter from Sequence options of Chemistry menu. Invoke File → Import → Sequences. Select Amino acid-/ DNA-/RNA-1 letter. The sequence (with bonds and leaving groups attached) should appear within the draw window. Mark the whole sequence with Select All from the Edit menu or by using Lasso tool. From the Chemistry menu, select Residue → Expand, the 1D (text string) sequence is transformed into the 2D molecular structure. Save it as struname.skc (*e.g.*, heptapeptide, STANLEY as stanley.skc). To draw 2D molecular structures, the users should refer to ISIS Draw Quick Start (ISIS Draw help) for operating instructions.

Draw the basic framework from template tools (horizontal template-tool icons and template pages from Template menu) and drawing tools (vertical drawing-tool icons). The small triangular sign on the drawing tool icons indicates additional tools available for selection. For example, pressing on the Single bond tool provides selection for drawing a double bond or a triple bond. Verify the chemistry of sketch by clicking run Chemistry inspector icon (or select Chemistry inspector from the Chemistry menu). To ensure uniform bond lengths and angles for the sketched molecule, select the molecule and then choose Object → Clean Molecule. Save the sketch as struname.skc. To place template, click one of the template-tool icons or an atom/ bond in the structural fragment/molecule on the template page and place the template anywhere inside the window by clicking an empty area.

The template can be fused/attached to an existing bond/atom by simply clicking the bond/ atom. To draw bonds, click a bond tool (Single/double/triple bond or Up wedge/down wedge/ either/up bond/ down bond), and then click the drawing area or drag the mouse from an existing atom to add a bond. To sprout a bond from an atom, click a bond tool and then click the atom. To draw chain in one direction/ring of specific shape, click chain/multibond tool. To draw atoms, click Atom tool and enter atom symbol or choose one from the drop-down list. Arrow tool provides options for drawing a variety of arrows (*e.g.*, unidirection, equilibrium, double-head, and electron-shift arrows, etc., after pressing arrow tool then choosing one of the arrows) for chemical reactions. Use Lasso select tool to select any structure/structural component for editing or relocating. To delete atom/bond/object one at a time, click Eraser tool and then the atom/bond/object. Text tool appends text description to the structures/reactions. To search for TOPS cartoons select Browse the Atlas of topology cartoons and Browse HTML page version to open the query form. Enter PDB ID on the Protein code query box (Chain query box can be left blank). The search may request a choice of the chain (if more than one chains are available) and returns TOPS atlas information listing the protein of your choice and representative protein in atlas. Click to view the TOPS cartoon(s) of the representative protein. Right click on the diagram to save the TOPS cartoon as cartoon.gif.

3D Structures with Molecular Graphics

For the 3D view of ISIS/Draw, ACD/3D Viewer Add-in can be installed. Retrieve ACD/3D Viewer for ISIS/Draw and installed it as an Add-in according to instructions. To view 3D structure which is opened/sketched on the ISIS/Draw window, select ACD/3D Viewer tool from Object menu to open ACD/3D Viewer window and subsequent display of the 3D structure. The 3D structure can be optimized and can be saved only in .s3d format, which is not recognizable by other modeling packages. The 2D structures of ISIS draw can be transformed into the 3D structures by WebLab Viewer Lite, which can be downloaded free for academic use from Accelrys Inc. Select Download Viewlite to register and download. Open the file by selecting MDL (*.skc); the 2D structure (*e.g.,* hexapeptide, stanley.skc) is converted into the 3D structure whose coordinate file can be saved as struname .pdb. Alternately, the commercial molecular modeling software programs such as ChemOffice can be used.

The ISIS draw in sketch format (struname.skc) is first converted to ChemDraw format (struname.cdx), which is then transformed into 3D structure (struname.c3d) with Chem 3D and saved as PDB format (struname.pdb). The common atomic coordinate files for 3D structure in biochemistry is PDB format. The pdb files of polysaccharides, proteins, and nucleic acids can be retrieved from the Protein Data Bank at RCSB. On the home page, enter PDB ID (check the box "query by PDB id only") or keywords (check the box "match exact word") and click Find a structure button. Alternatively, initiate search/retrieval by selecting SearchLite. On the query page, enter the keyword (*e.g.,* the name of ligand or biomacromolecule) and click Search button.

Select the desired entry from the list of hits to access Summary information of the selected molecule. From the Summary information, select Download/Display file and then PDB Text and PDB noncompression format to retrieve the pdb file. In order to display 3D structure online, choose View structure followed by selecting one of 3D display options. The display can be saved in .jpg or .gif image format. Most of molecular modeling software programs accept the pdb files (struname.pdb). RasMol, which is one of the most widely used molecular graphics freeware, can be downloaded.

In addition to PDB (struname.pdb or struname.ent) file, RasMol also read Alchemy, Sybtk MOL2, MDL mol, CHARMm, and MOPAC files. Launch RasMol (double click rswin.exe or rw32b2a.exe) to open the display window. Open the pdb file from File menu. The 3D structure can be displayed as wireframe, backbone, sticks, spacefill, ball and stick, ribbons, strands and cartoons. The display can be exported as bmp, gif, epsf, ppm, and rast graphics. KineMage (kinetic image) is an interactive 3D structure illustration software that can be downloaded. It is adapted for the structure representation of biological molecules by many biochemical textbooks. The program consists of two components: PREKIN and MAGE. The PREKIN program interprets struname.pdb file to kinemage struname.kin file that is then displayed and manipulated with the MAGE program. To start the PREKIN program, click Proceed to enter an output file name. This opens a dialog box, “Starting ranges.” Accepting the default (Backbone browsing script) saves the script producing Cα, disulfides for all subunits in the file to struname.kin. To start the MERGE program, click Proceed and select Open new file from File menu to open struname.kin with three windows (caption, display, and text). The 3D structure with connected series of alpha carbons is shown in the display window. To highlight the secondary structures, choose Selection of build-in scripts from the dialog box, “Starting ranges,” to open Build-in scripts box. Select ribbon: HELIX, SHEET from pdb to save as ribbon.kin. The MERGE program opens ribbon.kin.

Cn3D is a molecular graphics program that interprets structure files in MMDB (ASN.l) format (struname.val or struname.cgi) of Entrez/MMDB.

Cn3D can be accessed online from Entrez at http://www.ncbi.nlm.nih.gov/ Entrez or downloaded and used locally. This program accepts the coordinate file in MMDB ASN.1 format (*.val or *.cgi) but can be saved as PDB format (*.pdb) or KineMage format (*.kin). It is a structure-sequence interactive program.

In addition to structure view in the Graphic window, Cn3D provides the sequence view in the Sequence window which can be activated via View → Sequence Window. This enables the user to view the structure and sequence interactively. Select the region of protein molecule in the structure view by double clicking of the mouse, and both the region in the Graphic window and the amino acid residue(s) in the Sequence window are highlighted (yellow) and vice versa. Using this tool, it is possible to map the interaction sites between structure and sequence. The view menu of the Graphic window also provides an option for Animation, and the Sequence window offers options for alignment (Align menu). The style menu enables the user to display the structure in secondary structure, wireframe, neighbor, tabular, spacefill or ball-and-stick modes.

WORKSHOPS

1. Write SMILES strings for fumaric acid, D-gluconic acid, cholesterol, histidine, and AMP.
2. Use ISIS/Draw to sketch the above biochemical compounds.
3. Use ISIS/Draw to sketch maltotriose (malt3) and an octapeptide, GRAPHICS.
4. Convert the 2D sketch of matlS and GRAPHICS into 3D and save them as pdb files. Compare these files with the pdb file you retrieve from the Protein Data Bank at RCSB.
5. Retrieve protein topology cartoons (TOPS) for alcohol dehydrogenase (3BTO), concanavalin (2CNA), lysozyme (1LZ1), papain (1PPN), phosphofructokinase (1PFK), and rhodopsin (1BRD) or their representative TOPS. Compare their domain structures.
6. Retrieve a nucleic acid pdb file and visualize its 3D structure with RasMol. Save the structure as a graphic file in GIF format for printing.

7. Retrieve a protein pdb file and visualize its 3D structure with RasMol in different representations (Display menu). Identify the structural features or characteristics for which each display is best illustrative.
8. Retrieve a pdb file of an enzyme-substrate complex and visualize its 3D structure with KineMage. Save the structure as a graphic file in GIF format for printing.
9. Retrieve a structure file of a protein-DNA complex from Molecular Modeling Database at NCBI and visualize its 3D structure with Cn3D. Identify the interaction between the protein and DNA molecules.
10. Retrieve atomic coordinate files of two metalloenzymes, alcohol dehydrogenase (ADH) and Fe-superoxide dismutase (SOD), from PDB or MMDB. The subunit structure of ADH displays two metal ions, one catalytic and the other structural. The catalytic metal atom is chelated to Cys46, His67, Cys174, and a water molecule. Identify the catalytic metal atom and measure the approximate geometry of its chelation. The dimeric Fe-SOD similarly contains two Fe atoms per monomer. Search the literature to supplement the 3D structure view and present your findings regarding the function and geometry of the Fe atoms of Fe-SOD.

DYNAMICS IN MEMBRANES

Biological membranes hold cells together and divide them into compartments. They are also home to the lipid-soluble proteins that comprise roughly one-third of the genome and perform a variety of functions, including energy production and storage, signal transduction, and the formation of channels for the transport of substances into and out of cells and between their compartments. In 1972, Singer and Nicolso assimilated available data into a model for the molecular organization of biological membranes that is now taken for granted. According to this "fluid mosaic" model, a lipid bilayer forms the membrane matrix in which proteins are embedded. This matrix is fluid at physiological temperatures, and both lipids and proteins are free to diffuse in the plane of the membrane.

Biological membranes contain a complex mixture of several different types of lipids. Membrane lipids are amphiphilic molecules, with both polar and nonpolar substituents. A common membrane lipid that has been the subject of numerous experimental and theoretical studies, dipalmitoylphosphatidylcholine (DPPC), is diagrammed in Fig. 1. Like other membrane lipids containing two chains, DPPC is classified as an insoluble, swelling amphiphile. These lipids exhibit a variety of liquid crystalline phases that may be lamellar, cubic, or hexagonal, depending on the temperature and water content. Lamellar phases consist of stacks of lipid bilayers separated by layers of water (Fig. 4.2). The structures and mechanical properties of lipid bilayers depend on the amount of water present. As water is added to a lamellar phase, the spacing between the bilayers increases until the "fully hydrated" state is reached. At this point, which in DPPC is about 28 water molecules per lipid, additional water molecules do not go between the bilayers but rather go into a bulk water phase that is distinct from the lamellar bilayer water phase.

Hydrated bilayers containing one or more lipid components are commonly employed as models for biological membranes. These model systems exhibit a multiplicity of structural "phases" that are not observed in biological membranes. In the state that is analogous to fluid biological membranes, the "liquid crystal" or L_α bilayer phase present above the main bilayer phase transition temperature, T_m, the lipid hydrocarbon chains are conformationally disordered and fluid ("melted"), and the lipids diffuse in the plane of the bilayer. At temperatures well below T_m, hydrated bilayers exist in the "gel," or L_β, state in which the mostly all-trans chains are collectively tilted and pack in a regular two-dimensional array. Because the biologically relevant liquid crystalline phase is highly disordered, it is not amenable to atomic resolution structure determination. However, electron density and scattering length density profiles as a function of depth in the membrane have been mapped out by X-ray and neutron diffraction, and deuterium NMR has been used to characterize the average order in the hydrocarbon chains. The dynamics of lipids and water molecules in membranes on time scales presently accessible to simulations have been characterized by various experimental techniques, including neutron scattering and NMR.

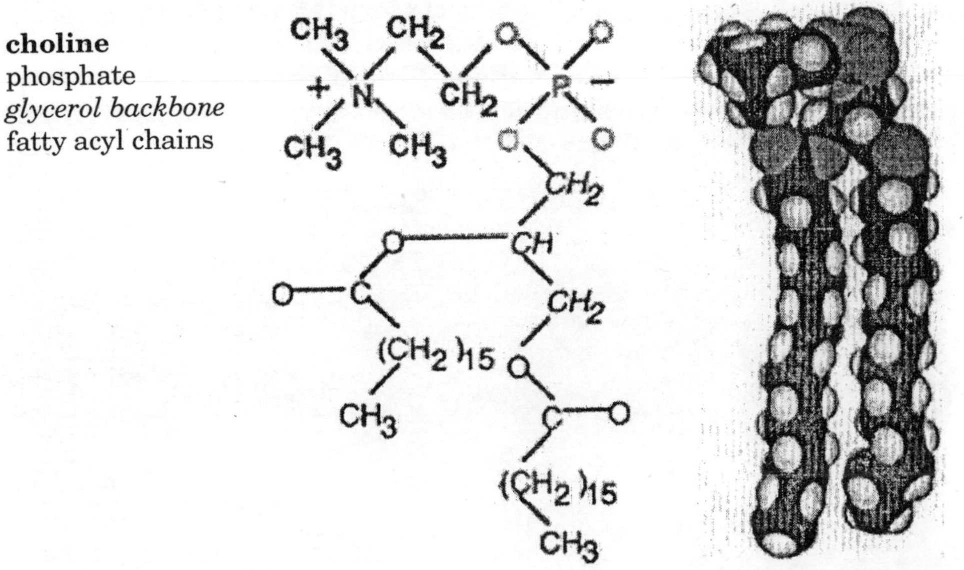

choline
phosphate
glycerol backbone
fatty acyl chains

Fig. 4.1. Chemical structure and space-filling representation of a phosphatidylcholine, DPPC. Different parts of the molecule are referred to by the labels at the left; together the choline and phosphate are referred to as the headgroup, which is zwitterionic. In the space-filling model, H atoms are white, O and P gray, and C black.

In recent years molecular dynamics (MD) simulations have emerged as a useful tool for filling in some of the details on the molecular structure and dynamics of membranes that are not available from experiments and for providing feedback on models used to interpret experimental data. Computer hardware and simulation methodology have matured to the point where stable simulations of membranes can now be routinely performed for at least a few nanoseconds, and many of the results of membrane simulations agree remarkably well with available experimental data. With the help of simulations, Singer and Nicolson's cartoon of a membrane drawn as a flat bilayer of lollypop-like lipids is evolving into images such as the one shown in Fig. 4.2.

This chapter describes some of the technical aspects of simulating membranes, presents results that illustrate the novel insight into membrane structure and dynamics that can be provided by simulation, and discusses the correspondence of the emerging atomic scale picture with the results of NMR and X-ray and neutron scattering experiments. We restrict our attention to pure lipid bilayers, which have largely been the focus of the field of membrane simulations to date (recent reviews can be found in with an emphasis on dynamics, which has been somewhat neglected compared to structure. However, we should mention that although the machinery for simulating membranes has been developed and tested on pure lipid bilayers, the number of applications to more complicated systems has been growing rapidly. A by no means exhaustive list includes the transport of small molecules (water, benzene, anesthetics and ions within and across bilayers, cholesterol effects on bilayers, and membrane-bound peptides and proteins (surface-bound tripeptides, transmembrane polyalanine, an amphipathic helical peptide, an amphipathic segment from the human corticotrophin releasing factor, the lytic peptide mellitin, the antimicrobial peptide dermaseptin B, bacteriophage Pf1 coat protein, the transmembrane domain of ErbB-2 tyrosine kinase receptor, and bacteriorhodopsin, including pores and ion channels (gramicidin, *E. coli* OmpFporin, alamethicin, and the influenza A M2 channel).

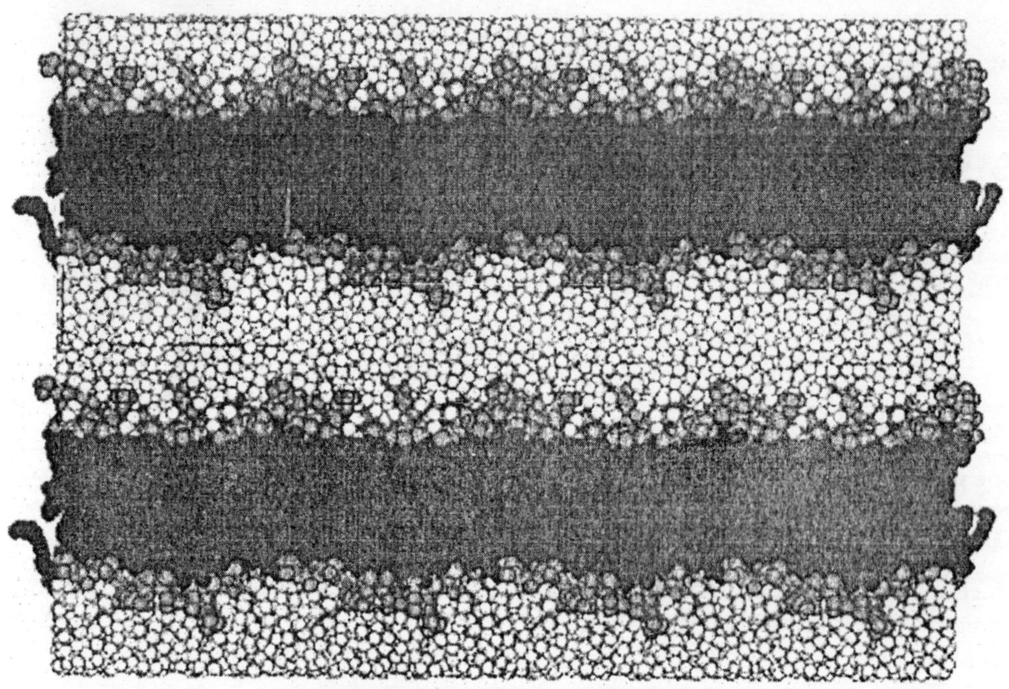

Fig. 4.2. Snapshot from an MD simulation of a multilamellar liquid crystalline phase DPPC bilayer. Water molecules are coloured white, lipid polar groups gray, and lipid hydrocarbon chains black. The central simulation cell containing 64 DPPC and 1792 water molecules, outlined in the upper left portion of the figure, is shown along with seven replicas generated by the periodic boundary conditions.

II. MOLECULAR DYNAMICS SIMULATIONS OF MEMBRANES

A. System Size and Construction

Algorithms and computer hardware have limited the size of most of the all-atom membrane simulations performed to date to 50-100 lipids plus water. However, with the increasing availability of parallel computers, larger systems containing up to 1000 lipids are starting to simulated, and systems containing more than 100 lipids will soon be routine. In the typical setup, the lipids are arranged in a bilayer, with water molecules on both sides, in a central simulation cell, or "box," which is then replicated by using three-dimensional periodic boundary conditions to produce an infinite multilamellar system (Fig. 4.2). It is important to note that the size of the central cell places an upper bound on the wavelength of fluctuations that can be supported by the system.

Because there are no crystal structures of fully hydrated lipid bilayers to start from, the initial configurations for the first membrane simulations had to be constructed "by hand." Many of the early simulations were based on the crystal structure of dimyristoylphosphatidylcholine dihydrate, in which the DMPC molecules form a bilayer with slightly tilted all-trans hydrocarbon chains. The unit cell was replicated to give the desired number of molecules, and the lipid center-of-mass positions were scaled to give the desired area per lipid. Then the system was hydrated and equilibrated to allow hydrocarbon chains to melt and the water molecules to penetrate through the headgroup regions. A less used approach that reduced the equilibration time was to pack together lipids randomly chosen from a library of configurations generated by simulating a single lipid in a mean field. Woolf and Roux extended this approach by using prehydrated lipids. Presently, most simulations are initiated from the end of a previous simulation, and lipid bilayer coordinates are generally provided upon request by authors of simulation papers. Incorporating solutes into bilayers can be tedious. The simplest approach, which consists of simply deleting enough lipids to create ample space, works well in some cases (*e.g.,* for inserting cholesterol molecules or transmembrane helices), but more sophisticated approaches are more generally useful. The latter include using radial forces to create spherical or cylindrical holes in the bilayer, and growing groups of soft spheres at locations that create a cavity in the membrane with the desired size and shape.

B. Force Fields

The force fields used in atomistic membrane simulations have the same form as the analytical, empirical molecular mechanics potentials used in classical simulations of proteins and nucleic acids. These potential functions contain harmonic terms for deforming bonds and bond angles, periodic and harmonic potentials for torsions, and van der Waals and Coulomb nonbonded interactions. The atomic charges are often obtained from quantum chemical calculations on lipid fragments. Most of the other potential parameters are taken from existing force fields for proteins and nucleic acids, which are generally built up from parametrizations of model compounds. Details on biomolecular force field parametrization are given in. The quality of lipid force fields may be evaluated with simulations of crystals of lipid fragments and by checking their ability to reproduce well-established experimental results in simulations of hydrated lipid bilayers. The level of accuracy that we have been able to achieve is typified by the densities of phospholipid and cholesterol crystals from constant-pressure MD simulations with fully flexible unit cells, plotted in Fig. 4.3. versus the experimental values.

Calculation of the energies and forces due to the long-range Coulomb interactions between charged atoms is a major problem in simulations of biological molecules. In an isolated system the number of these interactions is proportional to N^2, where N is the number of charged atoms, and the evaluation of the electrostatic interactions quickly becomes intractable as the system size is increased. Moreover, when periodic boundary conditions are used, the direct sum over all the Coulomb interactions in the periodic system is conditionally convergent.

There are two common solutions to these problems. The first, which was used in most biomolecular simulations until recently, is to employ a spherical truncation ("cutoff") scheme where the electrostatic interactions are smoothly switched off at a cutoff distance (typically around 10 Å). When used in conjunction with a neighber list, the computational work is proportional to *N*. An alternative approach, which is essentially exact for crystals, is to include all of the electrostatic interactions by using the Ewald method (or a related technique) for summing long-range interactions in a rapidly convergent fashion in an infinite periodic system. The disadvantages of the Ewald method are that it might enhance the artificial periodicity of a finite system replicated by periodic boundary conditions and that more computational effort is necessary compared to cutoffs (the straightforward implementation scales as N^{312}). The latter deterred most biomolecular simulators from using the Ewald method until recently, when the particle mesh Ewald method (PME) was introduced. In PME, the reciprocal space part of the Ewald sum is evaluated using fast Fourier transforms, resulting in an overall *N* log *N* scaling. The potential deleterious effects of the strict periodicity that results when lattice sums are used have not been assessed in membrane simulations. However, it has been explicitly demonstrated that the use of spherical truncation introduces serious artifacts into a variety of structural, thermodynamic, and transport properties of interfacial systems and lipid bilayers even when large cutoff radii are used.

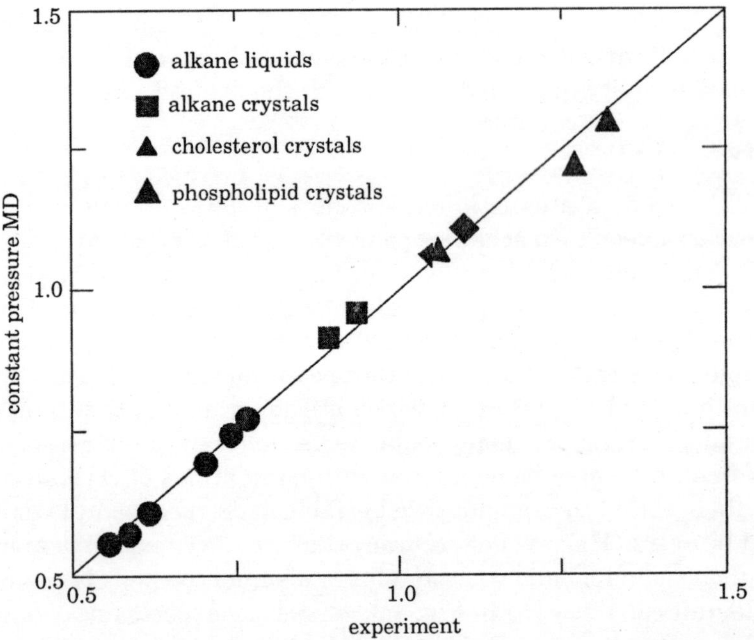

Fig. 4.3. Comparison of the densities (in g/cm³) of model compounds for membrane lipids computed from constant-pressure MD simulations with the corresponding experimental values. The model compounds include solid octane and tricosane, liquid butane, octane, tetradecane, and eico-sane, and the glycerylphosphorylcholine, cyclopentylphosphorylcholine monohydrate, dilauroly-glycerol, anhydrous cholesterol, cholesterol monohydrate, and cholesterol acetate crystals.

C. Ensembles

A molecular dynamics simulation consists of numerically solving the equations of motion of a set of particles (atoms), given the forces on the particles. Classical MD simulations that solve Newton's equations of motion generate trajectories belonging to the microcanonical [constant number of particles, volume, and energy (NVE)] statistical mechanical ensemble. It is generally desirable to perform simulations in other ensembles such as the isobaric-isothermal ensemble (constant NPT). In addition to being the natural choice for correspondence with typical experimental conditions, the NPT ensemble is useful for validating force fields by checking their ability to reproduce important structural parameters known from experimental measurements, such as the surface area per lipid, the interlamellar spacing, and the membrane thickness, and for predicting these quantities when they are not known (*e.g.*, in membrane-protein systems). Constant pressure and temperature are enforced in simulations by controlling the fluctuations of the particle kinetic energy and system volume, respectively, and there are various ways to do this. The best algorithms, in terms of their ability to rigorously generate the NPT ensemble, are based on the "extended system" approach in which additional dynamic variables are introduced, for example a time-dependent friction coefficient ("thermostat") to control the temperature and a piston to control the pressure. The equations of motion and a conserved energy are consistently formulated so that the microcanonical distribution function for the extended phase space gives the isobaric-isothermal distribution function for the particles after integration over the additional dynamic variables.

It is now widely accepted that it is best to simulate membranes at constant pressure, but there is some disagreement concerning the assumed form of the pressure tensor, *i.e.*, whether the pressure should be the same in all directions (isotropic) or whether the pressure in the plane of the membrane should be different from the pressure perpendicular to the membrane. The latter is appropriate for a true interfacial system, such as the air/ water and air/hydrocarbon interfaces, with a significant surface tension. Because membrane/water interfaces have either a vanishing or very small surface tension, in principle membranes should be simulated with an isotropic pressure tensor. However, some simulators have argued that to correct for finite size effects in simulations of small membrane patches, a modest surface tension should be imposed. As larger and larger bilayer patches are being simulated (≈1000 lipids), it is becoming evident that there are finite-size effects in systems containing ≈100 lipids. For example, in a recent bilayers simulated at constant isotropic pressure, the area per lipid in a system of 100 lipids was about 3% less than in a system of 1000 lipids. However, the explanation of the origin of this contraction and the best remedy for avoiding it in small systems (1000 lipids is still out of reach for most simulators) are debatable. On the one hand, one could argue that a surface tension should be applied to stretch a small membrane patch to counteract the shrinkage due to finite-size effects. However, an imposed surface tension can be expected to change the spectrum of fluctuations of the interface (in effect, stiffening the interface). On the other hand, one could impose constant isotropic pressure, keeping in mind the systematic error that results from finite-size effects. The results presented in this chapter were obtained by constant isotropic pressure calculations on bilayers containing 64 lipids, and hence the areas per lipid reported here are likely a couple of percent below the large system limit for the potential function employed.

D. Time Scales

Until recently, membrane simulations were generally limited to a couple of nanoseconds duration at most, and many of the simulations reported in the literature are less than a nanosecond long. As a result of this time scale limit, along with the size limit of roughly 100 lipids, many interesting phenomena occurring in membranes, such as lateral diffusion and complete rotation of lipids, collective undulations, bilayer phase transitions, and lateral phase separation in multicomponent membranes, to name a few, were well beyond the reach of the first generation of membrane simulations. Nonetheless, the early membrane simulations were useful for studying interactions between lipids, water, and membrane proteins at the atomic level, the complicated rearrangements of individual lipids, and the diffusion of water and other small molecules in and near membranes. In addition to permitting larger systems to be studied, parallel computing is presently extending the accessible time scale by an order of magnitude. Indeed, simulations of 10 ns duration are appearing, and these are enabling additional phenomena (*e.g.*, the early events in long-range lateral diffusion, undulations) to be characterized in unprecedented detail.

III. LIPID BILAYER STRUCTURE

A. Overall Bilayer Structure

We compare experimental results on DPPC bilayer dimensions and organization with our simulation results in Table 1 and the average locations of individual methyl and methylene groups along the bilayer normal, to demonstrate the level of accuracy that we have been able to achieve, which is considered quite high by present standards. Our simulations, which contain 64 lipids and numbers of water molecules corresponding to full hydration under the specified conditions.

B. Density Profiles

The snapshot from a fluid bilayer simulation shown in Fig. 4.2 reveals that the bilayer/water interface is quite rough and broad on the scale of the diameter of a water molecule.

Table 4.1. Comparison of MD and X-Ray Diffraction Results for Structural Parameters of Fully Hydrated DPPC Bilayers

	Gel phase (19°C, 12 water Molecules / lipid)		*Liquid crystal phase (50°C, 28 water molecules / lipid)*	
Quantity	*MD [51]*	*X-ray*	*MD*	*X-ray*
Area/lipid (Å^2)	45.8	47.2	61.8	62.9
Interlamellar spacing (Å)	65.2	63.4	67.3	67.2
Bilayer thickness (Å)	45.6	45.0	37.2	36.4
Chain tilt angle (0)	33.6	32.0		
Chain lattice parameters a, b (Å)	8.6, 5.5	8.5, 5.6		

The first atomic scale picture of the average structure of the interface was produced by measurement of the density distributions of different types of atoms along the bilayer normal by a combined neutron and X-ray diffraction study of a phosphatidylcholine bilayer at low hydration. The corresponding picture for fully hydrated bilayers has been provided by MD simulations and is exemplified by our results for the electron density profiles of the liquid crystalline phase of DPPC shown in Fig. 5. Defining the bilayer thickness as the distance between the peaks in the total electron density, we obtain 37.2 Å, which is in reasonable agreement with the values determined by X-ray diffraction analysis. In Fig. 4.4 we show the decomposition of the electron density profile into contributions from the lipid polar groups. The chemical heterogeneity of the membrane/water interface is clearly evident in the overlapping distributions of the lipid polar groups and water molecules in the broad interfacial region connecting the bulk water in the middle of the interlamellar space to the hydrocarbon in the middle of the bilayer. The total contributions from the water and the lipid show that the water density decays smoothly from the bulk value roughly 30 Å from the bilayer center and penetrates deeply into the bilayer. This is in contrast to the water density oscillations observed next to flat hydrophobic surfaces and the relatively narrow, noninterpenetrating air/water and hydrocarbon/water interfaces. Defining the interface as the range over which the water density goes from 90% to 10% of its bulk value, we find that the two interfaces occupy 23 Å, which is more than half of the total bilayer thickness. Thus, the membrane should be thought of as a broad hydrophilic interface, with only a thin slab of pure fluid hydrocarbon in the middle.

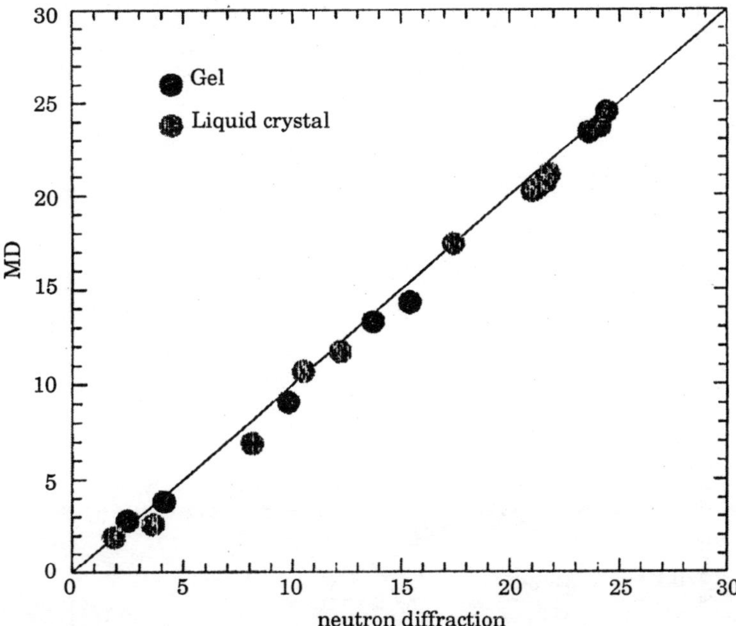

Fig. 4.4. Comparison of average distances from the bilayer center along the bilayer normal for deuterated methyl and methylene groups distributed throughout the DPPC molecule computed from constant-pressure MD calculations and neutron diffraction measurements on gel and liquid crystalline phase DPPC bilayers.

C. Solvation of the Lipid Polar Groups

To discuss the interactions between water molecules and specific lipid polar groups, we consider radial distribution functions for the water oxygen atoms surrounding the phosphate P, choline N, and carbonyl C atoms. The P—O *g(r)* has a sharp first peak at 3.8 Å, indicating tight solvation of the negatively charged phosphate by an average of four water molecules [obtained by integrating *g(r)* to the first minimum]. Inspection of the individual phosphate O-water O *g(r)* functions (not shown) reveals that the water molecules are almost exclusively associated with the unesterified phosphate oxygens. The N—O *g(r)* displays a broader first peak at 4.8 Å corresponding to 15 water molecules in the first solvation shell of the positively charged choline ammonium group. The relatively inaccessible carbonyl groups at the glycerol-fatty acyl ester linkages are solvated by an average of only 1.5 water molecules each.

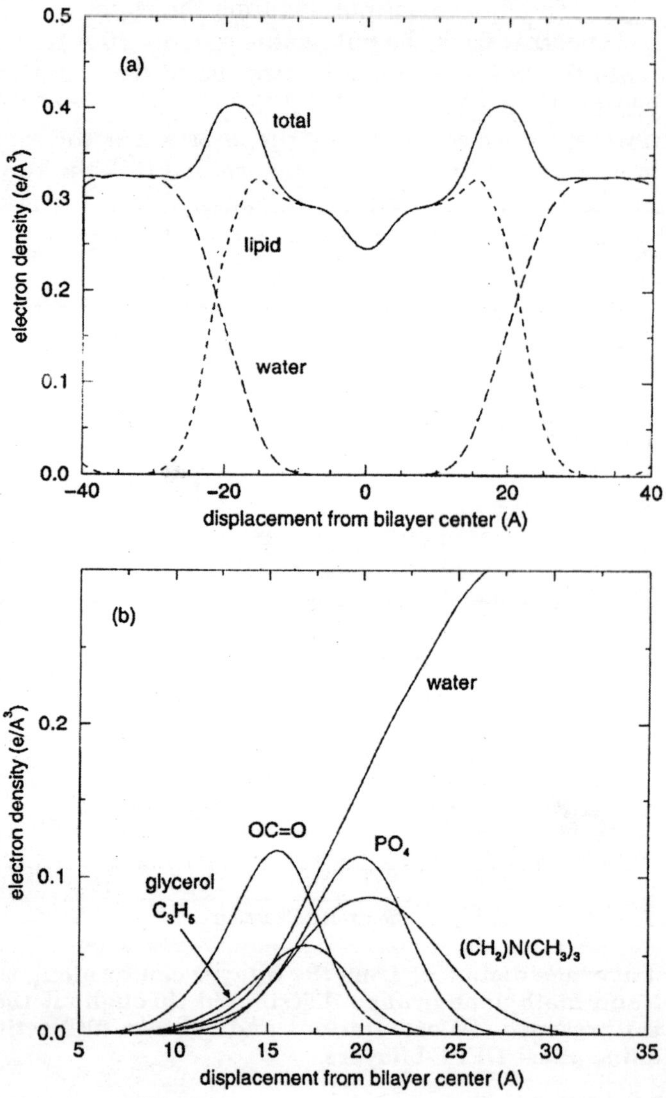

Fig. 4.5. Electron density distributions along the bilayer normal from an MD simulation of a fully hydrated liquid crystalline phase DPPC bilayer. (*a*) Total, lipid, and water contributions; (*b*) contributions of lipid components in the interfacial region.

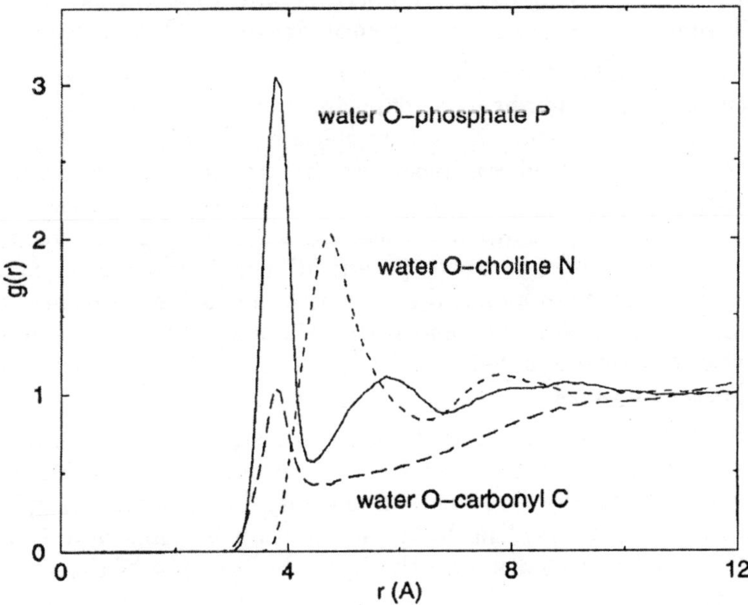

Fig. 4.6. Radial distributions of water oxygen atoms around sites in the polar groups in a DPPC bilayer.

The radial distributions of water around the lipid polar groups all contain well-defined first peaks signaling the existence of well-defined solvation shells. In subsequent analyses, we distinguish the water molecules that are closely associated with lipids by defining "bound" waters as those that fall within the first peaks of the water-phosphate, water-choline, or water-carbonyl *g(r)* functions. Thus, on average, roughly four waters are bound to the phosphate ("P-bound"), 15 to the choline ("N-bound"), and 1.5 to each carbonyl group ("CO-bound"). Some of these bound waters are shared between the polar groups, and overall roughly half of the 28 waters per lipid in the fully hydrated fluid phase bilayer are considered bound to lipids by our definition. We distinguish these bound waters from the "bulk" waters, a slab of water molecules roughly two to three water molecules thick in the middle of the interlamellar space, which we define as being more than 30 Å from the bilayer center. Of course, our definitions of bound and bulk water molecules are somewhat arbitrary, as are alternative definitions based on interaction energies, residence times, etc. However, we see in the next section that these definitions are useful for showing that the dynamics of water molecules near membrane surfaces depend on the strength of their interactions with specific lipid polar groups.

D. Water Orientational Polarization and the Membrane Dipole Potential

In membranes containing phospholipids such as DPPC, the negatively charged phosphate groups exert a strong influence on the structure of the water molecules. As the unesterified P—O bonds are preferentially oriented, on average pointing away from the membrane and into the water, the P-bound water molecules tend to orient their dipoles with their positive ends pointing toward the negative phosphates, resulting in a net orientational "polarization" [55]. The orienting power of the phosphate becomes clear when one observes that the orientational polarization is much less pronounced at water interfaces with lipid components that do not contain phosphate groups, *e.g.*, glycerol, decane, and decyl-β-glycoside.

These orientationally polarized water molecules appear to have a profound influence on the electrostatic properties of membranes. Experimental measurements on lipid monolayers and bilayers have demonstrated that there is an electric potential difference across lipid/water interfaces, typically a few hundred millielectronvolts, negative on the water side relative to the hydrocarbon. Thus, for the purpose of describing the electrostatics, the membrane can be thought of as a planar array of dipoles whose negative ends point toward the water. The molecular origins of the resulting "dipole potential" are of interest because the electrical properties of the bilayer surface influence the binding and passive transport of charged species. Experiments on phospholipid bilayers suggest that the primary negative contributions to the dipole potential arise from oriented water molecules and, to a lesser extent, from the carbonyl groups in the acyl ester linkages.

The dipole potential can be easily calculated from a simulation as a double integral of the average charge density. The total dipole potential profile from our DPPC bilayer simulation plotted in Fig. 4.7 monotonically decreases in the membrane/water interface to a value in the bulk water of about –500 mV. The lipid contribution is nonzero only in the interfacial region. The negative lipid contribution on the hydrocarbon side of the interface is canceled by a positive contribution on the water side, while the water contribution is monotonically decreasing throughout the interface. Thus, we conclude that the net dipole potential is due primarily to an excess of water molecules oriented with their dipoles pointing toward the membrane surface. Qualitatively similar results have been obtained from other simulations of DPPC bilayers, but different simulations do not agree on the extent of the lipid contribution.

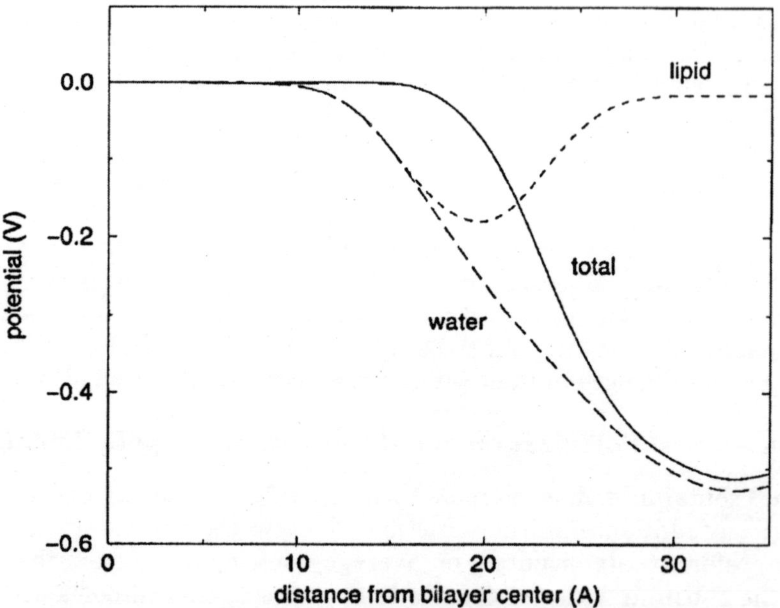

Fig. 4.7. The electric potential relative to the hydrocarbon ("dipole" potential) as a function of distance from the center of a fully hydrated DPPC bilayer.

IV. MOLECULAR DYNAMICS IN MEMBRANES

A. Overview of Dynamic Processes in Membranes

As in other biomolecular assemblies, functionally relevant motions in membranes span a wide range of length scales and time scales. Motions that have been detected experimentally include isomerizations in the hydrocarbon chains (10-100 ps), single lipid protrusions (10 ps to 1 ns), reorientation of the headgroups (1 ns), rotations of the lipid molecules about their long axes (10 ns), collective bilayer undulations (> 10 ns), and long-range lipid diffusion in the plane of the bilayer (10 ns to 1 s, depending on the length scale). Some of these are presently accessible to MD simulations, and more will become accessible as membrane simulations are extended to longer times. Detailed analyses of dynamics in membrane simulations have been relatively scarce compared to structural analyses. Pastor and Feller used short simulations and model calculations to analyze motions over a range of time scales, including chain, headgroup, and glycerol isomerizations, lateral diffusion, and whole lipid wobbling motion. More recently, Essmann and Berkowitz studied lipid center-of-mass diffusion and rotational dynamics using a 10 ns MD simulation, and Feller et al. used a 10 ns simulation to elucidate NOESY cross-relaxation observed in two-dimensional NMR experiments on lipid bilayers. In the remainder of this chapter, we analyze lipid and water motions on time scales up to 100 ps from our simulation of a liquid crystalline phase DPPC bilayer. Our discussion emphasizes a detailed comparison with experimental results, including the first close comparison with, and interpretation of, neutron scattering experiments that probe motions on the same time scale as the simulation.

B. Qualitative Picture on the 100 ps Time Scale

A qualitative picture of lipid and water dynamics in a membrane is given in Fig. 4.8, where we show configurations of water and lipid molecules at intervals spanning a period of 100 ps. On this time scale it is clear that the lipid centers of mass are not freely diffusing laterally, nor are they undergoing large excursions in the direction normal to the bilayer due to long-wavelength undulations; rather, they are "rattling in a cage" formed by their neighbours, with roughly equal amplitudes parallel and perpendicular to the bilayer plane. In addition to the isotropic center-of-mass motion, there is a considerable amount of motion in the internal degrees of freedom, especially in the acyl chains and, to a lesser degree, in the choline groups, involving the formation and disappearance of gauche bonds, gauche-trans-gauche kinks, and other chain defects. Although the details vary considerably among the individual water molecules, it is evident from Fig. 4.8 that, overall, the "bound" waters display less motion than the "bulk" waters and that the waters bound to the lipid phosphate and carbonyl groups are less mobile than those bound to the choline groups. Moreover, the trajectory of the bulk water molecule shown suggests that the motion of the bulk waters is anisotropic. In the remainder of this chapter, we expose these qualitative observations concerning the lipid and water dynamics on the 100 ps time scale in greater detail and discuss our results in relation to neutron scattering and NMR experiments.

C. Incoherent Neutron Scattering Measurements of Lipid Dynamics

Neutrons produced by present research-oriented sources typically have wavelengths on the order of angstroms and energies of a few millielectronvolts (1 me V ≈ 8 cm^{-1}) and hence

Fig. 4.8. Configurations of lipid and water molecules spanning a 100 ps interval during an MD simulation of a DPPC bilayer. The two left-hand panels show 10 configurations of two different lipids and three of their associated water molecules (one N-bound, one P-bound, and one CO-bound). The right-hand panel shows 20 configurations of a "bulk" water molecule in the interlamellar space of a bilayer stack.

are well suited for probing the structure and dynamics of molecules. Neutrons scatter from nuclei, and the total scattering is the sum of "coherent" and "incoherent" contributions. Coherent scattering depends on the relative positions and motions of the nuclei and thus conveys information on the structure and collective dynamics, whereas incoherent scattering reflects the motion of individual nuclei, *i.e.,* self-correlations or single-particle dynamics. The incoherent scattering cross section of hydrogen is much larger than all the other cross sections in organic molecules. Therefore, because hydrogen atoms are uniformly distributed in biological molecules, incoherent neutron scattering (INS) is particularly useful for studying the global molecular dynamics in biological systems. Moreover, selective deuteration provides a powerful mask for isolating the dynamics of selected parts of the system (*e.g.,* to separate water and biomolecule contributions). An energy-resolved spectrum typically has three regions: an elastic peak arising from atoms that move slowly compared to the resolution of the experiment, inelastic peaks due to scattering by normal modes, and quasielastic scattering, which appears as a broadening of the elastic peak and is due to diffusive motions. The range of energy and momentum transfers accessible on presently available neutron spectrometers overlaps well with the duration and size of MD simulations that are presently routine for biological molecules. Thus, INS data are a valuable resource for testing the dynamics produced by simulations, and simulations constitute a potentially valuable tool for interpreting INS data.

Most neutron spectroscopic experiments essentially measure the total dynamic structure factor, $S_{tot}^{meas}(\mathbf{W}, \omega)$, in which $\mathbf{Q}$ and ω are the momentum and energy transfers, respectively. The measured structure factor is the sum of coherent and incoherent contributions. However, because the incoherent scattering length of hydrogen is an order of magnitude larger than the scattering lengths of all the other atoms in lipids and proteins, we will assume that the coherent contribution is negligible, so that $S_{tot}^{meas}(Q, \omega) = S_{inc}^{meas}(\mathbf{Q}, \omega)$. In practice, the unavoidable 'spread in energies of the neutrons incident on the sample results in a finite energy resolution, and the measured spectrum is a convolution of the true spectrum, $5_{inc}(\mathbf{Q}, \omega)$, and the instrumental resolution function, $R(\omega)$:

$$S_{inc}^{meas}(\mathbf{Q}, \omega) = S_{inc}(\mathbf{Q}, \omega) \otimes R(\omega) \quad ...(4.1)$$

where ⊗ denotes a convolution product. The width of the resolution function determines the time scale of the dynamics probed by the instrument in a nontrivial way, with narrower widths (higher resolution) corresponding to longer observation times.

From the theory of neutron scattering [62], $S_{inc}(Q, \omega)$ may be written as the Fourier transform of a time correlation function, the "intermediate scattering function," $I_{inc}(\mathbf{Q}, t)$:

$$S_{inc}(\mathbf{Q}, \omega) = \frac{1}{2\pi}\int_{-\infty}^{\infty} I_{inc}(\mathbf{Q}, t)e^{-i\omega t}dt \quad ...(4.2)$$

$$I_{inc}(\mathbf{Q}, t) = \frac{1}{N}\sum_{j=1}^{N}\left\langle e^{iQ \bullet r_j(t)}e^{-iQ \bullet rj(0)}\right\rangle \quad ...(4.3)$$

Here r_j is the position operator of atom *j*, or, if the correlation function is calculated classically as in an MD simulation, r_j is a position vector; N is the number of scatterers (*i.e.*, H atoms); and the angular brackets denote an ensemble average. Note that in Eq. (4.3) we left out a factor equal to the square of the scattering length. This is convenient in the case of a single dominant scatterer because it gives $I(\mathbf{Q}, 0) = 1$ and $S_{inc}(\mathbf{Q}, \omega)$ normalized to unity.

The intermediate scattering function, $I_{inc}(\mathbf{Q}, t)$, is readily computed from an MD trajectory by using Eq. (4.3), and the result may be numerically Fourier transformed to give $S_{inc}(\mathbf{Q}, \omega)$. For the purpose of quantitatively comparing the result to neutron scattering data obtained on a given spectrometer, the instrumental resolution should be taken into account. Instrumental resolution functions are generally represented by a Gaussian or other peaked function centered at $\omega = 0$, with width $\Delta\omega$ (or $\Delta E = \Delta\omega$). Noting that a convolution in the frequency domain is equivalent to a product in the time domain, a resolution-broadened spectrum, $S_{inc}^{meas}(\mathbf{Q}, \omega)$, is obtained by Fourier transforming the product, $I_{inc}(\mathbf{Q}, t)\, R(t)$, where $R(t)$ is the Fourier transform of $R(\omega)$. It is instructive to look at $R(t)$ itself because it provides a direct measure of the longest time scale probed by a measurement at a given resolution. For example, the time-of-flight spectrometer IN5 at the Institut Laue Langevin in Grenoble has a resolution function that is well represented by a Gaussian with a full width at half-maximum (FWHM) ΔE value of 0.050 meV. Thus, $R(t)$ is also a Gaussian, with an FWHM of approximately 30 ps, and the product $I_{inc}(\mathbf{Q}, t)\, R(t)$ does not become negligible until $t \approx 100$ ps. Because $R(t)$ is generally a peaked function, there is no well-defined averaging time corresponding to a particular energy resolution, and motions with short correlation times will be weighted more heavily that those with long correlation times in the product $I_{inc}(\mathbf{Q}, t)\, R(t)$.

The **Q** and ω dependence of neutron scattering structure factors contains information on the geometry, amplitudes, and time scales of all the motions in which the scatterers participate that are resolved by the instrument. Motions that are slow relative to the time scale of the measurement give rise to a δ-function "elastic" peak at $\omega = 0$, whereas diffusive motions lead to "quasielastic" broadening of the central peak and vibrational motions attenuate the intensity of the spectrum. It is useful to express the structure factors in a form that permits the contributions from vibrational and diffusive motions to be iso-lated. Assuming that vibrational and diffusive motions are decoupled, we can write the measured structure factor as

$$S_{\text{nic}}^{\text{diff}}(Q, \omega) = \exp[-Q^2 \langle u^2 \rangle]\, S_{\text{nic}}^{\text{diff}}(Q, \omega) \otimes R(\omega) + B \quad \text{...(4.4)}$$

where the Debye-Waller exponential factor represents the attenuation of the intensity due to vibrational motions with mean-square amplitudes $\langle u^2 \rangle$, $S_{\text{nic}}^{\text{diff}}(Q, \omega)$ is the structure factor corresponding to the diffusive motions, and B is a constant background that may arise from coherent scattering and other, extraneous, scattering particular to the experimental setup. In Eq. (4.4) and the remainder of this chapter, we write the structure factor as a function of the magnitude of the momentum transfer Q, because we are interested in experiments in which the spectra are "powder averaged" (as opposed to single crystals).

Specification of $S_{\text{nic}}^{\text{diff}}(Q, \omega)$ requires models for the diffusive motions. Neutron scattering experiments on lipid bilayers and other disordered, condensed phase systems are often interpreted in terms of diffusive motions that give rise to an elastic line with a Q-dependent amplitude and a series of Lorentzian quasielastic lines with Q-dependent amplitudes and widths, *i.e.*,

$$(Q, \omega) = A_0(Q)\, \delta(\omega) + \sum_{i=1}^{n} A_i(Q)\, L_i(\Gamma_i(Q), \omega) \quad \text{...(4.5)}$$

where $L_i(\Gamma_i(Q), \omega)$ is a Lorentzian centered at $\omega = 0$ with half- width at half–maximum $\Gamma_i(Q)$:

$$L_i(\Gamma_i(Q), \omega) = \frac{1}{\pi}\left(\frac{\Gamma_i(Q)}{\Gamma_i(Q)^2 + \omega^2}\right) \quad \text{...(4.6)}$$

The amplitude of the elastic scattering, $A_0(Q)$, is called the elastic incoherent structure factor (EISF) and is determined experimentally as the ratio of the elastic intensity to the total integrated intensity. The EISF provides information on the geometry of the motions, and the linewidths are related to the time scales (broader lines correspond to shorter times). The Q and ω dependences of these spectral parameters are commonly fitted to dynamic models for which analytical expressions for (Q, ω) have been derived, affording diffusion constants, jump lengths, residence times, and so on that characterize the motion described by the models.

D. Comparison of MD and Neutron Scattering Results on Lipid Dynamics

Clearly, the best way to assess the ability of MD simulations to reproduce neutron scattering results is to compare measured and computed spectra directly, one on top of the other.

However, this is generally not possible because measured spectra are not available to simulators. The spectral parameters (EISFs and line widths) along with the parameters of the dynamic models used to fit their Q and ω dependence are what are generally reported in the literature. Quantities such as diffusion constants are often computed from MD simulations and their values compared to those extracted from neutron data. For the comparison to be meaningful, the same models must be used to analyze the simulations and the experiments.

However, as we discuss in more detail later, this is often not the case. It is actually easier, and certainly more appropriate, to directly compare the spectral parameters when they are available. Then, once the ability of the simulations to reproduce the data has been assessed, the simulations can be used to discuss the validity of the dynamic models used to fit the spectra and/or to inspire new models for interpreting experimental data. In the remainder of this section we show how the spectral parameters can be derived from simulations, and we compare our simulations to the results of neutron scattering experiments published by Konig et al. [63,64] on DPPC bilayers. In subsequent sections we discuss models used to describe lipid dynamics in membranes on time scales up to 100 ps.

For the comparison to be meaningful, it is essential to compute spectra from the MD trajectory that are broadened by the resolution of the experiment to which they will be compared and to process them by the same procedure as that used in the experimental data reduction. The processing consists of determining $\langle u^2 \rangle$, dividing the resolution-broadened spectra by the Debye-Waller factor, and fitting the result to an elastic line plus a sum of Lorentzians. Following Konig et al., we use two Lorentzians in our fits. The quality of the fits is illustrated in Fig. 4.9, where we show a typical spectrum computed for the lipid H atoms in our simulation of a fluid-phase DPPC bilayer. The value of $\langle u^2 \rangle$ is determined, as in the experimental analysis, as the slope of a plot of integrated intensity versus Q^2, according to the relationship.

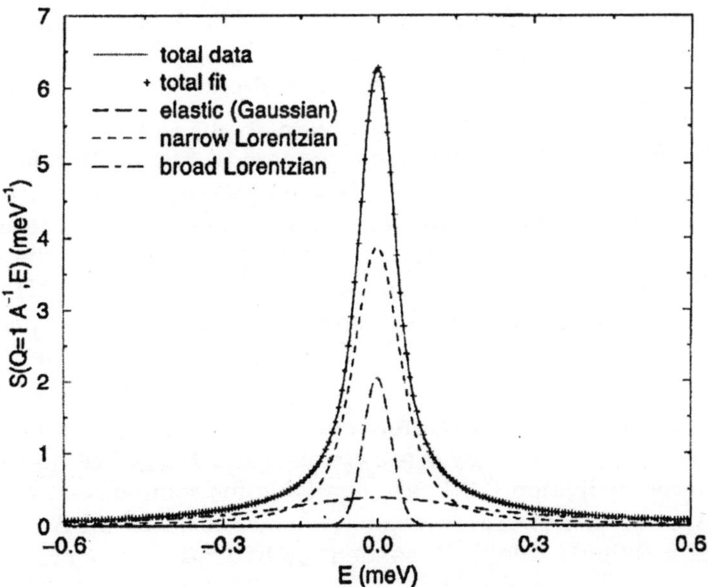

Fig. 4.9. Fit of an incoherent neutron scattering structure factor, $S(Q, \omega)$, computed for lipid H atom motion in the plane of the bilayer in a simulation of a DPPC bilayer, by the sum of an elastic line, a narrow Lorentzian with width Γ_1, and a broad Lorentzian with width G_1, convoluted with a Gaussian resolution function with $\Delta E = 0.050$ meV.

$$\int_{\omega_{min}}^{\omega_{max}} (Q, \omega)\, d\omega = \exp[-Q^2 \langle u^2 \rangle] \qquad \text{...(4.7)}$$

For lipid H atom motions in the plane of the bilayer, using an integration range -1.5 meV

$< E < 1.5$ meV, we obtain $\langle u^2 \rangle = 0.24$ Å^2, which is about twice as large as the 0.11 Å^2 reported by Konig.

In the remainder of this section, we compare EISFs and Lorentzian line widths from our simulation of a fully hydrated liquid crystalline phase DPPC bilayer at 50°C with experiments by Konig et al. on oriented bilayers that, in order to achieve high degrees of orientation, were not fully hydrated. We consider two sets of measurements at 60°C on the INS time-of-flight spectrometer at the ILL: one in which the bilayer preparations contained 23% (w/w) pure D_2O and another in which bilayer orientation was preserved at 30% D_2O by adding NaCl. The measurements were made on samples with two different orientations with respect to the incident neutron beam to probe motions either in the plane of the bilayers or perpendicular to that plane.

The EISFs for motions in the plane and perpendicular to the plane of the bilayer are compared in Fig. 4.10a and 4.10b, respectively. The EISFs from the simulations at 50°C and 43% hydration agree well with those measured at 60°C and 30% hydration. The experimental EISFs at the lower hydration decay more slowly with Q, indicating more restricted motion. This is reasonable, as we expect the surface area per lipid, and hence the range of lipid mobility, to decrease with dehydration. Carefully comparing Fig. 4.10a and 4.10b, we can see that the experimental EISFs at a given hydration level are very similar for motions in and perpendicular to the plane of the bilayer, indicating that the motion is isotropic. In contrast, the MD EISF for in-plane motion decays slightly faster with Q than that for motion perpendicular to the bilayer, suggesting slightly more mobility in the plane of the bilayer. Unfortunately, we cannot tell at this time if this is a minor fault of the simulation or if it is a correct prediction that there is a slight anisotropy to the motion at full hydration.

The widths of the narrow Lorentzians representing slow motions in the plane and perpendicular to the plane of the bilayer are compared in Fig. 4.11a and 4.11b, respectively. For the in-plane motion, the MD values for $Q = 0.5$ Å^{-1} agree well with the experimental results, but the increase with Q is significantly overestimated in the simulation compared to the experimental values. This suggests that the slower component of the in-plane motion in the simulation is too fast at short distances. On the other hand, the MD line widths for the slower component of the out-of-plane motion agree well with the experimental results at 30% hydration. As in the case of the EISF, the simulation predicts a slight anisotropy not seen in the experimental data.

The widths of the broad Lorentzians representing fast motions in the plane and perpendicular to the plane of the bilayer are compared in Fig. 4.12a and 4.12b, respectively. Only data at the lower hydration (23%) are available for comparison, and these agree well with the MD results, which show a slow, monotonic increase with Q. Although we expect the fast process to be at most only weakly dependent on hydration, it is not clear to what extent the comparison validates the simulation.

Overall the MD results at 50°C and 43% hydration agree well with the neutron results at 60°C and 30% hydration. Although it is desirable to compare simulation and experimental results under identical conditions, this is the most direct comparison we can make given the available data. We conclude from this comparison that the lipid dynamics reflected in H atom motions on time scales up to 100 ps are reasonably well reproduced by the simulation. The most significant discrepancy is that the time scale of the slower diffusive process decreases too quickly with Q for lipid motions in the plane of the bilayer.

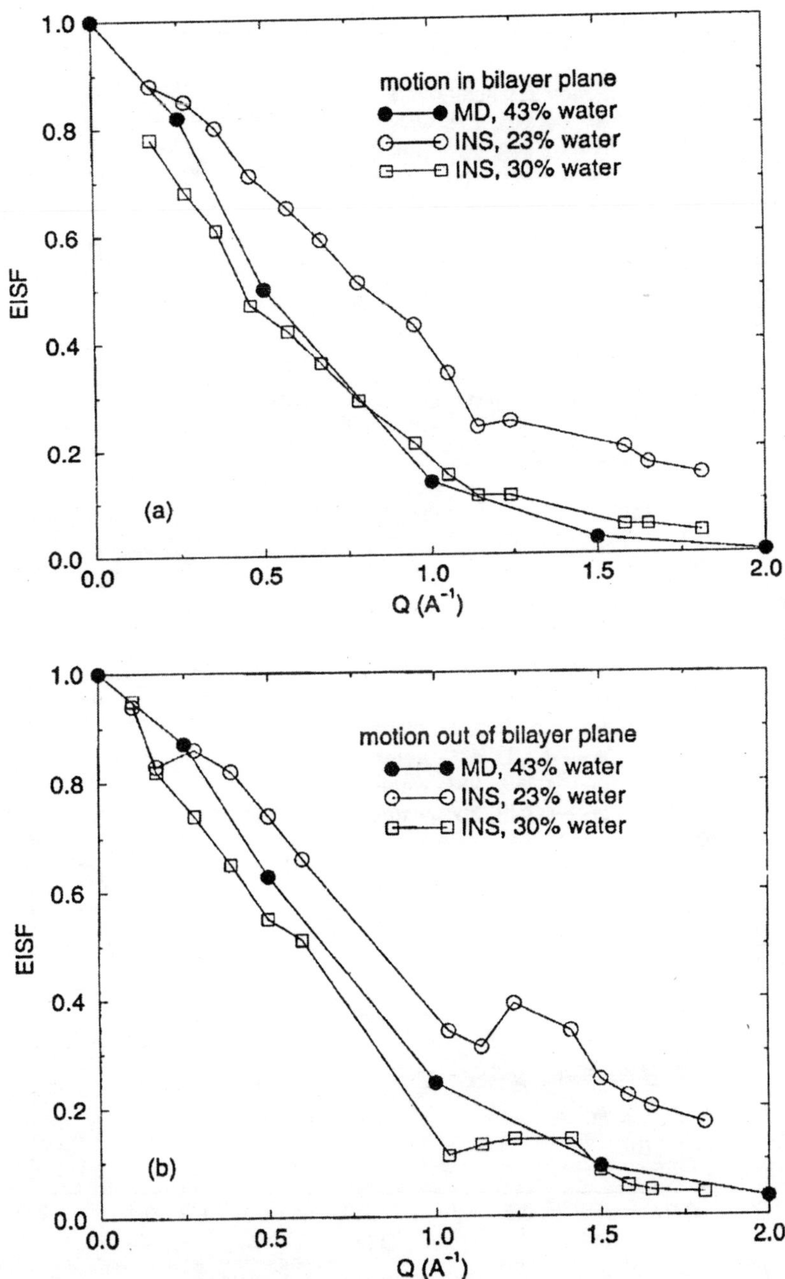

Fig. 4.10. Elastic incoherent structure factors for lipid H atoms obtained from an MD simulation of a fully hydrated DPPC bilayer, and quasielastic neutron scattering experiments on DPPC bilayers at two hydration levels for (a) motion in the plane of the bilayer and (b) motion in the direction of the bilayer normal.

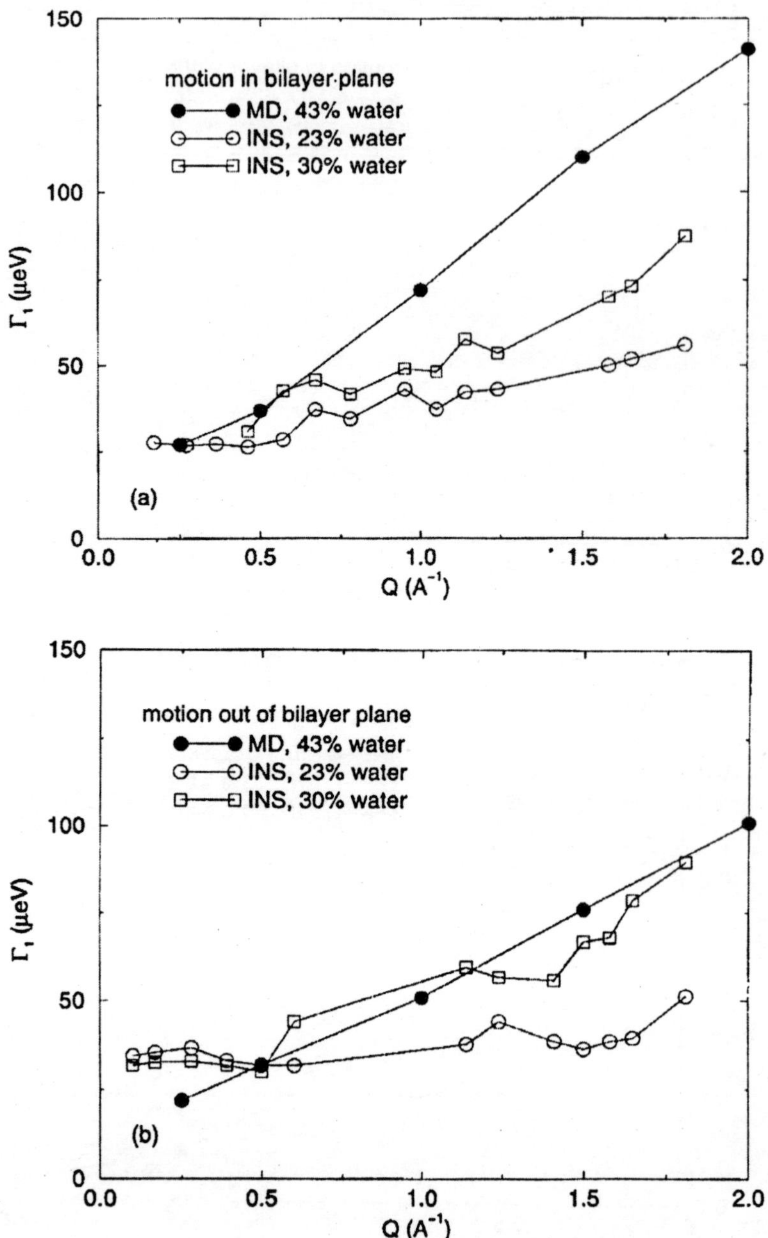

Fig. 4.11. Widths of the narrow Lorentzian components fit to structure factors for lipid H atoms obtained from an MD simulation of a fully hydrated DPPC bilayer and quasielastic neutron scattering experiments on DPPC bilayers at two hydration levels for *(a)* motion in the plane of the bilayer and *(b)* motion in the direction of the bilayer normal. The error bars on the experimental points are approximately ±5 μV.

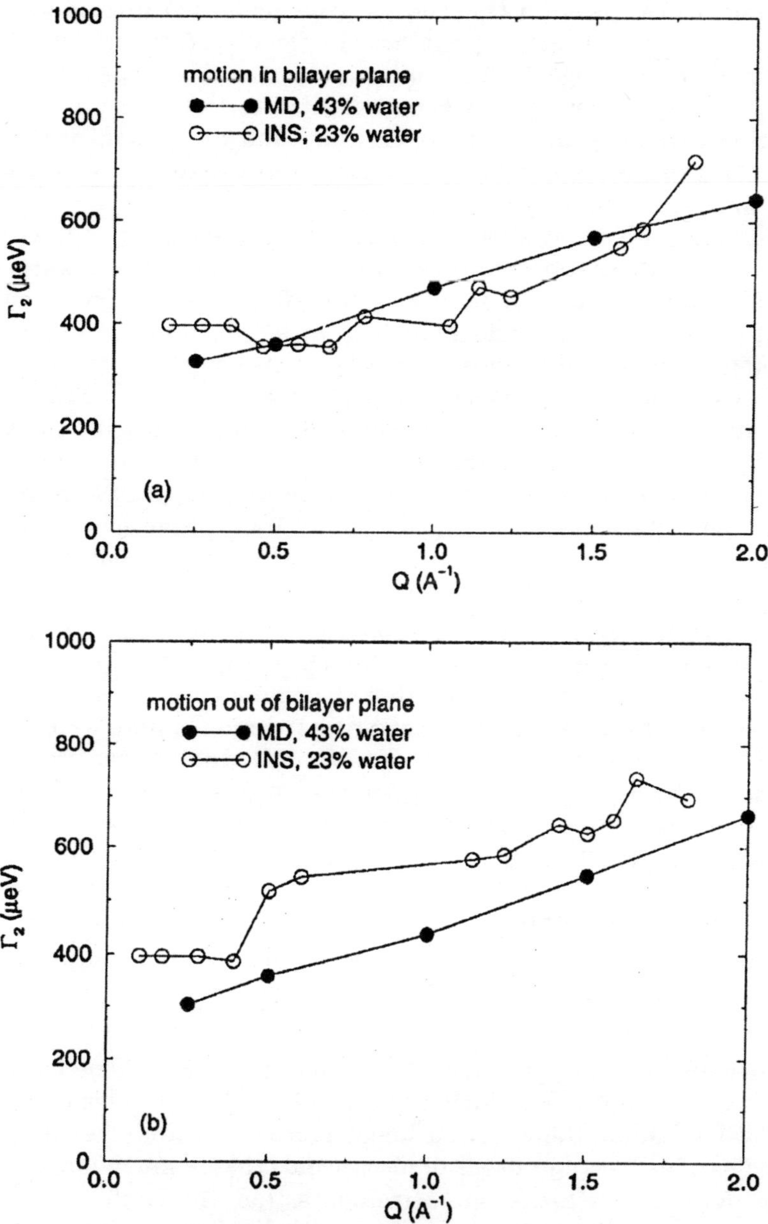

Fig. 4.12. Widths of the broad Lorentzian components fit to structure factors for lipid H atoms obtained from an MD simulation of a fully hydrated DPPC bilayer and quasielastic neutron scattering experiments on DPPC bilayers containing 23% water for (*a*) motion in the plane of the bilayer and (*b*) motion in the direction of the bilayer normal. The error bars on the experimental points are approximately ±150 μV.

In the stroboscopic picture in Fig. 4.8, the lipid dynamics appears to consist of two predominant dynamical processes on the 100 ps time scale: (1) intramolecular motions (librations, conformational transitions) superimposed upon (2) a rattling motion of the whole molecules in a confined space ("cage"). The center-of-mass motion appears to be essentially isotropic, with similar amplitudes in and perpendicular to the plane of the bilayer. The intramolecular motion primarily involves the creation and disappearance of a variety of conformational defects (gauche conformers, kinks, etc.) in the hydrocarbon chains. It is clear that on the 100 ps time scale there is no long-range diffusion and no significant rotational motion of the lipids around their long axes. This qualitative description of the lipid dynamics had been proposed, before simulation pictures such as those in Fig. 4.8 were available, by attempting to fit four dynamic models to neutron time-of-flight data : (I) rotation of the whole molecule plus out-of-plane diffusion inside a box (*i.e.,* with a fixed maximum amplitude); (II) diffusion of each proton inside a sphere (with different protons having different sphere volumes) superimposed on diffusion of the center of mass inside a cylinder; (III) diffusion of kinks in the hydrocarbon chains superimposed on diffusion of the whole molecule inside a cylinder; (IV) same as (III), except that rather than diffusing along the chains, the kinks stochastically appear and disappear. Model I was ruled out, but models II, III, and IV were all consistent with the data. Thus, the picture of lipid dynamics on the 100 ps time scale that has emerged from the neutron time-of-flight measurements is somewhat ambiguous, in the sense that physically distinct dynamic models fit the data equally well.

Having demonstrated that our simulation reproduces the neutron data reasonably well, we may critically evaluate the models used to interpret the data. For the models to be analytically tractable, it is generally assumed that the center-of-mass and internal motions are decoupled so that the total intermediate scattering function can be written as a product of the expression for the center-of-mass motion and that for the internal motions. We have confirmed the validity of the decoupling assumption over a wide range of Q (data not shown). In the next two sections we take a closer look at our simulation to see to what extent the dynamics is consistent with models used to describe the dynamics.

E. Lipid Center-of-Mass "Diffusion"

Center-of-mass translational motion in MD simulations is often quantified in terms of diffusion constants, D, computed from the Einstein relation

$$\langle r^2(t)\rangle = 2dDt \quad \text{...(4.8)}$$

where $\langle r^2(t)\rangle$ is the mean-square displacement (MSD) in d dimensions. The application of this relation is certainly valid for liquids in which, at long times, the molecules undergo Brownian dynamics and the MSDs display the required linear increase in time, but it is questionable for lipids on time scales of at least hundreds of picoseconds (likely much longer), where it has been deduced from neutron scattering measurements that the center-of-mass motion is confined. The anomalous (non-Brownian) diffusion of the lipid centers of mass is evident in the MSDs plotted in Fig. 4.13. At first glance it appears that the MSDs shown in Figs. 4.13a and 13b approach a linear time dependence on the 10 ps and 100 ps time scales, respectively. However, upon close inspection it is evident that the slopes of the MSDs are changing in time. In fact, it is clear from Fig. 4.13c that the MSD shows a power law, or fractal, time dependence :

$$\langle r^2(t)\rangle \sim t^{\alpha} \quad \text{...(4.9)}$$

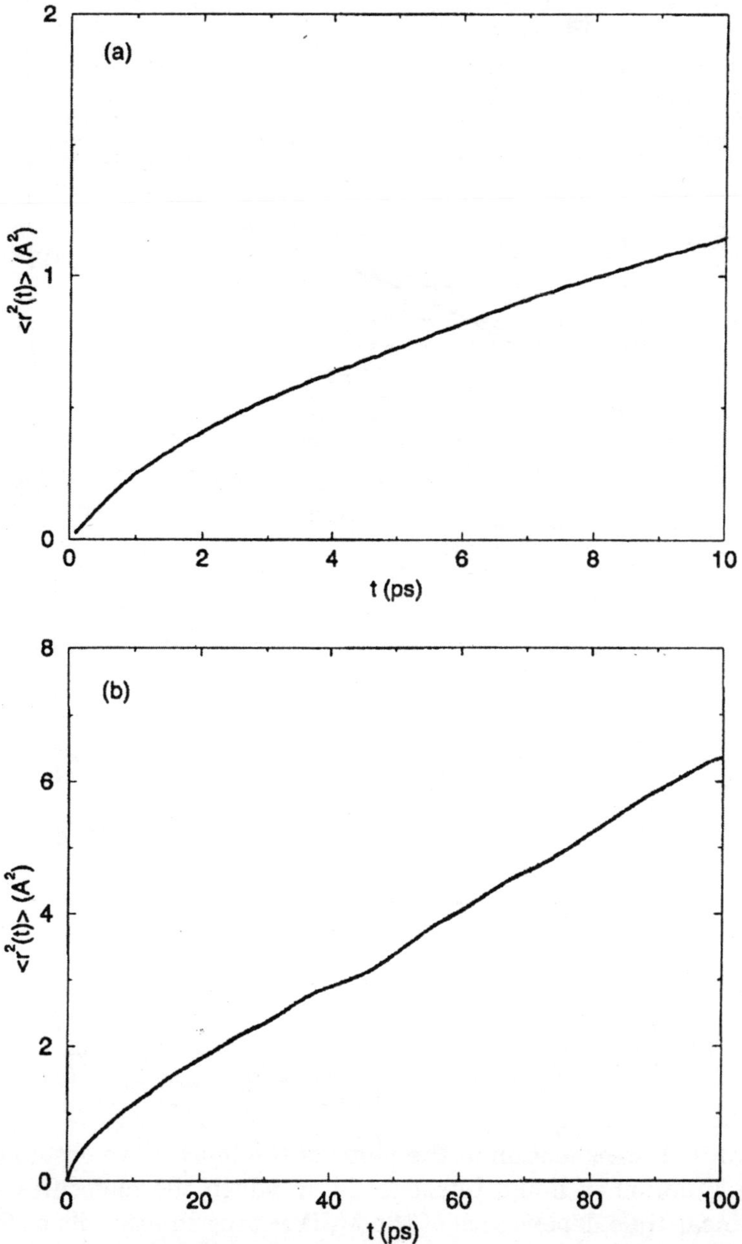

Fig. 4.13. Center-of-mass mean-square displacements computed from MD simulations at 323 K. (a) DPPC motion in the plane of a lipid bilayer averaged over 10 ps; (b) DPPC motion in the plane of a lipid bilayer averaged over 100 ps; (c) comparison of the DPPC in-plane mean-square displacement to linear and power law functions of time; (d) comparison of the center-of-mass mean-square displacement from an MD simulation of liquid tetradecane to a linear function of time.

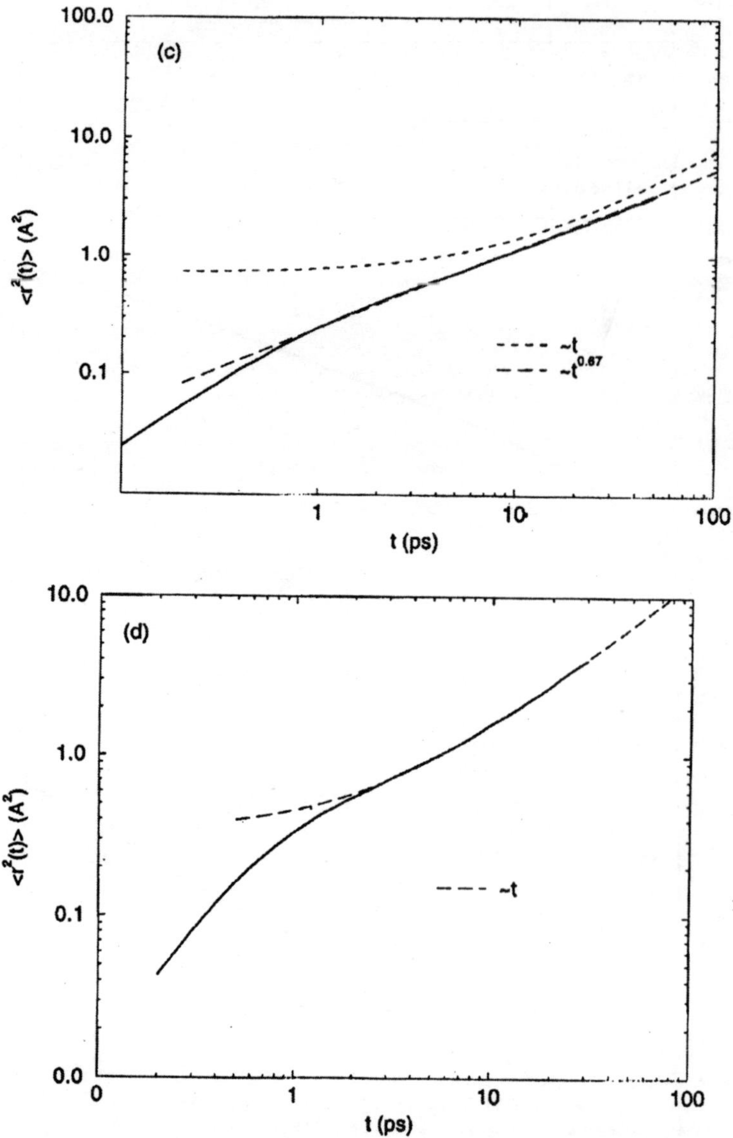

where for lipid center-of-mass motion in the plane of the bilayer, we obtain $\alpha = 0.67$. This is in contrast to the situation in liquid tetradecane, in which the molecules undergo normal diffusion, with a linear time dependence of the MSD setting in after about 3 ps (Fig. 4.13d). Application of the Einstein relation to the DPPC MSD in Fig. 4.13c would give a diffusion constant that depends on the length of the MSD used to calculate it and hence is ill-defined. The anomalous lipid diffusion is reminiscent of the cage effect that is universal in supercooled liquids and glasses. In light of the observation that lipids begin to change place only after several nanoseconds in fluid phase bilayers, it is likely that Brownian lateral motion of lipids will not be observed until well beyond 10 ns.

Analysis of neutron data in terms of models that include lipid center-of-mass diffusion in a cylinder has led to estimates of the amplitudes of the lateral and out-of-plane motion and their corresponding diffusion constants. It is important to keep in mind that these diffusion constants are not derived from a Brownian dynamics model and are therefore not comparable to diffusion constants computed from simulations via the Einstein relation. Our comparison in the previous section of the Lorentzian line widths from simulation and neutron data has provided a direct, model-independent assessment of the integrity of the time scales of the dynamic processes predicted by the simulation. We estimate the amplitudes within the cylindrical diffusion model, *i.e.,* the length (twice the out-of-plane amplitude) L and the radius (in-plane amplitude) R of the cylinder, respectively, as follows:

$$\mathrm{L} = 2\langle \Delta z^2 \rangle \qquad ...(4.10)$$

$$\mathrm{R} = (\langle \Delta x^2 \rangle + \langle \Delta y^2 \rangle)^{1/2} \qquad ...(4.11)$$

where $\langle \Delta x^2 \rangle$, $\langle \Delta y^2 \rangle$, and $\langle \Delta z^2 \rangle$ are the mean-square fluctuations in the x, y, and z directions, respectively. Averaging over 100 ps, we find $L = 1.3$ Å and $R = 0.96$ Å. These may be compared to the values $L = 2.3$ Å and $R = 1.1$ Å obtained in the model III/IV fits to neutron time-of-flight data taken on IN5 by Konig et al. on a DPPC bilayer at 60°C and 23% water. The cylinder radius derived from the simulation is slightly smaller than that derived from the neutron data, and the cylinder length is substantially shorter. This in contrast to our expectation that the simulation values should, if anything, be larger because the simulation was carried out at full hydration. Moreover, the simulation predicts an appreciable anisotropy that it absent from the experimental data. There are two possible explanations for the discrepancies. The first is that simulation models the lipid center-of-mass motion poorly. The second, which is probably more correct, is that the combined diffusion in a cylinder-and-chain defect model used to determine the L and R values quoted here attributes too little motion to the chains and too much to the center of mass. We examine models for the motion of the lipid internal degrees of freedom in the next section.

F. Hydrocarbon Chain Dynamics

Two physically reasonable but quite different models have been used to describe the internal motions of lipid molecules observed by neutron scattering. In the first the protons are assumed to undergo diffusion in a sphere. The radius of the sphere is allowed to be different for different protons. Although the results do not seem to be sensitive to the details of the variation in the sphere radii, it is necessary to have a range of sphere volumes, with the largest volume for methylene groups near the ends of the hydrocarbon chains in the middle of the bilayer and the smallest for the methylenes at the tops of the chains, closest to the bilayer surface. This is consistent with the behaviour of the carbon-deuterium order parameters, S_{CD}, measured by deuterium NMR :

$$S_{CD} = \frac{1}{2}\left\langle 3\cos^2\theta - 1 \right\rangle \qquad ...(4.12)$$

where θ is the angle between the C—D bond vector and the bilayer normal. We show the negative of the usual order parameters, which may be thought of as "disorder parameters," averaged over both chains in all the molecules in our bilayer simulation, in Fig. 4.14a.

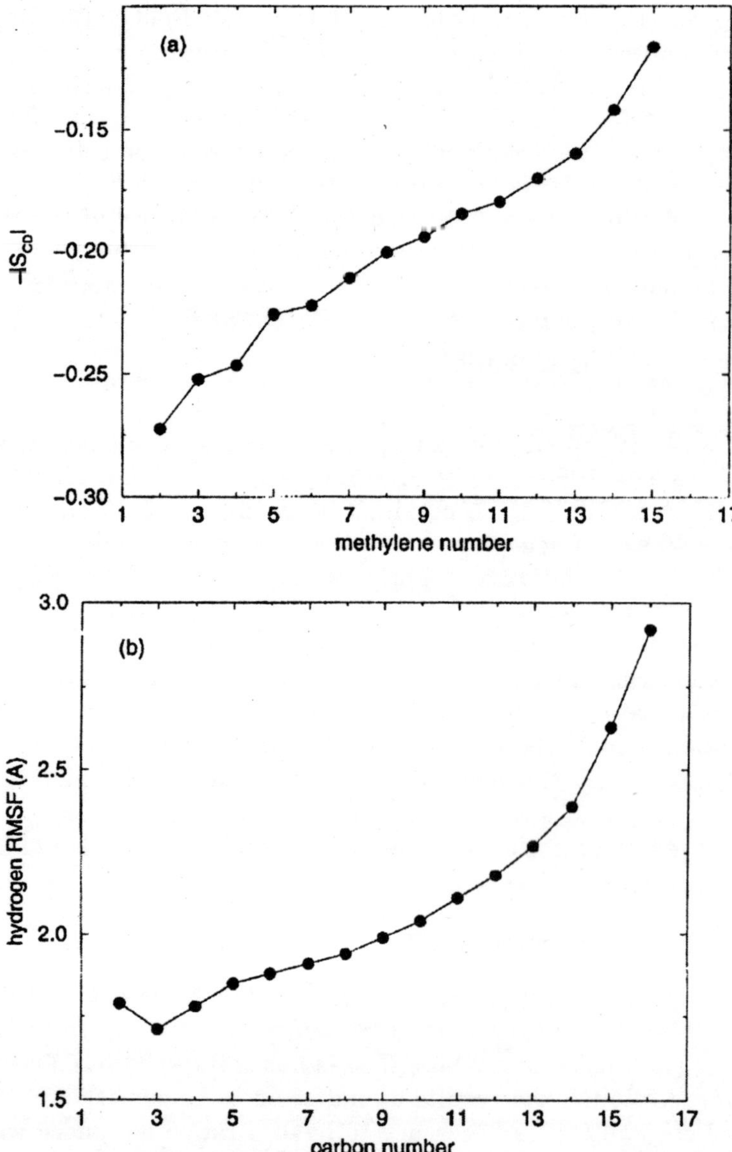

Fig. 4.14. Measures of disorder in the acyl chains from an MD simulation of a fluid phase DPPC bilayer, (*a*) Order parameter profile of the C—H bonds; (*b*) root-mean-square fluctuation of the H atoms averaged over 100 ps.

Although our values are not in quantitative agreement with the experimental values, they faithfully reproduce the well-known observation that the disorder increases from the tops of the chains towards the middle of the bilayer, with a more pronounced increase in the last few carbons of the chain. The S_{CD} represent different types of disorder, including chain conformational defects and whole molecule "wobbling".

To make contact with the diffusion-in-a-sphere model, we have defined the spherical radius as the root-mean-square fluctuation of the protons averaged over 100 ps. The variation of the spherical radius as a function of position in the chains is plotted in Fig. 4.14b. As expected, the radius increases monotonically from a minimum valuc of ≈1.8 Å at the tops of the chains to a maximum of ≈3 Å in the middle of the bilayer, with a profile closely resembling that of the disorder parameter. Our range of spherical radii is significantly narrower than the ranges reported by Konig et al. for the bilayer at 30% hydration, assuming a linear increase from the minimum to maximum values: 0.8-7.6 Å from measurements with Q in the plane of the bilayer, and 1.1-5.8 Å from measurements with Q along the bilayer normal. The origin of this discrepancy is difficult to ascertain without knowing the sensitivity of the fits to the data on the range and shape of the radius profile. To gain more insight it would be useful to see if the neutron data could be fit well using a variation of spherical radii resembling the profile in Fig. 4.14b.

Although the diffusion-in-a-sphere model is illuminating in the sense that it conveys the notion of increasing dynamic disorder moving toward the bilayer interior, as a model for hydrocarbon chain dynamics it is not completely satisfying because it assumes indepen-dent motion for each proton and therefore lacks an explicit connection to conformational transitions. In this regard perhaps a more appealing model is the chain defect model introduced subsequently by Konig et al. as an alternative to the diffusion-in-a-sphere model. In this model it is assumed that the primary mechanism for the proton jumps in the hydrocarbon chains detected by neutron time-of-flight measurements is the creation/annihilation of a *gtg'*(g^-tg^+ or g^+tg^-) kink from/to an all-trans conformation. Two types of kink dynamics were considered. In the first, the kink is assumed to diffuse along the chains, whereas in the second a kink forms and disappears randomly at different points in the chains. Both models fit the data equally well, but the stochastic kink model is more consistent with the picture from MD simulations. The resulting model contains two parameters: the average number of gtg^1 kinks per chain, 0.9, and the transition rate, $r_2 = 7.5 \times 10^{-10}\ S^{-1}$, which corresponds to a kink lifetime of $1/r_2 = 13$ ps.

The chain defect model is useful because it incorporates a realistic mechanism for proton jumps into a model that leads to analytical expressions that can be used to fit neutron data. However, as pointed out by Konig et al., this model is an oversimplification, because clearly there are other types of chain defects (*g*, *gg*, g^-tg^-, and g^+tg^+) that are forming and disappearing on the time scale of the experiment. Indeed, we find in our simulation that *gtg'* kinks account for less than 20% of the gauche bonds in a fluid phase DPPC bilayer. Moreover, we find that there are as many *gtg* (g^-tg^- or g^+tg^+) kinks as *gtg'* kinks. This is remarkable because a *gtg* kink produces a chain defect that significantly alters the path of the chain, whereas a *gtg'* kink leaves the chain path largely unaltered. Consequently, a *ttt* to *gtg* transition is expected to lead to a larger displacement of more protons than a *ttt* to *gtg'* transition. Thus, if our simulation results are trustworthy, we may conclude that the chain defect model including only *gtg'* kinks significantly underestimates the proton motions arising from conformational transitions in the chains. The average total number of *gauche* bonds per chain in our simulation is in good agreement with estimates from FTIR spectra, but the numbers of kinks are significantly lower than the experimental estimates. The seriousness of the discrepancy in the number of kinks is hard to assess because of uncertainties in assigning the spectra and the considerable disagreement between independent experimental measurements. We finish this section by noting that our simulation predicts a kink lifetime of 6 ps, which is about half that assumed in the chain defect model.

Table 4.2. Conformational Defects in the Hydrocarbon Chains in Fluid Phase Lipid Bilayers

	Average number per chain	
Type of defect	*MD*	*Expt*
g	3.6	3.7 [72], 3.6-4.2
gtg	0.35	1.2
gtg + *gtg'*	0.61	1.0

G. Water Dynamics

In contrast to the lipids, which can exhibit a wide variety of motions over a wide range of time scales, water dynamics in membranes on the time scale of tens of picoseconds are relatively uncomplicated, consisting of (ignoring biologically uninteresting high frequency vibrations) rigid-body translation and rotation. In this section we analyze the translational and rotational motion of water in multilamellar bilayers, highlighting the distinction between the bound and bulk waters identified earlier.

The center-of-mass mean-square displacements of the bulk and the three different classes of bound water molecules are plotted in Fig. 4.15a. In all cases the MSDs display a linear time dependence beyond about 5 ps. Thus, the water translation at times longer than a few picoseconds can be described as Brownian motion, and the mobility can be quantified by a diffusion constant, *D,* proportional to the slope of the linear part of the MSD. The values obtained for motion in the plane and perpendicular to the plane of the bilayer, from the average of five 20 ps MSD increments, with the slopes calculated from 10 ps to 20 ps, are listed in Table 4.3. As expected, the bulk water molecules have the greatest translational mobility, followed by the choline-bound, phosphate-bound, and carbonyl-bound water molecules. The tightly associated P-bound and CO-bound water molecules have roughly equal diffusion constants for in-plane and out-of-plane translational motion. The structural organization of the "bulk" waters that occupy a thin slab in the middle of the interlamellar space is quite similar to that of pure water at the same temperature, but the dynamics are different. As one might expect for water confined between two slabs; we find that the diffusion constant in the plane of the bilayer is significantly larger (~50%) than in the direction normal to the bilayer.

We discuss the rotational dynamics of water molecules in terms of the time correla-tion functions, $C_l(t) = \langle P_1[\cos\theta_l(t)]\rangle$ (l - 1, 2) where P_l is the lth Legendre polynomial, $\cos\theta_l(t) = \mathbf{u}_l(0)\cdot\mathbf{u}_l(0)\cdot\mathbf{u}_l(t)$, $\mathbf{u}_1$, is a unit vector along the water dipole (HOH bisector), and $\mathbf{u}_2$ is a unit vector along an OH bond. Infrared spectroscopy probes $C_2\ (t)$, and deuterium NMR probes $C_2(t)$. According to the Debye model (Brownian rotational motion), both correlation functions are exponential, and $C_1(t)$ decays three times as fast as $C_1(t)$. The $C_1(t)$ for the different classes of water molecules, plotted in Fig. 4.15b, display a rapid initial decay (<10 ps) decay followed by a slower relaxation. The $C_2(t)$ (not shown) look similar but show a more rapid initial decay, as expected. Although the $C_1(t)$ are better described by multiexponential or stretched exponential functions, we have estimated rotational correlation times by fitting the short-time (first 10 ps) decays to single exponentials, $\exp(-t/\tau_l)$. The trend in the rotational rates (l/τ_l) is the same as that in the translational diffusion constants, *i.e.*, bulk > bound, and the ratios τ_1/τ_2 are approximately 1.5 for all classes of water molecules, suggesting that the Debye model is not appropriate for describing the rotational motion of water molecules in the vicinity of membranes.

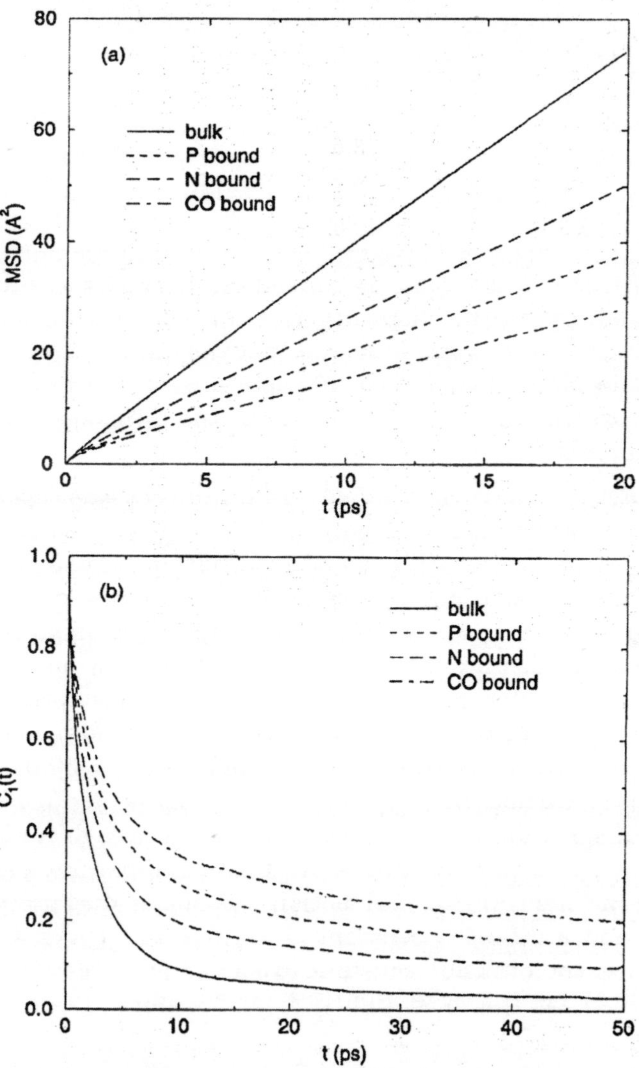

Fig. 4.15(*a*). Mean-square displacements of water molecules in three dimensions (see text for definitions of bound and bulk waters). (*b*) Time correlation functions for reorientation of the water O—H bonds.

We finish this section by comparing our results with NMR and incoherent neutron scattering experiments on water dynamics. Self-diffusion constants on the millisecond time scale have been measured by NMR with the pulsed field gradient spin echo (PFGSE) method. Applying this technique to oriented egg phosphatidylcholine bilayers, Wassall demonstrated that the water motion was highly anisotropic, with diffusion in the plane of the bilayers hundreds of times greater than out of the plane. The anisotropy of the long-range diffusion observed on the microsecond time scale is consistent with our results but much more pronounced, as expected, given the disparity in time and length scales probed in the experiment and simulation. Wassall also observed that the diffusion constant increased with the number of water molecules per lipid molecule according to the Finer model of hydration shells and consistent with the differences we found in the diffusion constants between the different classes of bound and bulk water.

Table 4.3. Diffusion Constants and Rotational Correlation Times of Water Molecules from an MD Simulation of a Fully Hydrated Fluid Phase DPPC Bilayer[a]

Water class	*In-plane D (10^{-5} cm²/s)*	*Out-of-plane D (10^{-5} cm²/s)*	*Dipole τ_1 (ps)*	*OH bond τ_2 (ps)*
Bulk	6.6	4.5	3.6	2.2
N-bound	4.4	3.6	5.6	3.7
P-bound	3.2	2.7	7.1	5.0
CO-bound	2.3	2.0	9.1	5.9

An INS study of water dynamics in oriented DPPC bilayers was reported by Konig et al. Neutron time-of-flight measurements were performed on oriented samples at 44°C and two hydration levels, using a spectrometer with an energy resolution (0.015 meV) that corresponds to a maximum observable correlation time of a few hundred pico-seconds. The spectral parameters are not available for comparison, so we compare parameters of the model used to fit the data, which includes bound water molecules that are assumed to undergo rotation diffusion only, and quasi-free water molecules that undergo both rotational and translational diffusion. At low hydration (three to four water molecules per lipid molecule) the data could be modeled as rotation only, with a rotational correlation time τ_R = 60 ps. Apparently, this $\tau_R = 1/D_R$, where D_R, is the rotational diffusion constant, should be divided by 6 for comparison with the simulation results in Table 4.2. The resulting 10 ps is close to our bound water values. To fit the data at higher hydration (11 waters per lipid), the parameters for the rotational motion were kept fixed and the translational motion was modeled as jump diffusion, giving a residence time of 2 ps and a diffusion constant $D = 1.6 \times 10^{-5}$ cm²/s. Considering the 6°C difference in temperature, the diffusion constant extracted from the scattering data is close to the values we obtained for the most tightly bound waters (P- and CO-bound) in our simulation.

V. SUMMARY AND CONCLUSIONS

This chapter has given an overview of the structure and dynamics of lipid and water molecules in membrane systems, viewed with atomic resolution by molecular dynamics simulations of fully hydrated phospholipid bilayers. The calculations have permitted a detailed picture of the solvation of the lipid polar groups to be developed, and this picture has been used to elucidate the molecular origins of the dipole potential. The solvation structure has been discussed in terms of a somewhat arbitrary, but useful, definition of bound and bulk water molecules.

The majority of the chapter was focused on an analysis of dynamics, which to date have been neglected relative to structure in simulation studies of membranes. We compared results from our simulation of a fluid phase bilayer with incoherent neutron scattering experiments that probed the motions of lipid H atoms on time scales of up to 100 ps. For the most part the simulation and experimental results were in good agreement, but the simulation predicted a slight anisotropy in the lipid motion that was not detected experimentally and overestimated the spatial dependence of the time scale of the slower of two dynamic processes resolved by the experiment. Having established a reasonable level of agreement between the simulation and neutron data, we examined the correspondence between the motions observed in the simulation and the dynamic models used to fit the experimental data. The center-of-mass motion and internal rearrangements are decoupled, and the former is well described as diffusion in a confined space (cylinder) on the 100 ps time scale, but not as Brownian motion. There were some significant discrepancies between the picture that emerged from the simulation and the models used to describe the internal motions. In particular, the simulation predicted a much weaker variation in the radii of the diffusion-in-a-sphere model and the involvement of more conformations in the acyl chain dynamics than the single kink assumed in the chain defect model.

Our simulation suggested that both the translational and rotational dynamics of water molecules in a fully hydrated, multilamellar lipid bilayer system depend on where the water molecules are located. As expected, both the translational and rotational mobilities of the "bulk" water molecules located in the middle of the interlamellar space are significantly greater than those of the "bound" water molecules located in the first solvation shell of the lipid polar groups (carbonyl, phosphate, and choline). The translational diffusion constants and the rotational rates of the bound water molecules increase in the order carbonyl-bound < phosphate-bound < choline-bound. On the time scale of tens of pico-seconds, the bound water molecules exhibit largely isotropic translational motion, whereas the bulk water molecules diffuse approximately 50% faster in the plane of the membrane than out of the plane. None of the water molecules in the membrane-water system obey the Debye rotational diffusion model. The detailed picture of water dynamics near membranes derived from the simulation is in qualitative agreement with available data from NMR and neutron scattering experiments, but a more quantitative, model-independent comparison with neutron scattering would be useful for a better assessment.

DYNAMICS OF OLIGONUCLEOTIDES

The biological functions of DNA and RNA were initially assumed to involve only their primary sequence as required for storage of the genetic code. Consistent with this view was the helical B structure of DNA initially proposed by Watson and Crick. While initial experimental work based on fiber diffraction indicated heterogeneity in DNA structure, such as the A and B forms of DNA, it propagated the idea that the structure of DNA was that of a regular helix. This view started to change when the first structures of DNA based on single-molecule X-ray crystallography were obtained, which showed local conformational heterogeneity to be present in DNA while the overall structures still assumed canonical forms. Later, structural studies of RNA, particularly transfer RNA (tRNA), revealed the structure of RNA to have significant tertiary characteristics beyond the helical structures dominating DNA. More recently, X-ray crystallographic studies of DNA-protein complexes revealed DNA structures that are significantly distorted from the helical conformations traditionally envisioned for DNA. Furthermore, it has become evident that the structural distortion of DNA and the wide variety of tertiary structures of RNA are essential for their biological activity.

Although experimental studies of DNA and RNA structure have revealed the significant structural diversity of oligonucleotides, there are limitations to these approaches. X-ray crystallographic structures are limited to relatively small DNA duplexes, and the crystal lattice can impact the three-dimensional conformation. NMR-based structural studies allow for the determination of structures in solution; however, the limited amount of nuclear overhauser effect (NOE) data between nonadjacent stacked basepairs makes the determination of the overall structure of DNA difficult. In addition, nanotechnology-based experiments, such as the use of optical tweezers and atomic force microscopy have revealed that the forces required to distort DNA are relatively small, consistent with the structural heterogeneity observed in both DNA and RNA.

Computational studies of nucleic acids offer the possibility to enhance and extend the information available from experimental work. Computational approaches can facilitate the experimental determination of DNA and RNA structures. Dynamic information, although often isotropic in nature from experimental studies, can be obtained from computations at an atomic level of detail. Of particular interest is a detailed knowledge of the influence of base sequence, base composition, and environment on DNA structure. Finally, computational approaches can reveal the subtle relationship between structure and energetics, yielding an understanding of the properties of oligonucleotides that allow for the conformational changes required for their biological function.

In the coming lines we provide an overview of the current status of nucleic acid simulations, including studies on small oligonucleotides, DNA, RNA, and their complexes with proteins. This is followed a presentation of computational methods that are currently being applied for the study of nucleic acids. The final section of the chapter includes a number of practical considerations that may be useful in preparing, performing, and analyzing MD simulation based studies of nucleic acids.

OVERVIEW OF COMPUTATIONAL STUDIES ON OLIGONUCLEOTIDES

A. DNA

Computational studies of nucleic acids initially lagged behind protein-based calculations. Nucleic acids, being extended polyanions, require a more rigorous treatment of the solvent environment, whereas the globular structure of many proteins allows for greater tolerance for the vacuum environment applied in early MD simulations due to computational limitations. Early attempts at simulating nucleic acids in vacuum involved decreasing the phos-phate charges or setting them to zero. Alternative approaches included the inclusion of "hydrated" counterions, or solvatons, that mimicked a sodium ion hydrated by six water molecules and the use of distance-dependent dielectrics for buffering the electrostatic interactions. Whereas results from these calculations produced some insights into the properties of DNA, the quality of the results were generally in poor agreement with experiment, emphasizing the need for better treatment of the solvent environment.

Inclusion of explicit solvent in calculations on DNA involved simulations in which the DNA was both held rigid and allowed to evolve along with the solvent molecules. Application of the former approach allowed for a better understanding of the solvation of DNA to be obtained. For example, the hydration of AT and polyA-polyT B-form tracts was studied via Monte Carlo calculations, and MD simulations were used to investigate differences in hydration of the B and Z forms of DNA. Although these works contributed to the understanding of the solvation of oligonucleotides, the local conformational heterogeneity of DNA structure observed in crystal structures of DNA emphasized the need to include both the DNA and solvent as flexible degrees of freedom in the simulations. One of the earlier calculations on DNA with an explicit solvent representation was performed on a d(CGCGA) duplex in a sphere of water that included neutralizing counter-ions. The structure resulting from this simulation was shown to be similar to the B form of DNA; however, the total simulation time was only 114 ps, not long enough to allow for significant relaxation of the DNA, which has been more recently shown to require 1 ns or longer. Though limited, this work strongly indicated that MD simulations of DNA duplexes with an explicit solvent representation were both feasible and a useful method to better our understanding of DNA structure.

Over the next decade a number of efforts were made to apply MD simulations using explicit solvent representations to DNA. A number of these calculations were performed on the d(CGCGAATTCGCG) or "Drew" dodecamer, making this sequence the benchmark for DNA duplex calculations. Calculations performed on this molecule using explicit solvent included a 140 ps simulation with harmonic constraints on the hydrogens involved in Watson-Crick basepairs and a 1 ns simulation with reduced charges on the phosphates. In both cases relatively stable structures were obtained, although significant distortions of the structures from the canonical B form were evident. Similar results were obtained on the Drew dodecamer in a 150 ps simulation that included a 9 Å solvation shell of explicit water molecules, where the structure continued to drift from the canonical B form over the course of the simulation. Although these and other efforts produced reasonably stable simulations, calculations up to approximately 1995 generally yielded distorted DNA structures if performed long enough to allow for significant relaxation of the system.

During 1995 several groups showed that stable simulations were possible. A simulation on the Drew dodecamer in the crystal environment was performed for 2.2 ns, yielding root-mean-square (RMS) differences in the range of 1.0-1.5Å from the experimental structure [18]. Similarly good results were obtained on a crystal simulation of a Z-DNA d(CGCGCG) duplex. These successes were reproduced for DNA in solution, with simulations of the d(CCAACGTTGG) duplex and a DNA triplex both performed for 1 ns. Initially, these successes were attributed to the use of the Ewald method, in some cases via the particle mesh Ewald (PME) approach. This assertion seemed appropriate given the highly charged nature of oligonucleotides and the ability of the Ewald methods to accurately treat the long-range electrostatic interactions that should dominate in these systems. Subsequent studies, however, using atom-based truncation methods for simulations of the Drew dodecamer and deoxy and ribo GCGCGCGCGCGC duplexes in solution showed that stable structures could be obtained without the use of Ewald-based methods. Furthermore, simulations of the d(CGCGCG) DNA duplex in aqueous solution with a nonbond cutoff of 12 Å with smooth shifting to zero at the cutoff distance of just the electrostatic energy or both the electrostatic energy and force gives simulation results indistinguishable from those obtained using Ewald summation in terms of the RMS difference from the initial structure and RMS fluctuations around the final average. These results indicated that improvements in the AMBER and CHARMM force fields used in the calculations contributed significantly to the ability to perform stable simulations.

Since those initial successful simulations, several simulations of DNA in solution of more than 1 ns have been performed. Two different simulations have been performed on the Drew dodecamer. One simulation, extended for 3.5 ns using the CHARMM force field, showed a change in the overall conformation from the canonical B to the A form of DNA. The second simulation, using AMBER, was initially performed for 5 ns, over which time the structure was shown to fluctuate approximately 2.5-3.5 Å from the canonical B form of DNA; that simulation has since been extended to 14 ns. Two 10 ns simulations have been performed on the DNA duplex d(C_5T_5) using the AMBER and CHARMM force fields. This system was selected because of the suggestion that it assumes a B-type structure in the AT region and a A-like structure in the GC region. Results revealed that both force fields yield reasonable results, although disagreements in both force fields with experiment were identified. Of note are the relaxation times of the overall structures indicated from these simulations. Initial relaxation times of 1 ns or more are reported, with significant conformational fluctuations occurring for the remainder of the simulations.

The methodological advances just presented have brought the field of nucleic acid force field calculations to a point where results from the calculations can be used with reasonable confidence to aid in the interpretation of experimental data as well as to be used for scientific investigations that are not accessible to experiment. Accordingly, a number of studies based on MD simulations, as well as other methods, have been undertaken to study a wide array of biologically relevant events associated with DNA. A brief overview of some of these efforts follows.

1. Environmental and Base Sequence Influences on Duplex DNA

Alterations of DNA structure associated with changes in water activity have long been known [2], although the exact mechanisms associated with these phenomena are still in question. Further, it is known that base sequence as well as base composition also have a central role in dictating both the local and overall structure of duplex DNA. Transitions from the A to B and from the B to A forms of DNA have been observed, indicating the lack of significant energetic barriers between these two forms of DNA. The presence of ethanol, hexamminecobalt (III), and 4 M NaCl have been shown to stabilize the A form of DNA, consistent with experiment. Studies combining results from MD simulations with entropy estimates from harmonic analysis and contin-uum models to estimate the free energies of solvation indicate that internal energies favour the B form and solvation contributions favour the A form; however, the approach does not fully account for the switch to the A form of DNA at high salt concentrations. This discrepancy is consistent with calculations indicating that changes in the hydration of the phosphodiester backbone of DNA lead to changes in the conformational preferences of the backbone that influence the equilibrium between the A and B forms. A similar combined MD-continuum study, however, properly predicted stabilization of the A and B forms in low and high water activity, respectively, when ethanol is used to alter the water activity.

Concerning the influence of sequence on the structure and dynamics of DNA, a number of interesting studies have been performed on the TATA box. The TATA box is a consensus 7mer that is essential for the initiation of transcription in eukaryotes. Crystal structures have been determined for the TATA box DNA bound to the TATA box binding protein (TBP). In these structures the DNA is observed to be significantly distorted from the B form to a form closer to the A form of DNA that has been dubbed the TA form. On the basis of these results it was suggested that the inherent conformational preference of TATA box DNA may be similar to the TA form, thereby facilitating binding with the TBP. Two separate MD studies of the TATA box, using different force fields and simulation methodologies, both indicated that sequence to indeed assume a more A-like conformation than other forms of DNA. These results are an example of how MD simulations can provide information that is difficult to obtain or inaccessible via experimental approaches.

2. DNA-Protein Interactions

Protein-DNA complexes present demanding challenges to computational biophysics: The delicate balance of forces within and between the protein, DNA, and solvent has to be faithfully reproduced by the force field, and the systems are generally very large owing to the use of explicit solvation, which so far seems to be necessary for detailed simulations. Simulations of such systems, however, are feasible on a nanosecond time scale and yield structural, dynamic, and thermodynamic results that agree well with available experimental data. Some aspects common to the various systems are briefly summarized in the following paragraphs.

Only a handful of MD simulations of protein-DNA complexes have been reported. All but one, a model of the chromosomal HMG-D system, deal with sequence-dependent DNA binding either to a restriction endonuclease (EcoRI) or to transcription factors. Studies on transcription factors include the repressers, the antennapedia homeodomain, the TATA box binding protein, and the DNA binding domains (DBDs) of hormone receptors. This set of systems contains representative proteins of several of the known DNA-binding structural motifs: helix-turn-helix (*e.g.*, homeodomain and the lac repressor), Zn finger proteins (*e.g.*, hormone receptors), ribbon proteins (TBPs), and the HMG box (HMG-D, SRY). These examples contain DNA conformers of the canonical B type as well as DNA with bends and kinks present. All of the listed studies are on solvated systems that contain several thousand atoms. For example, the first simulation of the lac repressor headpiece with 51 amino acids bound to a 14 basepair DNA duplex in water contained 12,889 atoms and lasted for 0.125 ns, in part with NOE restraints. Simulations up to 2 ns and of systems as large as 36,000 atoms have been performed. Comparisons of similar systems, *e.g.*, wild-type versus mutants or cognate versus noncognate complexes, have also been made in some cases.

Some proteins bind to DNA of any sequence as part of their biological function, such as in the tight packing of DNA in chromosomes. The structures of at least two HMG-box-containing proteins that are important for chromatin structure have been experimentally determined in complexes with the proteins specifically bound to DNA, but no complex between a sequence-independent HMG box protein-DNA complex has been determined. In the study by Balaeff et al., three models of a complex between the HMG-D protein (a nonspecific HMG protein) and DNA were constructed, and the model complexes were subjected to 160 ps of MD simulation. The quality of these models was assessed on the basis of a number of criteria, including the stability of the structure and the geometry of the protein and the DNA. The model based on docking HMG-D to a DNA model similar to the bent DNA conformer observed in TBP-DNA complexes was chosen for a final 60 ps MD simulation that indicated that the protein adapted its conformation slightly to better fit the DNA. In addition to a number of contacts between basic amino acid residues with the DNA phosphodiester backbone, there were many hydrophobic interactions in the DNA minor groove formed by hydrophobic residues on the surface of the HMG box. Comparisons with the sequence-dependent HMG box protein-DNA complexes showed how nonspecific HMG domain proteins can bind in a similar way to many different DNA sequences by using nonpolar interactions instead of the polar interactions found at key sites in the specific complexes.

For the proteins that bind to a specific DNA sequence it is quite natural to compare cognate complexes with complexes in which either the protein or the DNA has been altered. Such comparisons are also undertaken in simulations of hormone receptors, including variations of both the protein and DNA, variation of the DNA bound to EcoRI, the SRY-DNA complex, and the TBP in which TBPs bound in different orientations to the same DNA sequence are compared. Here the nuclear hormone receptors have attracted the most attention, with several simulations presented for both the glucocorticoid and estrogen receptors. The overall picture emerging from these simulations is that the systems are well behaved; the DNA adapts its conformation to the protein, which holds on to the DNA with a number of well-defined hydrogen bonds to the phosphate backbone, allowing specific recognition by rather complex, and dynamic, networks of both direct and water-mediated hydrogen bonds between the protein and DNA. Point mutations can, to some degree, be accommodated by this network through side chain rearrangements and moving water molecules, but there can also be larger changes. In the glucocorticoid receptor DBD changes of a single amino acid at the protein/DNA interface lead not only to a slight change in the orientation of the protein on the DNA but also to significant conformational changes some 15 Å away, in the part of the protein where contacts are made with the other protein in the dimer that is bound to the DNA. Other recent simulation studies of protein-DNA complexes were performed on the TRP operator and the ZIF268-DNA complex. From all these studies it is clear that calculations of DNA or RNA in complexes with protein will greatly facilitate our understanding of a wide variety of processes associated with growth, differentiation, and signal transduction at an atomic level of detail.

3. DNA-Drug Interactions

Numerous drugs, including many antibiotics, function via direct interactions with DNA. In addition, a number of anticancer agents, including cisplatin, function through alkylation of DNA. Computational approaches offer the means to better understand the nature of the interactions between drugs and DNA as well as a rational approach for the optimization as well as identification of lead compounds. The mode of interaction of two antibiotics, Esperamicin and Dynamicin, both of which lead to the cleavage of DNA, were investigated via MD simulations. These studies yielded information on the mechanisms and cleavage patterns of DNA. In another study, the relative binding of daunomycin and 9-aminoacridine to B-DNA were studied via free energy perturbation calculations. Although the calculations reproduced the experimental trends, the agreement may have been fortuitous considering that the calculations were performed in vacuum. An interesting study was the application of QM/MM methods to investigate the cross-linking of guanine bases in DNA by nitrous acid. Although not a study of a drug per se, the work strongly indicates that details of the reactions of alkylating agents with DNA can be investigated.

B. RNA

RNA structures, compared to the helical motifs that dominate DNA, are quite diverse, assuming various loop conformations in addition to helical structures. This diversity allows RNA molecules to assume a wide variety of tertiary structures with many biological functions beyond the storage and propagation of the genetic code. Examples include transfer RNA, which is involved in the translation of mRNA into proteins, the RNA components of ribosomes, the translation machinery, and catalytic RNA molecules. In addition, it is now known that secondary and tertiary elements of mRNA can act to regulate the translation of its own primary sequence. Such diversity makes RNA a prime area for the study of structure-function relationships to which computational approaches can make a significant contribution.

To date, RNA calculations have been performed on a variety of systems of different topologies including helical duplexes, hairpin loops, and single strands from tRNA, rRNA, and ribozymes. In a simulation of an RNA tetraloop of the GRNA type, which is very common and known to be remarkably stable, it was found that without imposing any external information the simulation found the right conformation even when it started from the wrong one. Studies have used Ewald summation methods to handle the long-range electrostatic interactions in several simulations of tRNAAsp; both the whole molecule and the anticodon arm have been simulated. These simulations basically find that RNA molecules maintain the experimentally observed structures, and the authors proceed to analyze hydrogen bonding and hydration patterns; the latter, being more difficult to observe directly in experiments, provide new information from the simulations. In particular, several C—H...O "hydrogen bonds" are found. They are shown to be important for stabilizing the preferred nucleotide conformation in RNA through base-backbone interactions and also to stabilize the anticodon loop conformation.

Results from the anticodon studies are similar to what is found in the simulation of the whole tRNA, indicating that no serious artifacts were introduced in the fragment simulations. Simulation studies have also been performed on a ribozyme; once again the molecule remained structurally stable, and hydrogen bonding and hydration as well as specific interactions involving Mg ions were analyzed. The structural stability observed in all these simulations is attributed to the use of Ewald summation for long-range electrostatic interactions. As discussed earlier, this structural stability is not unique to MD simulations using Ewald summation. In quite a large number of studies, standard spherical truncation schemes were successfully used for DNA as well as RNA systems. Some spherical truncation schemes are known to cause problems, especially for charged systems, *e.g.,* "neutral group" switching or truncation, and these should thus be avoided. Furthermore, it is not clear that stable results can be obtained using a "mixed" force field, in which charges for different portions of the system (*e.g.,* DNA versus protein) are obtained from different sources without reparametrization.

C. Dynamics and Energetics of Oligonucleotides

One of the most powerful attributes of computational studies is the ability to obtain direct relationships between energetics and structure. As discussed earlier, the structures of both DNA and RNA are extremely sensitive to environmental conditions. In essence, alteration in the environment leads to changes in the conformational free energy surface of the molecules. Moreover, experimental studies have shown that DNA can be significantly distorted from the canonical forms by using very small forces. Such plasticity presumably allows for the opening of DNA required for transcription and replication, for formation of nucleosomes and other processes of central biological importance. Detailed knowledge of the phenomena that allow for this plasticity will, in addition to furthering our knowledge of biological processes, facilitate the use of oligonucleotides as mechanical devices as the field of nano-technology develops.

Initial applications of computational techniques have involved the use of potentials of mean force (PMFs) or umbrella sampling to investigate the energetics of ribo and deoxyribodinucleotides. From a series of PMF calculations on the 16 combinations of the dinucleotide X_pY (X,Y = A,C,G,U) and their deoxyribose counterparts, it was found, in accordance with the experimental data, that purine-purine pairs stack best, pyrimidine-pyrimidine pairs not at all, and the purine-pyrimidine heterodimers were in between. It is quite clear from these studies that the relative free energies are not dominated by direct base-base interactions, but that the driving force for stacking is of an enthalpic character [83]. Differences between DNA and RNA were small, with the methyl group in thymine stabilizing stacking and the 2'-hydroxyl group of RNA in some cases stabilizing and in other cases destabilizing. Reduced stabilization of the stacked conformation in nonaqueous solvents (DMSO, chloroform, methanol) was also observed.

PMF calculations have also been used to investigate the end-to-end extension of a duplex DNA 16mer, with the calculations designed to reproduce the conditions used in atomic force microscopy (AFM) experiments. Results from these calculations were consistent with the barrierless extension of DNA observed in the AFM experiments. Detailed analysis of the simulated results yielded a model in which the unfavourable intramo-lecular mechanical energy of DNA associated with extension is compensated for by DNA-solvent interactions to yield the barrierless extension of the DNA. Alternative computational studies of the extension of DNA have been performed in vacuo with both atom-based and internal coordinate based methods. Results from the internal coordinate based calculations, which have been developed to implicitly include solvent effects, yielded a qualitative picture of the structural changes in DNA as end-to-end extension is observed from which a new form of DNA, the S form, was identified. Atom-based calculations of the stretching DNA in vacuum indicated that Watson-Crick hydrogen bonds remained intact throughout the extension of the DNA; however, the omission of solvent could significantly influence this conclusion. In combination, the calculations to date on the energetics of oligonucleotides associated with structural perturbations strongly indicate that much information is to be gained from these types of computational studies.

D. DNA Phase Transitions

Molecular dynamics simulations have also been used to interpret phase behaviour of DNA as a function of temperature. From a series of simulations on a fully solvated DNA hexamer duplex at temperatures ranging from 20 to 340 K, a "glass transition" was observed at 220-230 K in the dynamics of the DNA, as reflected in the RMS positional fluctuations of all the DNA atoms. The effect was correlated with the number of hydrogen bonds between DNA and solvent, which had its maximum at the glass transition. Similar transitions have also been found in proteins.

E. Modified Oligonucleotides

Interest in chemically altered oligonucleotides has been generated by the possibility of using antisense technology in drug therapy as well as to exploit chemically modified species in oligonucleotide structure-function studies, including nucleic acid-protein interactions. Modifications studied to date via computational approaches include modifications of the phosphates, sugars, and bases. Modified phosphodiester backbones subjected to computational studies include phosphoramidate-modified species, peptide nucleic acids (PNAs), and 2′-5′ phosphodiester linkages. Results from these studies yield in-sights into the influence of the backbone on the overall conformation of the oligonucleotide, including stabilization of triple helical structures. In another study, calculations were performed on DNA duplexes and triplexes, with guanidine groups replacing the normal phosphodiester linkages, yielding a polycation. Results showed the triplex to be more stable than a polyanionic triplex. Examples of the study of modified sugar include calculations on oligonucleotides containing hexitol instead of the furanose rings. These calculations indicated that oligonucleotides with hexitol sugars may form more stable duplexes with RNA than DNA due to alterations in solvation of the minor groove. Some of the calculations involving modified bases included free energy perturbation calculations investigating the influence of bromination and methylation on the stability of Z-form oligonucleotides. Other studies of modified oligonucleotides have been performed on the thymine dimers caused by UV photodamage of DNA showing the presence of bent structures consistent with experiment along with information on possible mechanisms by which repair enzymes may identify damaged sites [100]. Although these studies yielded useful insights, it should be emphasized that the quality of the results will be strongly influenced by the force field parameters and methods used in the calculations. Special care is required to ensure that the new parameters introduced into a force field to treat the novel chemical structures are (1) consistent with the force field and (2) modeling the modified structures correctly.

F. Alternative Secondary and Tertiary Motifs of Oligonucleotides

Beyond the helical structures and the RNA folded structures already discussed, both DNA and RNA are known to assume alternative structures, such as quadraplexes, that are of biological relevance. In addition, the nature of the hydrogen bonding patterns in oligonucleotides allows the design and construction of structures with the potential to significantly advance the field of nanotechnology. To date the number of calculations on alternative oligonucleotide structures is not large; however, some good examples exist. MD simulations have been performed on four-stranded iDNA, which involves intercalated cytosines and have been shown to yield stable structures. Quadraplexes are structures suspected to exist at the end of telomeres, stabilizing the terminal DNA in chromosomes. The quadraplexes themselves are stabilized by the presence of cations. Free energy perturbation calculations have been applied to investigate the influence of cation type on the extent of stabilization. MD simulations have also been used to facilitate the determination of the structure and dynamics of quadraplexes based on NMR data. Although only a few studies on alternative oligonucleotide folds have been studied to date, the ability of computational methods to investigate the stability of alternative structural motifs should facilitate the use of oligonucleotides in nanotechnology.

METHODOLOGICAL CONSIDERATIONS

Biological functions of nucleic acids occur at the level of single nucleosides up to chromosomes, with the majority of the functions intimately involving proteins as well as other biological molecules. Accordingly, computational methods for the study of nucleic acids must be able to access these variously sized systems. In this volume, the focus is on calculations at the atomistic level, and the present chapter remains consistent with this goal. Such a limitation, however, confines us to a discussion of computational studies of oligonucleotides that contain less than 50 basepairs. To overcome this limitation, a brief section on the different computational methods for the study of oligonucleotides, including approaches to the study of larger structures, follows.

A. Atomistic Models

Initial atomistic calculations on nucleic acids were performed in the absence of an explicit solvent representation, as discussed earlier. To compensate for this omission, various changes in the energy functions were employed, including reduced charges in the phosphate groups, the inclusion of hydrated sodiums, and altered dielectric constants. These approaches did yield results that furthered our knowledge of the structure-function relationships of nucleic acids in a qualitative fashion. However, as computational methods matured, the demand for more quantitative results increased, requiring more realistic models in oligonucleotide calculations. These demands have been met by way of both improved algorithms for simulations and improved potential energy functions, not to mention the incredible increase in computational power that has occurred over the last two decades. These aspects are discussed in this section.

Central to the quality of any computational study is the mathematical model used to relate the structure of a system to its energy. General details of the empirical force fields used in the study of biologically relevant molecules are covered in and only particular information relevant to nucleic acids is discussed in this chapter.

Initial applications of computational techniques to study nucleic acids used models adapted from protein force fields. Other parameter work of note included early studies on the properties of the furanose ring in nucleic acids, which showed the importance of the proper treatment of the pseudorotation profile in potential energy functions. The first widely used force fields for the study of nucleic acids were associated with the programs CHARMM and AMBER. These force fields were developed for both extended atom and all-atom representations. Extended atom models, which have been used extensively for simulation studies of proteins, are of lesser value for calculations on nucleic acids because the smaller number of nonpolar hydrogens in nucleic acids makes the gain in computer efficiency for an extended atom model smaller for them than for proteins. Both force fields were also developed primarily for modeling and simulation studies in vacuum, based on the use of distance-dependent dielectric constants in the calculation of electrostatic interactions. Such an approach allows short-range electrostatic interactions to dominate while longer range interactions are damped in accord with the damping due to the dielectric constant of water. Both of these force fields were quite useful, and a number of the studies cited earlier were based on these works.

Computational studies of nucleic acids via atomistic models in the absence of solvent are expected to yield poor representations owing to the polyanionic nature of oligonucleotides. Although a number of efforts were made to circumvent the use of explicit solvent models, the results were generally unsatisfactory (see above). These failures motivated the inclusion of solvent models in MD calculations. Initial efforts, primarily based on the TIP3P water model, used the parameter sets developed for vacuum calculations; the only difference was the use of a dielectric constant of 1 for the electrostatic calculations rather than the distance-dependent dielectric. Results from these studies, along with work based on an earlier version of the GROMOS force field, generally led to improved agreement with experiment compared to the vacuum calculations. However, as MD simulations were significantly extending beyond 100 ps, it was observed that the structures deviated significantly from the canonical forms of DNA. It was not until the "second generation" force fields were developed that stable simulations of oligonucleotides in solution were achieved.

The second generation force fields for nucleic acids were designed to be used with an explicit solvent representation along with inclusion of the appropriate ions. In addition, efforts were made to improve the representation of the conformational energetics of selected model compounds. For example, the availability of high level ab initio calculations on the conformational energetics of the model compound dimethylphosphate yielded significant improvements in the models for the phosphodiester backbone. Similarly, new data on the interaction strengths for the Watson-Crick and Hoogsteen basepair interactions allowed for improvements in the nonbonded parameters. In the context of the nonbonded parameters is the need to properly balance the solute-solute interactions (*e.g.,* Watson-Crick basepair interactions) with solute-solvent and solvent-solvent interactions. In the development of the CHARMM force field, special emphasis was placed on this balance. On the basis of these force fields it was shown that stable MD simulations could be performed on a variety of sequences and structures; however, the results still had limitations.

Motivated by the need for further improvements, the AMBER and CHARMM force fields were optimized further. In addition, the BMS force field was developed, which incorporates features from the second generation CHARMM and AMBER force fields along with additional optimization of the parameters. In all of these force fields, emphasis was placed on further optimization of the conformational energetics, yielding improved agreement with data on the canonical structures of DNA, surveys of dihedral distributions, and helical parameters of crystal structures from the Nucleic Acid Database (NAD), along with a variety of other experimental and ab initio data. With the CHARMM force field, emphasis was placed on balancing local contributions, based on potential energy data for a series of model compounds, with global properties for duplexes in solution. This approach was designed to ensure that the proper contributions in the force field were combining to yield the desired properties of DNA and RNA in solution. It is expected that these improved energy functions will further facilitate the ability of MD simulations on nucleic acids to yield better quantitative agreement with experimental data.

Further extension of the atomistic models of nucleic acids will be achieved through additional optimization of the force fields using the present forms of the potential energy functions and extending the form of the potential energy function to include electronic polarizability along with other terms. Current work indicates that improvements in the revised CHARMM force field with respect to the treatment of sugar puckering are possible. Concerning electronic polarizability, the polyanionic nature of nucleic acids and the influence of salt on their structure strongly indicate that gains in the quality of results from MD simulations will be made via its inclusion. To date, no calculations on nucleic acids using models that include electronic polarizability have been performed; however, QM studies have suggested that the polarization contribution to the solvation free energy of DNA is only 1-3%. Thus, conclusions concerning the true gains to be made via the inclusion of electronic polarizability in nucleic acid simulations must wait until more detailed studies are performed. At this time one may speculate that some calculations addressing specific problems (*e.g.*, energetics of phosphate-counterion interactions) will require polarizable models whereas other phenomena (*e.g.*, conformational sampling of DNA and RNA duplexes) will see little improvement.

B. Alternative Models

Although the discussion thus far has concentrated on atomistic models of oligonucleotides in which the solvent is included explicitly, alternative models exist that allow for computational studies of larger oligonucleotides. For example, supercoiling of circular DNA has been studied using ribbonlike models. This approach allows for the generation of trajectories using time steps of 100 fs, from which folding from the circular state into supercoiled forms was observed. Another approach is the use of internal coordinates combined with the implicit treatment of solvent, as in the program JUMNA. This method is basically atomistic, but the movements of the system are described entirely in internal coordinates, greatly facilitating the locating of minima and the sampling of conformational space via Monte Carlo methods. The internal coordinate method has also been used with a minimal hydration model where a 5 Å shell of explicit water molecules was used to hydrate the DNA, allowing for an integration time step of 10 fs. Results showed the structures to be close to the B form of DNA; however, variations in RMS differences were smaller than occur in simulations using full solvent representations with periodic boundary conditions. Another "low resolution" method involves treating individual base-pairs as three-point representations. The degrees of freedom between the individual "base-pairs" can then be sampled to investigate the structural properties of extended DNA or RNA duplexes. This approach can be combined with atomistic models to allow for both the overall fold of the oligonucleotide and specific interactions in small portions of the structure to be modeled, an approach that has been used to study portions of the 16S rRNA. A method based on the use of a segmented rod model along with Brownian dynamics allows for studies of DNA molecules hundreds of basepairs in length. Although these methods sacrifice varying levels of detail, they extended computational approaches to significantly larger oligonucleotides, allowing for access to a wide variety of biological processes, such as the winding of DNA into supercoils and mechanisms associated with nucleosome and, ultimately, chromatin formation.

PRACTICAL CONSIDERATIONS

Performing successful calculations on nucleic acids requires selection of the appropriate models for the goals of the calculations followed by determination of the proper starting configuration. When designing a computational study one should carefully consider the type of information desired from the calculations along with the available resources. In many instances, atomic details of interactions between oligonucleotides and the environment or with a bound protein are desired, making the use of atomistic models appropriate. These methods, however, require significant computational resources for the generation, storage, and analysis of the MD simulations. Continuing increases in computational power with the simultaneous decrease in computer costs makes the required facilities accessible to most laboratories. An alternative is the use of supercomputing centers. For systems larger than about 50 basepairs or where atomistic details of interactions between the nucleic acid and the solvent are not required, the methods discussed in the preceding section are appropriate.

The remainder of this chapter focuses on practical aspects of the preparation and implementation of atomistically based computations of nucleic acids. A flow diagram of the steps involved in system preparation and the performance of MD studies of nucleic acids is presented. Additional details on many of the procedures described here may be found in books by Alien and Tildesly and Frenkel and Smit.

A. Starting Structures

A significant advantage of computational studies on nucleic acids is that reasonable guesses of the starting geometries can be made. When studying duplexes, these are typically based on the canonical forms of DNA and RNA. A number of available modeling and graphics packages have the ability to generate canonical structures for a given sequence. Alternatively, experimental structures from crystal or NMR studies, obtained from the nucleic acid or protein databanks, can be used. While with DNA and RNA duplexes, crystal and NMR structures generally do not differ significantly from canonical structures, in cases where there are loops, bulges, hairpins, or unstacked bases, as in tRNA, the use of experimental structures is helpful. Alternatively, if the helical and nonhelical regions are known, reasonable guesses for a starting geometry, followed by relaxation of the structure via MD simulations, can be applied. This approach is useful when low resolution data on a nucleic acid structure are available. A useful alternative is the program NAB, which generates structures of both helical and nonhelical regions of oligonucleotides and is accessible via the Internet. When creating starting models of RNA or DNA, efforts should be made to check that the model is consistent with available biophysical and biochemical experimental data.

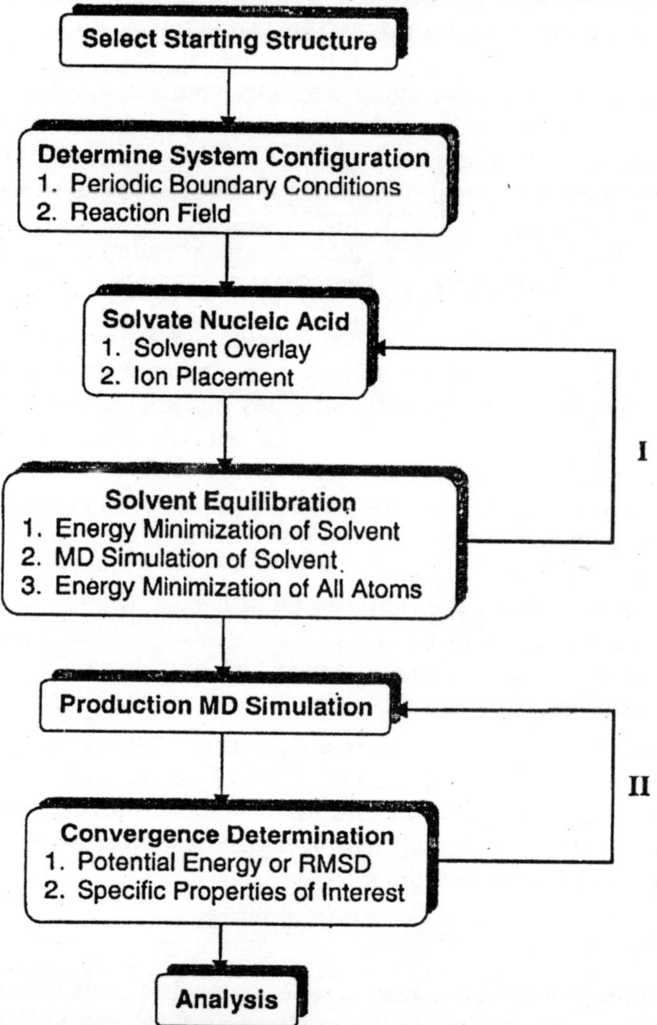

Fig. 5.1. Flow diagram of the parameter optimization process. Loops I-II represent iterative stages of the optimization process as discussed in the text.

B. System Configuration, Solvation, and Ion Placement

Essential for MD simulations of nucleic acids is a proper representation of the solvent environment. This typically requires the use of an explicit solvent representation that includes counterions. Examples exist of DNA simulations performed in the absence of counterions, but these are rare. In most cases neutralizing salt concentrations, in which only the number of counterions required to create an electrically neutral system are included, are used. In other cases excess salt is used, and both counterions and co-ions are included. Though this approach should allow for systematic studies of the influence of salt concentration on the properties of oligonucleotides, calculations have indicated that the time required for ion distributions around DNA to properly converge are on the order of 5 ns or more. This requires that preparation of nucleic acid MD simulation systems include careful consideration of both solvent placement and the addition of ions.

As a first step in setting up an MD study of nucleic acids in solution, the overall configuration of the system must be considered. This configuration is defined by the boundary conditions to be used in the solvent simulation. Boundary conditions are required to maintain the proper density of the system as well as to minimize edge effects if the system is set up so that the condensed phase environment is finite, thereby interacting directly with vacuum. The most commonly used and most rigorously correct are periodic boundary conditions (PBCs). In this approach the system (nucleic acid, surrounding solvent, and ions) is created in, typically, a cubic or rectangular shape. The edges of the system, however, do not see a vacuum, but the edge on the opposite side of the cube or rectangle, allowing for interaction of the solvent on each edge of the cube or rectangle with that on the opposing edge. In addition to cubic or rectangular systems, PBC simula-tions may also be performed using octahedral or rhombic dodecahedral symmetries, which are appropriate for spherical molecules and minimize the total number of solvent molecules required to properly solvate the spherical macromolecule. An advantage with the PBC approach is that it can be used with Ewald methods, which are currently considered the most rigorous methods for treating long-range electrostatic interactions. Alternative approaches for the treatment of boundaries are reaction field based methods, which include a potential energy barrier that keeps the solvent molecules from diffusing away from the simulation system and reaction field terms that account for the absence of water (or the presence of vacuum) outside the barrier. If a reaction field is not present, then the water molecules at the surface will tend to interact to a greater extent with each other than with interior water molecules, leading to problems with the solvent density and solvent transport problems at the surface of the system that may adversely effect the properties of the entire simulation system. These approaches are typically used on spherical systems, although cylinders and planes have also been used. For treatment of long-range electrostatics, reaction field methods can use atom truncation, extended electrostatic, or fast multipole methods. Reaction field methods are most useful for systems too large to treat via PBC, such as protein-nucleic acid complexes. In all cases it is important that care be taken to ensure that the boundary conditions being used do not adversely affect the properties of the systems under study. This can typically be checked by performing simulations on the system and comparing properties calculated from the system with available experimental data, the most accessible being structural properties from X-ray or NMR experiments.

When preparing a PBC or reaction field calculation, the total size of the system is important. In general, the larger the system, as judged by the amount of solvent, the better, in that there are *less* likely to be adverse contributions from the boundary condition on the nucleic acid. The minimum amount of solvent surrounding the nucleic acid is dictated by the treatment of long-range electrostatics. With PBC the solvent box should be of such a size that the nucleic acid molecules in adjacent cells are further apart than (1) the real space cutoff for Ewald-based methods or (2) the atom truncation interaction cutoff distance. For reaction field based calculations, the distance from the nucleic acid to the edge of the solvation shell should be greater than the atom truncation distance. Concerning these distances, for Ewald methods a real space cutoff of 10 Å or greater and for atom truncation a cutoff distance of 12 Å or more are suggested. In all cases the simulator should perform tests to ensure that the applied boundary methods and treatment of long-range electrostatic interactions do not adversely affect the calculated result.

Once the geometry and size of the system to be studied are determined, a pure solvent system (*i.e.,* no DNA or RNA) of those dimensions should be built. This can typically be done via standard procedures included with the various modeling packages. These systems should then be subjected to MD simulations using the identical methods for treatment of the nonbonded interactions to be used in the final calculation. This will (1) allow the solvent to properly equilibrate with respect to itself and any ions included at this stage and (2) offer a test of the proposed methodology by ensuring that water density and transport properties are in satisfactory agreement with experiment. Once the solvent is equilibrated, it is overlaid onto the nucleic acid molecule, and all solvent molecules with nonhydrogen atoms within a given distance of solute nonhydrogen atoms (typically 1.8 Å) are then deleted. At this stage ions can be added to the system as required.

Ion placement in simulations of nucleic acids can have a significant impact on the computed results, consistent with the role of water activity on the structure of DNA. A comparison of the influence of ion placement on MD simulations was reported by Young et al. Methods applied included a Monte Carlo based method that places counterions at low energy positions around DNA using a sigmoidal distance-dependent dielectric function for calculation of the interaction energy of the ion with the environment. The second method used was based on calculation of the electrostatic potential around the DNA followed by the placement of ions at the most favourable locations in the potential. This is performed in an iterative fashion such that subsequent ions take into account previously placed ions. The final method used involved the placement of sodium counterions "6 Å from the P atoms along the bisector of the backbone OPO groups." All three methods yielded similar results when the electrostatic jnteractions were treated with the Ewald method. Additional methods include the replacement of water oxygens in a previously solvated system with counterions, with the selection criteria based on the interaction with the surrounding water molecules and the oligonucleotide. This can also be done in an iterative fashion that allows for the positions of ions to be sensitive to the presence of other ions. In this approach, the omission of water hydrogens from the interaction energy calculation eliminates orientational problems that would require energy minimiza-tion or dynamics, thereby significantly decreasing the computational requirements. Other simulators simply replaced randomly selected water molecules with ions. A final approach is to initially overlay the DNA or RNA with a solvent box or sphere that already contains ions at the desired concentration and has been previously equilibrated. In this method all ions and solvent molecules that overlap the solute are removed and additional ions are added at random positions or deleted, based on those furthermost from the DNA or RNA, to obtain electrical neutrality. The last two methods are particularly well suited for the placement of excess salt (counter- and co-ions beyond those needed to achieve electrical neutrality). In all cases, after the addition of ions it is essential that the solvent and ions be adequately equilibrated around the nucleic acid, as discussed in the following paragraph. The simulator should also be aware that ions strongly interacting at specific sites on the nucleic acid can strongly influence both the structural and dynamic properties.

Once the DNA or RNA has been overlaid with solvent and ions, the energy of the system is minimized, and the system is subjected to an equilibration MD simulation. These calculations should be performed with the nucleic acid atoms fixed or harmonically constrained to allow for the solvent to relax around the nucleic acid. Properties such as the potential energy and the solvent-solute interaction energy should be monitored as a function of simulation time to ensure that the system has relaxed to a satisfactory extent. In cases where the MD studies will be performed at constant volume (*i.e.,* NVT or NVE simulations) it is suggested that the pressure of the system be monitored to ensure that it is in the vicinity of 1 atm. If the pressure differs significantly from 1 atm, the distances used for the deletion of solvent molecules can be increased or decreased accordingly. Note that in simulations containing fixed or harmonically constrained atoms the calculated pressures may be incorrect, requiring that short MD simulations without constraints be performed to obtain accurate pressures. Once the solvent has adequately equilibrated around the solute, the entire system, including the nucleic acid, should be energy minimized. At this stage of the system preparation it is important that checks be performed to ensure that the system is still properly solvated. If problems in the simulation system are evident, the system should be resolvated in appropriate fashion to correct the problem and the equilibration redone.

C. Production MD Simulation

At this stage the production MD simulation can be initiated. It is generally preferable to perform simulations in the isobaric, isothermal (NPT) ensemble, which yields thermodynamic properties that correspond to the experimentally accessible Gibbs free energy. In some cases initiation of the MD simulations can involve gradual heating of the system, although initiation of the simulation at the final desired temperature is often sufficient. Several algorithms for NPT simulations are available, including Berendsen's method, the Langevin piston, and an extended method from Klein and coworkers. It should be noted that Berendsen's method does not correspond to a true ensemble, and problems associated with systematic oscillations in the system volume can occur. When the simulation is initiated it is important to closely monitor both structural and energetic properties to ensure that significant perturbations of the solute do not initially occur due to the applied methodology. If such perturbations are present, the system preparation and equilibration approach should be evaluated for potential problems.

D. Convergence of MD Simulations

Rigorous proof of convergence of MD simulations cannot be performed. Efforts can be made, however, to ensure that various properties calculated from a simulation have reached satisfactory levels of convergence. First, the simulations should be analyzed to determine if global (1) energetic and (2) structural properties have stabilized. Energetic stability is typically investigated by monitoring the potential energy versus time. During most simulations the potential energy initially relaxes, after which it fluctuates around a constant value for the remainder of the simulation. In NVE simulations the ratio of RMS fluctuations of the potential energy and kinetic energy can be monitored, with $RMS_{PE}/RMS_{KE} = 0.01$ indicating proper energy conservation. Analysis of structural properties versus time (*e.g.*, RMS difference with respect to the starting structure) will also typically show an initial relaxation followed by fluctuations around an average value for the remainder of the simulation. In certain cases these fluctuations may be relatively large, indicating the sampling of alternative conformations that may be biologically relevant. From the analysis of structural and energetic properties versus time, the initial portion of the production simulation during which relaxation of these properties occurs is discarded. Additional convergence tests are performed only on the remainder of the production simulation.

Tests of convergence on the remaining portion of the simulation can most readily be performed by calculating the desired property (*i.e.*, the property that is of most interest to the simulator) over different simulation time lengths and monitoring the change in the average value as a function of simulation time. Convergence is indicated by the lack of significant change in the average value of the property as the simulation time increases. Another method involves separating the total trajectory into independent blocks (*e.g.*, a 500 ps production simulation may be separated, into five 100 ps blocks), calculating average values for each block, and comparing the average values from the individual blocks. These block averages can also be used to calculate overall averages and standard errors for individual properties. This is an excellent method for obtaining the statistical significance of results from a simulation.

For an entire MD simulation, as well as for separate blocks from a simulation, determination of convergence of a property can be most rigorously carried out by calculating the time series of a property and determining its autocorrelation function and accompanying relaxation time. For adequate convergence the total MD simulation time or the block time should be approximately four times as long as the relaxation time. This test is also appropriate for determining the length of blocks that can be considered independent. It should be emphasized that the relaxation time and, accordingly, the amount of simulation time required for convergence are dependent on the property being investigated. Thus, the total required simulation time is dependent on the type of information that is to be obtained from a simulation. If it is determined that the properties of interest have not converged, then the production simulation should be extended.

E. Analysis of MD Simulations

Molecular dynamics simulations at an atomic level of detail contain enormous amounts of information, making rational analysis of simulations essential. In cases where studies have been designed in close collaboration with experiment, the information to be obtained is that which corresponds to the experimental data. Alternatively, systematic analysis of the MD trajectory may be performed to identify specific properties of interest. Properties determined from simulations of nucleic acids can be separated into those associated with structure, hydration, and energetics. Structural properties include the dihedrals associated with the phosphodiester linkage, the sugar and the glycosidic linkage, Watson-Crick base pairing, and a variety of intramolecular distances such as the minor groove width based typically on interstrand P distances. Nucleic acid sugar puckering is typically not analyzed on the basis of the individual dihedrals, but rather with respect to the concept of pseudorotation, where the sugar conformation on is defined in terms of two variables, the pseudorotation angle and the amplitude. An important class of structural properties comprises the helical or helicoidal parameters. These terms define the orientations of the bases with respect to a global helical axis or locally with respect to adjacent bases or basepairs. Both methods have been used extensively, and a number of programs have been developed to perform this analysis, including CURVES, FREEHELIX, and a program by Babcock and Olson. The CURVES method has been incorporated into an analysis and plotting package, DIALS and WINDOWS, which analyzes MD trajectories as well as individual coordinate files and presents data, including time series, in a highly compact fashion. When using these programs the user should be aware of both the global and local definitions of the helicoidal parameters. Whereas in idealized helices the two definitions yield similar or identical results, the global method is preferable for standard helices and the local definition is more suited for distorted helices, such as those found in DNA-proteiri complexes, where a global helical axis cannot be rigorously defined. When performing structural analysis it is often best to initially calculate overall averages, use that information to identify terms that may be of interest, and then analyze those terms via time series or probability distributions. Probability distributions of DNA dihedrals and helicoidal parameters calculated via CURVES have been published, and those based on FREEHELIX are available.

Hydration properties, including interactions with ions, are strong determinants of DNA structure. Hydration numbers can be determined on the basis of a variety of criteria for comparison with individual crystal structures or with data from surveys of the NDB. This type of analysis is often performed on different portions of the DNA such as the minor groove and is typically limited to first shell hydration. Complicating such analysis is the overlap of waters hydrating different sites. For example, a water molecule hydrating the sugar O4' atom may also be hydrating atoms in the minor groove, and it has been shown that counterions can replace water molecules in the first shell of hydration. Hydration can also be studied with respect to the dynamic properties of the water molecules. Examples include changes in water dynamics around DNA leading to local alterations of the dielectric environment and results indicating that decreased water activity due to increased salt concentration is associated with changes in water mobility rather than hydration number. Dynamics properties that have been analyzed include diffusion constants and rotational correlation times. Rotational correlation times may be analyzed on the basis of the water dipole axis, the axis perpendicular to the plane of the water, and the H—H axis. These motions are associated with molecular twisting (dipole axis), rocking (perpendicular), and wagging (H—H vector), respectively. Difficulties in determining correlation times of individual waters occur with water molecules that undergo little motion on the time scale of the simulation, leading in some cases to undefined correlation times. A method of analysis based on the initial 4 ps of the decay of the autocorrelation function can overcome some, but not all, of these problems.

Supplementing direct analysis of structural and hydration properties is the use of energetic analysis. For example, instead of monitoring individual Watson-Crick hydrogen bonds, interaction energies between basepairs can be monitored. Similarly, interaction energies between water and the nucleic acid can be determined to supplement the information discussed in the preceding paragraph. The advantage of energetics over direct structural or hydration information is the ability to more readily take into account all possible contributions, not just those envisioned by the simulator. Furthermore, it may be possible to better gauge the relative impact of different types of interactions on the trajectory obtained from an MD simulation. Such analysis has been used to identify energetic and, subsequently, structural phenomena allowing for the barrierless extension of DNA beyond its canonical B form and contributions to the barrier to strand separation [86]. Typically, energetic analysis should be performed using the same truncation scheme as that used in the MD simulation. This can be problematic in Ewald simulations, as it is currently not possible to use Ewald summations to determine interaction energies. One suggestion is to calculate interaction energies between atoms with the real-space truncation distance used in the simulation. An interesting alternative to direct analysis of interaction energies is the recalculation of solvation free energies using continuum models on individual nucleic acid structures from a solvated MD simulation. This approach has been used to understand the equilibrium between the A and B forms of DNA as a function of water activity. Although the method successfully ordered the equilibrium when water activity was modified with ethanol, it was not successful at predicting the salt effect, which may be associated with atomic detail interactions not modeled in continuum approaches. Overall, energetic analysis of MD simulations offers an additional method to analyze simulations, often allowing for identification of structural contributions that are difficult to identify via direct structural analysis.

V. CONCLUSION

In this chapter we have attempted to give an overview of the types of nucleic acid systems that are accessible to computational study. These vary from nucleosides, through duplex DNA and RNA, up to nucleic acid-protein complexes. Accessibility to both longer time scales and larger systems is expected to increase as advances in both computational power and methods continue to occur. Although the work cited cannot be considered complete, it should allow the reader to access information required for moving into and obtaining a more general background of the field of nucleic acid simulations.

6 DNA EXPRESSION DATA

Short, conserved sequence elements, or DNA motifs, located upstream of the transcriptional start site are often binding sites for transcription factors that regulate a group of genes involved in similar cellular functions. Additional genes with this motif in their upstream regions are good candidates for involvement in the same cellular function. Thus, upstream regulatory motifs can provide powerful hypotheses about links in the genetic regulatory network. These upstream regulatory motifs can be discovered computationally by local alignment of up-stream regions from coregulated sets of genes. In order to discover a motif computationally by using local alignment methods, an accurate list of coregulated genes, or a regulon, is needed.

The AlignACE program can tolerate some noise in the alignment process from extra sequence that does not contain the motif of interest, but too much extra sequence will prevent the motif from being aligned. Thus, we must employ additional experimental and computational methods to obtain accurate regulon predictions. There are several methods that can be used for regulon prediction. Each of these methods will be discussed in more detail below. mRNA expression data have proved to be very useful for predicting coregulated sets of genes. Genes that are upregulated or downregulated in a single experimental condition are likely to be co-regulated, as are genes with similar mRNA expression profiles across many conditions or time points.

There are also several computational methods for predicting regulons that are based on comparative genomics. Here we will discuss several methods for predicting regulons, motif discovery algorithms, parameters for evaluating the significance of a proposed regulatory motif, and applications of motif finding using. AlignACE in bacterial genomes and in *Saccharomyces cerevisiae.*

REGULON PREDICTION

Groups Derived from mRNA Expression Data

A powerful method for predicting regulons is to look for patterns in mRNA expression data from DNA chips or microarrays. This has been done for several sets of experimental conditions in *S. cerevisiae.* One particularly effective method for obtaining coregulated sets of genes from mRNA expression data is to cluster genes

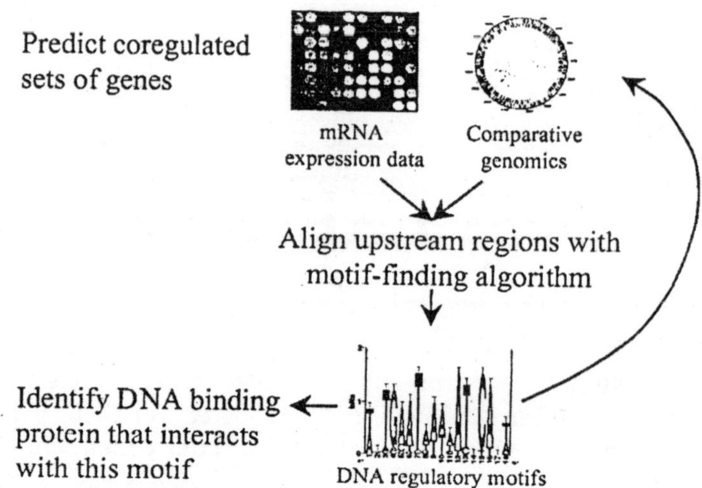

Fig. 6.1. Process of motif discovery. The first step is to predict coregulated groups of genes, or regulons. This can be done either using expression data from microarrays or by computational methods based on comparative genomics. Next, the upstream regions of genes within the predicted regulons are aligned, using a motif discovery algorithm. The resulting DNA regulatory motifs can be illustrated using a motif logo, as shown here (Schneider and Stevens, 1990). The height of a stack of letters is proportional to the information content, and the relative frequency of each base is given by its relative height. The presence of a significant regulatory motif found in this manner can be used to refine the contents of the predicted regulons, because the presence of a regulatory motif provides additional evidence that this set of genes is coregulated. Experiments can also be designed to test the function of the regulatory motif and to identify the transacting factor that binds to it.

on the basis of their expression profiles over a large number of experimental conditions, or over a large number of time points. Different algorithms for clustering such data are reviewed in Sherlock (2000). Many known motifs, as well as new motifs, have been found in the upstream regions of S. *cerevisiae* genes clustered by this method. A database of expression data in S. *cerevisiae* for 217 experimental conditions is available.

Groups Derived from Conserved Operons

Operons (tandem genes, usually functionally related, that are transcribed onto a single mRNA) are the main transcriptional regulatory units in bacteria. Often several bacterial operons that share an upstream binding site for a common transcription factor are coregulated to form a higher-order regulatory unit (a regulon). In different microbial organisms the genes that make up a regulon are often assorted into operons in different combinations. Two genes that are spatially separated on the chromosome in one organism, but spatially close in several other microbial species, are good candidates for being coregulated or functionally related. Such pairings between genes have been used to construct larger groups of genes that are predicted to be functionally coupled. Thus regulons can often be reconstructed by looking at conserved operon composition across other genomes. Fig. 6.3 shows a simple example illustrating this method.

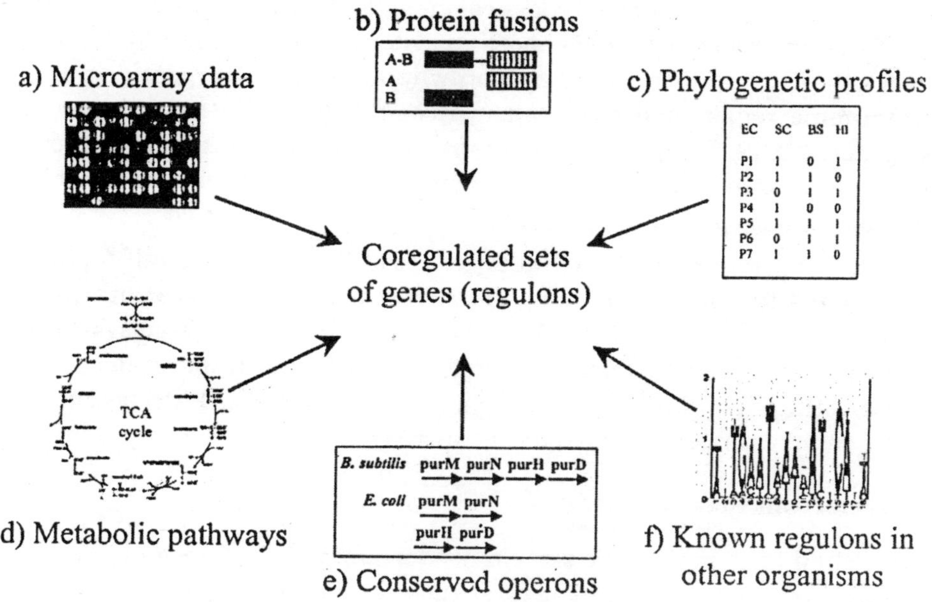

Fig. 6.2. Methods for regulon prediction. Method (*a*) relies on experimental mRNA expression data from microarrays. Methods (*b*), (*c*), and (*e*) are comparative genomics methods based purely on sequence information. Methods (*d*) and (*f*) incorporate known biology of well-studied organisms.

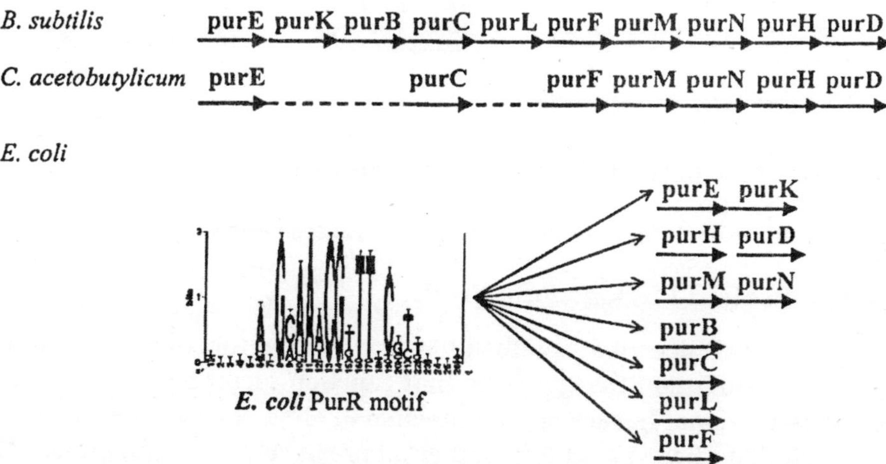

Fig. 6.3. Predicting the *E. coli* purine biosynthesis regulon based on conserved operons. In B. subtilis, C. acetobutylicum, and several other bacteria, a large group of genes involved in purine biosynthesis is found together in a single operon. In *E. coli* these genes are split into seven different operons that are coordinately regulated by binding of the transcription factor PurR to a specific DNA motif found in the upstream region of each operon. Each colour (not shown here) represents a separate but coregulated operon in *E. coli*.

In eukaryotes operon structures are less common, and each individual gene typically has an instance of the upstream regulatory motif in order to achieve coordinated transcription. However, two genes involved in the same pathway are often divergently transcribed from the same intergenic region and share an upstream regulatory site. Hence, divergently transcribed genes can be thought of as operons in a broader sense. Several of the groups constructed in this manner correspond to known metabolic pathways, including the purine and arginine biosynthesis pathways in *E. coli.*

By aligning the upstream regions of these groups of genes with a motif-finding algorithm, the regulatory motifs for the DNA-binding proteins that regulate these pathways (PurR and ArgR, respectively) can be found. Therefore, the purine and arginine biosynthesis regulons, as well as the binding motifs for PurR and ArgR, can be found "from scratch," with no input of biological data other than the genome sequence itself. This method will become increasingly powerful as the number of genomes sequenced increases.

Groups Derived from Protein Fusions

Two genes found to be part of the same multidomain protein fusion in one organism, but scattered in the genome in another organism, are good candidates for being functionally related. Fig. 6.4 shows a simple example illustrating this method. This method will become more powerful as the amount of genome sequence information increases.

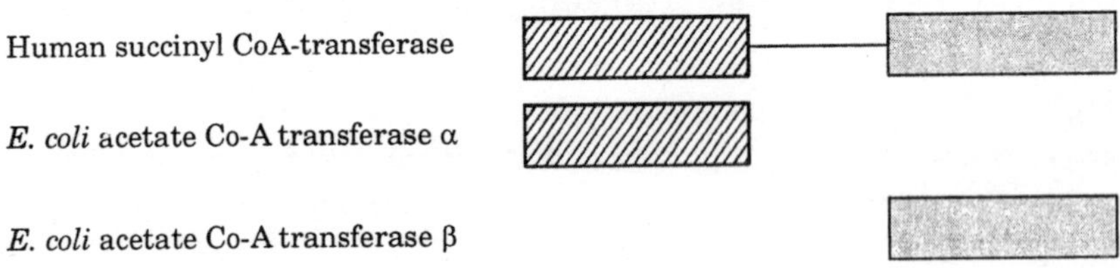

Fig. 6.4. Illustration of the protein fusion method of predicting functionally related genes. *E. coli* acetate Co-A transferases a and ft are pre-dicted to interact because their human homologues are fused into a single polypeptide chain. Boxes represent protein domains.

Groups Based on Protein Phylogenetic Profiles

Pellegrini et al. (1999) describe a method for determining functionally related groups of genes based on the assumption that proteins that function together in the cell are likely to evolve in a correlated fashion. In each species functionally linked genes (*i.e.,* all of the proteins that make up a certain pathway) tend to be either all preserved or all eliminated. Thus, genes present in the same set of organisms and absent in the same set of organisms are likely to be functionally linked. Pellegrini et al. constructed groups of evolutionarily correlated genes by constructing a phylogenetic profile for each protein (a vector representing whether an orthologue to this protein is present or not present in each organism) and clustering these vectors based on their similarity. Fig. 6.5 illustrates this method for predicting functional interactions between genes.

Groups Derived from Metabolic and Functional Pathways

Metabolic and functional group categories tabulated in existing data-bases are also useful data sets for motif finding. Some examples of such databases are KEGG (Kyoto Encyclopedia of Genes and Genomes; Ogata et al., 1999), MIPS (Munich Information Center for Protein Sequences; Heinemeyer et al., 1998), and YPD (yeast protein database; Hodges et al., 1999).

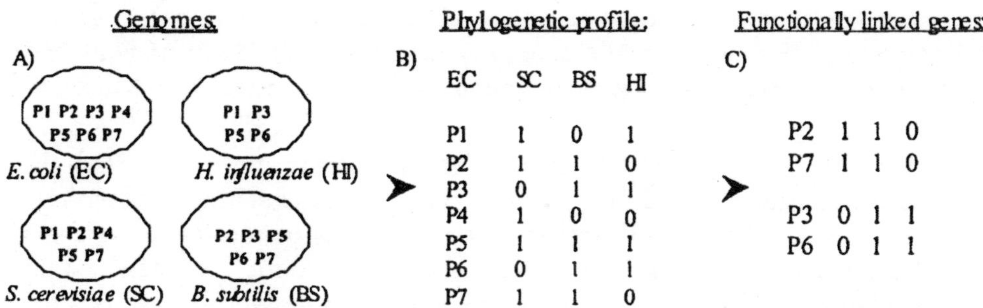

Fig. 6.5. Predicting functionally related genes based on conserved evolution. Panel (A) illustrates which of the seven *E. coli* proteins (P1-P7) are found in each of three additional bacterial genomes. In panel (B), a 1 or a 0 indicates that the protein is found or not found in this organism, respectively. The rows in panel (B) constitute the phylogenetic profile for this protein. By grouping genes with similar phylogenetic profiles, we obtain functionally linked groups of genes, illustrated in panel (C). P2 and P7 are predicted to be functionally linked, as are P3 and P6.

Since these groups are based on a compilation of experimental data in these organisms, the motifs found within these groups are biased toward known motifs. However, new motifs have also been found in the upstream regions of these groups. Hughes et al. (2000) used the MIPS and YPD groups to search for motifs in *S. cerevisiae,* and McGuire et al. (2000) used the KEGG groups to search for motifs in 17 complete prokaryotic genomes.

Groups Derived from Homologues to Footprinted Regulons in Other Species

Groups based on known regulons in other organisms, such as orthologues to the *E. coli* genes known to be regulated by 55 DNA-binding proteins, can also be used to search for upstream regulatory motifs. Such methods have been used to identify known regulatory motifs that are conserved in other organisms, as well as different motifs that could regulate orthologous sets of genes in different organisms.

Combining Several Methods

Information obtained from a combination of the above methods can be used to obtain better predicted regulons. For example, McGuire and Church (2000) combined the three methods described above that rely only on information derived from comparative genomics (methods based on conserved operons, protein fusions, and phylogenetic profiles) to predict regulons in 24 complete genomes. These regulons were then used to search for new regulatory motifs.

MOTIF-FINDING ALGORITHMS

Once a group of genes believed to be coregulated (a predicted regulon) has been obtained by one of the methods described above, DNA regulatory motifs can be discovered by local alignment of the upstream regions of this set of genes. Several different algorithms have been used for motif discovery within DNA sequences. Only a sampling of the large number of motif-finding algorithms available will be mentioned here. The Gibbs sampling algorithm is based on the statistical method of iterative sampling. This algorithm has been optimized for use with DNA sequence alignments in the program AlignACE. The AlignACE algorithm, along with applications of AlignACE in bacterial genomes and in *S. cerevisiae,* will be discussed in more detail below.

MEME (Multiple EM for Motif Elicitation) is a motif-finding algorithm that uses an expectation maximization algorithm to locate one or more repeated patterns in the input sequence. MEME searches by building statistical models of motifs (matrices of discrete probability distributions) and maximizing the posterior probability of these models, given the data. MACAW (Multiple Alignment Construction and Analysis Workbench) is a motif-finding algorithm that allows the user to construct multiple alignments by locating, analyzing, editing, and combining blocks of aligned sequence segments.

Several methods based on detection of over-represented oligonucleotide frequencies have been proposed, including a method by van Helden et al. (1998). This simple and fast motif-finding method defines the statistical significance of a site based on tables of oligonucleotide frequencies in noncoding regions. However, this method is limited to short, highly conserved motifs.

AlignACE Algorithm

For clarity in describing the Gibbs sampling algorithm used in AlignACE, we will present the simple case of searching for a single pattern of fixed width W within each of N input sequences (S_I... S_N). However, the AlignACE algorithm allows for automatic detection of variable pattern widths as well as multiple motif instances per input sequence. The pattern is described by a probabilistic model of residue frequencies at each position, q_{ij}, where i ranges from 1 to W and j ranges from 1 to 4 (A, C, T, and G). We also consider the background frequencies for each base, p_j, where j ranges from 1 to 4. The p_{ij} are different for each organism. The start site for the motif within each input sequence is designated by a set of positions a_k, where k ranges from 1 to N. To choose the best pattern, the algorithm must choose values of a_k that maximize the ratio of the pattern probability to the background probability.

The first step is to choose random start positions (a_k). The next step (the predictive update step) is to choose one of the N sequences at random (z), and calculate the q_{ij} for the current positions, excluding z. The q_{ij} are calculated using the equation

$$q_{ij} = \frac{c_{ij} + b_j}{N - 1 + B}, \qquad ...(6.1)$$

where c_{ij} is the count of nucleotide j at position i, b_j is a residue-dependent "pseudocount" suggested by Bayesian statistical analysis for the purpose of pattern estimation, and B is the sum of the Bj.

The next step (the sampling step) is to consider every possible segment x of length W within z, and to calculate the probability Q_x of generating each such segment x according to the current pattern probabilities q_{ij}, and the probability. P_x of generating each segment according to the background probabilities p_j. A weight equal to $\Lambda x = Q_x/P_x$ is assigned to segment x, and with each segment so weighted, a random position is selected. This new position becomes the new value for a_z, and the algorithm is iterated. The most probable alignment is the one that maximizes the equation

$$F = \sum_{i=1}^{W} \sum_{j=1}^{4} c_{ij} \log \frac{q_{ij}}{p_j}, \qquad ...(6.2)$$

where c_{ij} is the count of nucleotide j at position i. As the pattern description calculated in the predictive update step becomes more accurate, the determination of the location in the sampling step also becomes more accurate.

Several optimizations of the Gibbs sampler algorithm were added in the AlignACE program for use in finding DNA motifs : both strands of DNA are now considered; the near-optimum sampling method was improved in order to obtain higher scoring alignments; simultaneous multiple motif searching was replaced with an iterative masking approach; and the model for base background frequencies was set to equal the background nucleotide frequencies in the genome being considered.

INDICES FOR EVALUATING THE SIGNIFICANCE OF A MOTIF

When using a motif discovery algorithm such as AlignACE, often a large number of motifs are identified. Additional indices are necessary to assess these motifs and identify those most likely to correspond to specific regulatory motifs that are biologically relevant. Highly conserved and common genomic elements, such as repetitive elements and ribosome binding sites, are also found by AlignACE and must be separated from potential binding sites for specific transcription factors. Several indices used in assessing newly discovered motifs are described below.

MAP Score

The MAP score is the index used by AlignACE to determine the statistical significance of alignments sampled, given the composition of the input sequence. MAP scores are normalized so that the score for an alignment of zero sites is assigned a score of 0. The MAP score is higher for similar motifs with greater numbers of aligned sites and for more tightly conserved motifs, and lower for an identical alignment of sites derived from a larger set of input sequences, motifs with more dispersed information content, and motifs rich in nucleotides more prevalent in the genome. The MAP score is useful for ruling out alignments that are not statistically significant, but this score is not very useful in discriminating real regulatory motifs from other statistically significant motifs found by AlignACE because many nonspecific chromosomal features have high MAP scores (i.e., ribosome binding sites and repetitive elements).

Specificity Score

The specificity score is a measure of how specific a motif is for the sequence in which it was aligned, compared to the genome as a whole. This is the most useful index for selecting functional motifs. Two different specificity score measures have been described: the specificity score (S) is more useful for yeast and eukaryotes, while the site specificity score (S_{site}) is more useful for bacterial sequences and genomes containing operons and multiple copies of a motif within a single upstream-region. The main difference between these two specificity scores is that S is a measure of how many of the top predicted ORF targets in the genome were used to align the motif originally, whereas S_{site} measures the probability of obtaining the observed fraction of the top motif instances in the genome within the sequence input to AlignACE.

For each motif a position-specific weight matrix is constructed and the whole genome is searched for additional instances of the motif. The Berg and von Hippel weight matrix, implemented in the ScanACE program, is used to perform this search. To calculate S_{site}, the top 200 ScanACE hits are considered, and the number of these sites located within the upstream regions used to align the motif is calculated. The probability of obtaining this number of sites within this subset of the genomic sequence by random means is calculated using the following equation (the hypergeometric distribution):

$$S_{site} = \sum_{i=x}^{\min(s_1,s_2} \frac{\binom{s_1}{i}\binom{N-s_1}{s_2-i}}{\binom{N}{s_2}} \qquad \text{...(6.3)}$$

N is the total number of possible sites (equal to the number of base pairs in the genome), s_1 is the number of ScanACE hits considered (200), s_2 is the total number of possible sites in the set of upstream regions used to align the motif (equal to the number of base pairs in the sequence input to AlignACE), and i is the intersection of s_1 and s_2. To calculate S, the 100 ORFs that are the most likely targets for the factor binding the motif are selected (the ORFs with the strongest sites in the region between -100 and 500 bp upstream of the translational start). This list of gene targets is compared against the list of genes in the group used to find the motif, using the same equation as above; however, now N is the total number of ORFs in the genome, s_1 and s_2 are the numbers of ORFs in the group used to find the motif and in the list of target genes, respectively, and x is the number of ORFs in the intersection of the two lists.

Positional Bias

This is a measure of bias in the locations of the top ScanACE hits in the genome relative to the start codon of the closest gene. The subset of the top 200 ScanACE hits that fall within a certain distance s of the start codon (300 bp for bacteria and 600 bp for *S. cerevisiae*) are considered. A smaller window w (30 bp for bacteria and 50 bp for *S. cerevisiae*] is then chosen such that it contains the maximum number of hits within s. The probability of seeing the observed number of sites or greater in a window of size w, out of a sample space of size s, is calculated as follows :

$$P = \sum_{i=m}^{t} \binom{s_1}{i} \left(\frac{w}{x}\right)^{i} \left(1-\frac{w}{s}\right)^{t-i} \qquad \text{...(6.4)}$$

Similarities Between Motifs

The CompareACE program, used to compare two motifs by finding the best alignment, calculates the Pearson correlation between position-specific scoring matrices. Two motifs with a CompareACE score greater than 0.7 are considered similar. New motifs can be compared against databases of known motifs using CompareACE, including a database of footprinted *E. coli* motifs. CompareACE scores can also be used as a metric to cluster motifs. After calculating a matrix of pairwise CompareACE scores, motifs can be clustered using a simple joining algorithm. Often when a large set of motifs is analyzed, multiple instances of similar motifs will be present, and clustering will reduce the set of motifs to unique clusters.

Palindromicity

Many known bacterial motifs are palindromic. To identify palindromic motifs, the CompareACE program can be used to compare a motif and its reverse complement. The same CompareACE cutoff score as for comparing motifs to one other (0.7) has been used.

ALIGNACE APPLICATIONS

Applications of AlignACE in Complete Bacterial Genomes

AlignACE has been applied to several kinds of potentially coregulated groups of genes in prokaryotic genomes. McGuire et al. (2000) used groups derived from conserved operons in the WIT database, groups derived from metabolic pathways in the KEGG database, and groups derived from homologues in other species to members of known *E. coli* regulons to search for regulatory motifs in 17 prokaryotic genomes. Specific, new motifs were found using each grouping method and in each organism.

McGuire and Church (2000) combined and compared three techniques for predicting functional interactions based on comparative genomics (methods based on conserved operons, protein fusions, and correlated evolution) in order to construct predicted regulons in 22 complete prokaryotic genomes. These predicted regulons were then used to find a large number of new motifs using AlignACE. Identifying motifs in bacterial genomes is complicated by the presence of operons. It is difficult to locate the regulatory region for a gene within an operon.

If the gene of interest lies within an operon, its promoter could lie several genes upstream. Thus, we must include several possible intergenic regions to ensure that we have included the segment containing the regulatory region. This is possible in bacteria because regulatory regions are shorter; hence, we can include several intergenic regions without adding so much extra sequence that the motif-finding algorithm will no longer be able to align the regulatory motif. An additional problem associated with motif finding in bacteria is that there are fewer instances of most regulatory motifs because there is usually only one instance per operon instead of one instance per gene.

Motif conservation can used to aid in finding new motifs by grouping upstream regions from closely related bacteria, thus increasing the number of instances of each motif in the sequence to be aligned. The effectiveness of AlignACE in finding known *E. coli* motifs was analyzed. Of the 32 *E. coli* footprinted regulons in our database with 5-100 known binding sites (Robison et .al., 1998), 26 have motifs that can be found by AlignACE (81%). The groups obtained from homologues in other species to members of known *E. coli* regulons were used to study conservation of *E. coli* DNA motifs in other organisms, as well as to identify potential new mechanisms for regulating the same cellular process in more distantly related organisms. At least 30% of the known *E. coli* DNA regulatory motifs were found to be conserved in closely related bacteria. A new motif can indicate either a different mechanism for regulating a similar cellular process or divergence of binding site residues in a conserved DNA-binding protein.

Applications of AlignACE in Saccharomyces Cerevisiae

AlignACE has been used for motif finding in *S. cerevisiae* with mRNA expression data, as well as with functional group categories (Hughes et al., 2000). Selecting upstream regions is more straightforward in yeast because there are no operons. Because intergenic regions are longer in yeast, and regulatory elements are often found farther upstream than in bacteria, longer stretches of noncoding sequence have been used for motif finding (up to 600 bp). Roth et al. (1998) first used AlignACE to find motifs, employing regulons derived from experiments employing Affymetrix GeneChips in the following three systems: galactose utilization, mating type, and heat shock response.

By aligning the groups of genes with the largest fold changes in these conditions, using AlignACE, they were able to find the binding sites for many of the factors known to be involved in these processes. Tavazoie et. al., (1999) looked at a time series of mRNA abundance data measured over two synchronized *S. cerevisiae* cell cycles. These data were used to group 3000 yeast genes into 30 clusters with similar temporal expression profiles. These groups were highly enriched with genes involved in the same cellular processes.

By aligning the genes making up these clusters with AlignACE, they were able to find many known motifs as well as several new ones. Hughes et al. (2000) used AlignACE to search for motifs upstream of the genes making up the MIPS and YPD (yeast protein database) functional groups. Many known and new motifs were found in this analysis, including a new motif found upstream of genes involved in proteasome formation. This motif has been independently identified as the binding site for Rpn4 in Church's lab as well as by Mannhaupt et al. (1999). Several other potentially functional new *S. cerevisiae* motifs are being experimentally tested.

CONCLUDING REMARKS

The use of motif-finding algorithms to discover new regulatory motifs is a powerful method for understanding the interconnections in genetic regulatory networks. Groups of genes (predicted regulons) obtained from mRNA expression data can be extremely powerful in this regard, though such groups of genes have the drawback that they are dependent on the experimental conditions chosen. These groups can be complemented by theoretically obtained predicted regulons that are less dependent on the experimental conditions chosen.

The presence of a significant regulatory motif shared by the members of a predicted regulon provides additional evidence that this group of genes is actually coregulated. The motif-finding program AlignACE has been used successfully to predict new motifs in *S. cerevisiae*, as well as in over 20 complete bacterial genomes. A new motif discovered by AlignACE upstream of proteasome subunits in *S. cerevisiae* has been confirmed experimentally, and numerous other new motifs should be verified as well. AlignACE has also been used to analyze conservation and divergence of regulatory mechanisms between microbial genomes. The methods that we have described here for predicting regulons and their cis-regulatory motifs will become increasingly powerful as the size of the sequence database and the number of complete genomes grows.

The discovery of DNA regulatory motifs will be facilitated by having complete genomes for a greater number of closely related organisms, since sequences from closely related organisms can be pooled to find conserved motifs that are found in only a few copies in each genome. In addition, regulon-prediction methods based on comparative genomics (including the methods described here that involve conserved operons, protein fusions, and phylogenetic profiles) will yield more information about each organism as more complete and partial genome sequences become available. Several complete genomes have recently been made available, including *Drosophila, C. elegans*, and several human chromosomes. As many as 50 additional bacterial genomes could become available by 2002, along with the human genome sequence. The framework described here for predicting regulons and their cis-regulatory motifs will soon become increasingly powerful.

COMPUTATIONAL MODELS OF CELL

In modeling whole cells we try to understand complex properties of cells by combining interlocking transport and regulatory mechanisms. We use a modular approach and develop models of each individual process separately using available experimental data. We then construct progressively more complete models by combining components to understand how they work together. Sometimes, we proceed in the opposite order, beginning with a comprehensive model, which we simplify in order to determine the minimal essential elements. One particularly useful simplification technique is to exploit separation of time scales to set fast processes to equilibrium. Three model systems are investigated in this chapter; the bullfrog sympathetic ganglion neuron the pituitary gonadotroph and the pancreatic β-cell.

The order chosen is pedagogical and is in fact opposite to the historical order in which the models were developed. These models focus on the consequences of integrating plasma membrane (PM) and endoplasmic reticulum (ER) membrane fluxes into a coordinated system for control of membrane potential and Ca^{2+} concentrations. The concentration of cytoplasmic Ca^{2+}, $[Ca^{2+}]_i$, is of great importance for the life and death of cells. It is a key regulator of many cell processes, such as secretion, gene transcription, and apoptosis.

We will restrict our attention to models with a small number of discrete, well-mixed compartments; diffusion in spatially extended compartments will be discussed elsewhere in the chapter. Other models for these particular systems can be found. These simplified models incorporate ion pumps and voltage and ligand gated channels like those discussed in previous chapters, and provide illustrations of the modular approach. A common dynamical theme in all three models is oscillations, some driven by the ER, some by the plasma membrane, and some by interactions between the two. We will see, that the combination of fast positive feedback with slow negative feedback is a ubiquitous way for cells to produce oscillations.

Models of ER and PM Calcium Handling

The most striking feature of Ca^{2+} in cells is its low concentration, approximately 0.1 μM at rest, and only 1 μM or so at its peak. In contrast, Na^+ and K^+ are found in millimolar concentrations. Cells need to keep $[Ca^{2+}]_i$ so low because Ca^{2+} binds to many proteins and modifies their enzymatic properties. Thus, rises in Ca^{2+} must be kept brief and highly localized to avoid runaway activation of enzymatic cascades. These needs are met by two basic mechanisms, buffering and sequestration. Buffers are specialized Ca^{2+}-binding proteins that soak up 95-99% of the Ca^{2+} in the cytosol. Ca^{2+} is sequestered in internal stores—the

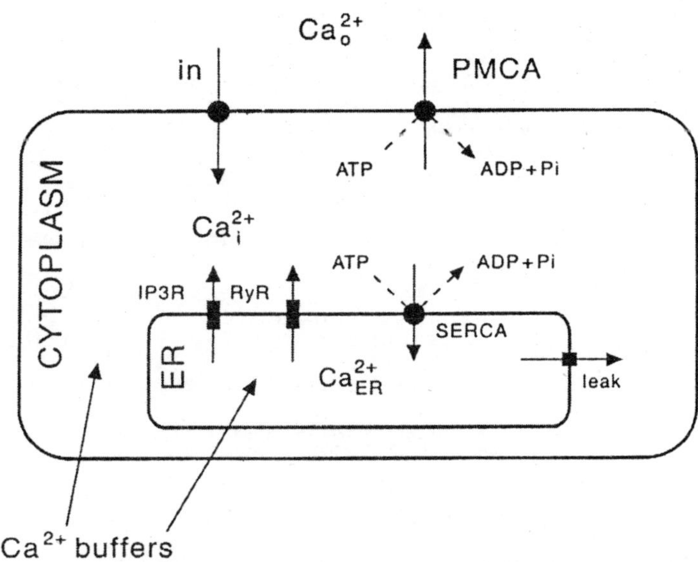

Fig. 7.1. Basic elements involved in Ca^{2+} signaling in cells.

sarcoplasmic reticulum (SR) in muscle cells, and the endoplasmic reticulurn (ER) in other cells—by proteins that hydrolyze ATP to move Ca^{2+} against steep concentration gradients. These proteins are called sarco/endoplasmic Ca^{2+}-ATPase (SERCA) pumps. Other pumps, plasma membrane Ca^{2+}-ATPases (PMCA), remove Ca^{2+} from the cell. ER and SR membranes contain ion channels, analogous to but different from plasma membrane channels, that allow Ca^{2+} back out of the stores into the cytoplasm when needed. Whereas all cells have Ca^{2+} pumps for negative feedback and homeostasis, some cells have evolved ion channels such as the ryanodine receptor (RyR) and the IP_3-Receptor, which are both activated and inhibited by Ca^{2+} and provide both positive and negative feedback.

The large ratios of bound to free Ca^{2+} and sequestered to cytoplasmic Ca^{2+} mean that brief channel openings can lead to large excursions of free cytoplasmic Ca^{2+} or oscillations that can be exploited for signaling. Mitochondria and vesicles may also act as Ca^{2+} reservoirs and have their own specialized uptake and release mechanisms for Ca^{2+} and other ions. These have not been modeled as extensively, and will be treated only in passing here. Ca^{2+} indicator dyes, which are exogenous buffers that fluoresce differentially depending on $[Ca^{2+}]_i$ are the most common way to measure the concentration of free Ca^{2+}. As we will see below, buffers change (generally, slow down) the dynamics of Ca^{2+} in cells.

The optical measurement of Ca^{2+} is discussed further elsewhere. Some cells possess Ca^{2+}-dependent K^+ or Cl- channels, and one can use the currents generated by these channels as an alternative measurement. A third approach is to use aequorin, a naturally occurring protein found in jellyfish, which luminesces when bound to Ca^{2+}. Aequorin can be targeted to Ca^{2+} stores in cells of other organisms by genetic techniques.

Flux Balance Equations with Rapid Buffering

We begin with a general description applicable to any cell with a cytosolic compartment (subscripted by i, originally for "intracellular") and an ER (subscripted by ER) and one species of Ca^{2+} buffer in each compartment. That gives a total of four species, with bound and free Ca^{2+} in each of the two compartments.

We define N_i = total number of Ca^{2+} ions, bound and free, in the cytoplasm and N_{ER} = total number of Ca^{2+} ions in the ER. Then

$$[Ca^{2+}]_i^{tot} = \frac{N_i}{\overline{V}_i}, [Ca^{2+}]_{ER}^{tot} = \frac{N_{ER}}{\overline{V}_{ER}}, \quad ...(7.1)$$

where the $\overline{V}$'s representing the volumes of the cytosol and ER have overbars to distinguish them from membrane potentials.

The fundamental physical principle used in constructing equations for the four Ca^{2+} concentrations is conservation of Ca^{2+} ions. The number of ions in each compartment changes due to fluxes across each of the membranes separating the compartments:

$$\frac{dNi}{dt} = J_{PM}^{in} - J_{PM}^{out} - J_{ER}^{in} + J_{ER}^{out} \quad ...(7.2)$$

$$\frac{dN_{ER}}{dt} = J_{ER}^{in} - J_{ER}^{out} \quad ...(7.3)$$

The J's have units of μmoles/s and represent common fluxes found in cells. In terms of concentrations we have here we have absorbed the volumes into the j's, giving them units of concentration/time, typically μM/s.

$$\frac{d[Ca^{2+}]_i^{tot}}{dt} = \frac{1}{\overline{V}_i}\left(J_{PM}^{in} - J_{PM}^{out} - J_{ER}^{in} + J_{ER}^{out}\right) = j_{PM}^{in} - j_{PM}^{out} - j_{ER}^{in} + j_{ER}^{out}, \quad ...(7.4)$$

$$\frac{d[Ca^{2+}]_{ER}^{tot}}{dt} = \frac{1}{\overline{V}_{ER}}\left(J_{ER}^{in} - J_{ER}^{out}\right) = \frac{\overline{V}_i}{\overline{V}_{ER}}\left(j_{ER}^{in} - j_{ER}^{out}\right) \quad ...(7.5)$$

Table 7.1. Key to Ca^{2+} Levels and Fluxes.

Ca^{2+} levels	$[Ca^{2+}]_i^{tot}$	bound (buffered) + free cytosolic Ca^{2+}
	$[Ca^{2+}]_{ER}^{tot}$	bound + free ER Ca^{2+}
	$[Ca^{2+}]_i$	free cytosolic Ca^{2+}
	$[Ca^{2+}.B]_i$	bound cytosolic Ca^{2+}
Fluxes	J_{PM}^{in}	influx through PM Ca^{2+} channel
	j_{PM}^{in}	j_{PM}^{in} divided by cytosolic volume $\overline{V}_i$
	J_{PM}^{out}	extrusion by plasma membrane Ca^{2+} ATP-ase
	j_{PM}^{out}	J_{PM}^{out} divided by cytosolic volume $\overline{V}_i$
	J_{ER}^{in}	pumping into ER by SERCAs
	J_{ER}^{out}	flux from ER to cytosol through IP3R or RYR

Now we come to grips with the buffering equation. The term $[Ca^{2+}]^{tot}$ consists of free Ca^{2+} plus Ca^{2+} bound to buffer B. In the cytosol,

$$[Ca^{2+}]_i + [Ca^{2+}.B]_i = [Ca^{2+}]_i^{tot}, \quad ...(7.6)$$

$$[B]_i + [Ca^{2+}.B]_i = [B]_i^{tot} \quad ...(7.7)$$

with similar expressions for the ER. We avoid adding a differential equation to describe these reactions by assuming rapid equilibrium between Ca^{2+} and buffer;

$$[Ca^{2+}]_i = \frac{K_i[Ca^{2+}.B]_i}{[B]_i} \quad ...(7.8)$$

Methods for treating Ca^{2+} buffering are discussed in more detail elsewhere. Combining (7.6 and (7.8) gives

$$[Ca^{2+}]_i^{tot} = [Ca^{2+}]_i\left(1+\frac{[B]_i}{K_i}\right) \quad ...(7.9)$$

Combining (7.7) and (7.8) gives

$$[B]_i^{tot} = [B]_i\left(1+\frac{[Ca^{2+}]_i}{K_i}\right). \quad ...(7.10)$$

Finally, solving (7.10) for [B]i and substituting into (7.9) gives the desired algebraic relation between $[Ca^{2+}]_i^{tot}$ and $[Ca^{2+}]_i$:

$$[Ca^{2+}]_i^{tot} = [Ca^{2+}]_i\left(1+\frac{[B]_i^{tot}}{K_i+[Ca^{2+}]_i}\right). \quad ...(7.11)$$

To get the balance equation for $[Ca^{2+}]_i$, we apply the chain rule and (7.11) to obtain

$$\frac{d[Ca^{2+}]_i^{tot}}{dt} = \frac{d[Ca^{2+}]_i^{tot}}{d[Ca^{2+}]_i}\frac{d[Ca^{2+}]_i}{dt} = \frac{1}{f_i([Ca^{2+}]_i)}\frac{d[Ca^{2+}]i}{dt}, \quad ...(7.12)$$

where

$$f_i\left([Ca^{2+}]_i\right) = \left(1+\frac{K_i[B]_i^{tot}}{(K_i+[Ca^{2+}]_i)^2}\right)^{-1}. \quad ...(7.13)$$

This is a form of the rapid buffering approximation (RBA) (Wagner and Keizer 1994), specialized to cells that may be considered spatially homogeneous. The extension to situations in which diffusion of $[Ca^{2+}]_i$ and buffer cannot be ignored is taken up elsewhere.

In general, f_i is a function of $[Ca^{2+}]_i$, but for low-affinity buffers $K_i \gg [Ca^{2+}]_i$, and f_i can be taken as constant:

$$f_i \approx \frac{1}{1+([B]_i^{tot}/K_i)} \quad ...(7.14)$$

In this case, f_i can be interpreted as the fraction of $[Ca^{2+}]_i^{tot}$ that is free. Typical measured values for f_i are 0.01-0.05.

Combining (7.12) and the corresponding equation for the ER with (7.4) and (7.5) gives

$$\frac{d[Ca^{2+}]_i}{dt} = f_i\left(j_{PM}^{in} - j_{PM}^{out} - j_{ER}^{in} + j_{ER}^{out}\right) \quad ...(7.15)$$

$$\frac{d[Ca^{2+}]_{ER}}{dt} = \frac{\overline{V}_{f_i ER}}{\overline{V}_{ER}}\left(j_{ER}^{in} - j_{ER}^{out}\right) \quad ...(7.16)$$

Note that as $[B]_i^{tot}$ increases, f_i decreases and $[Ca^{2+}]_i$ becomes slower. This can be a problem when measuring $[Ca^{2+}]i$ using fluorescent dyes, which are themselves Ca^{2+} buffers. On the other hand, changing the cell's buffering power with dyes or other exogenous buffers can also be a useful experimental tool.

Equation (7.16) is often rewritten

$$\frac{d[Ca^{2+}]_{ER}}{dt} = \frac{f_i}{\sigma}\left(j_{ER}^{in} - j_{ER}^{out}\right) \qquad ...(7.17)$$

where

$$\sigma = \frac{\overline{V}_{ER} f_i}{V_i f_{ER}} \qquad ...(7.18)$$

We can interpret $\overline{V}_{ER/f_{ER}}$ and $\overline{V}_{i/f_i}$ as the "effective volumes" of the ER and cytosol, respectively, that is, the volumes taking into account the fraction of Ca^{2+} that is free in each compartment. Then σ is the ratio of effective volumes, an important dimensionless quantity that captures the combined effects of volume and buffering differences between the ER and the cytosol. Note that (7.15) and (7.17) have two fewer parameters than the original set (7.15) and (7.16). Thus, two of the parameters have been revealed to be redundant by this simple substitution. A more extensive example of the use of dimensionless parameters is given in Section 7.3 for the gonadotroph model.

If we multiply (7.17) by σ and add it to (7.15), the ER flux terms cancel, and we obtain

$$\frac{d[Ca^{2+}]_T}{dt} = f_i\left(j_{PM}^{in} - j_{PM}^{out}\right), \qquad ...(7.19)$$

where

$$[Ca^{2+}]_T = [Ca^{2+}]_i + \sigma[Ca^{2+}]_{ER} \qquad ...(7.20)$$

The reader may check that

$$[Ca^{2+}]_T = \frac{f_i(N_i + N_{ER})}{\overline{V}_i} \qquad ...(7.21)$$

and $[Ca^{2+}]_T$ may thus be interpreted as the "total free Ca^{2+}" of the cell with the cytosolic volume as the reference volume.

In several examples to follow we will find it convenient to replace $[Ca^{2+}]_{ER}$ by $[Ca^{2+}]_T$. For example, if the plasma membrane is impermeable to Ca^{2+} (the closed-cell model), then $j_{PM}^{in} = j_{PM}^{out} = 0$ and $[Ca^{2+}]_T$ is constant, even if $[Ca^{2+}]_i$ and $[Ca^{2+}]_{ER}$ fluctuate. This allows us to eliminate one differential equation. Because Ca^{2+} is conserved, no information is lost, and we can recover $[Ca^{2+}]_{ER}$ from the algebraic relation

$$[Ca^{2+}]_{ER} = ([Ca^{2+}]_T - [Ca^{2+}]_i)/\sigma \qquad ...(7.22)$$

Although the ER is necessarily smaller in volume than the cytosol, it can have greater surface area because it is convoluted. If pump density is the same, the ER fluxes will then be larger than the plasma membrane fluxes. One can study the influence of ER surface area by multiplying the ER fluxes by a common factor, but it may be more convenient to factor out the ratio and make it explicit in the equations. Equations (7.15) and (7.17) can be rearranged to give

$$\frac{d[Ca^{2+}]_i}{dt} = f_i\left(j_{PM}^{in} - j_{PM}^{out} - \frac{1}{\lambda_{ER}}\left(J_{ER}^{\sim in} - J_{ER}^{\sim out}\right)\right) \quad ...(7.23)$$

$$\frac{d[Ca^{2+}]_{ER}}{dt} = \frac{f_i}{\sigma\lambda_{ER}}\left(J_{ER}^{\sim in} - J_{ER}^{\sim out}\right), \quad ...(7.24)$$

where $j_{ER}^{\sim in} = \lambda_{ER} j_{ER}^{in}$, $j_{ER}^{\sim out} = \lambda_{ER} j_{ER}^{out}$ and $\lambda_{ER} = A_{PM} / A_{ER}$ is the ratio of plasma membrane to ER membrane surface area. This ratio will be important in understanding the β-cell and gonadotroph models below, and will be derived by nondimensionalizing the latter model.

Expressions for the Fluxes

The final step in specifying a particular model is to replace the general terms j_{PM}^{in}, j_{PM}^{out}, j_{ER}^{in} and j_{ER}^{out} by appropriate biophysical expressions. Efflux from the cell is generally mediated by a pump flux, j_{PMCA}, influx into the ER from the cytosol by j_{SERCA}. In whole cell models we generally use empirical, Hill-type formulas for pump rate, of the form

$$j_{SERCA} = \frac{v_{SERCA} \cdot [Ca^{2+}]_i^2}{K^2_{SERCA} + [Ca^{2+}]_i^2}, \quad ...(7.25)$$

rather than detailed kinetic models. The maximal pump rate $V_{inferior}$ is proportional to the total number of SERCA or PMCA pumps and to the rate of a single pump, and is typically expressed here in μM/s.

Efflux from the ER to the cytosol often includes a constant, unregulated leak conductance of the form

$$j_{LEAK} = v_{LEAK} \cdot ([Ca^{2+}]_{ER} - [Ca^{2+}]_i), \quad ...(7.26)$$

where v_{LEAK} is the leak permeability (also expressed in μM/s) and is the thermodynamic driving force for a symmetric channel. Note that there is no contribution of membrane potential to the ER driving force. It is generally, though not universally, believed that there is no significant ER membrane potential because of counterions that balance the fluxes of Ca^{2+}.

Interesting dynamics such as oscillations generally require nonlinear feedback on the efflux rate by Ca^{2+}. This can be mediated by the RyR, for which a detailed model is presented in Section 7.2, or the IP_3R.

Ca^{2+} generally enters cells through ion channels. Ionic currents are measured in pA (10^{-12} C/s) or nA (10^{-9} C/s), so an additional factor is needed to convert to μmoles:

$$j_{PM}^{in} = -\alpha I_{Ca} \quad ...(7.27)$$

where $\alpha = 1/(2F\overline{V}_i)$, F = 96480 C/mol is Faraday's constant (F = *eA* = the elementary charge × Avogadro's number), and the factor 2 accounts for the two positive charges of a Ca^{2+} ion. As an example, if I_{Ca} is measured in pA, and cytosolic volume is measured in μm^3, then j_{PM}^{in}

will have units of M/ms. Note that by the Hodgkin-Huxley convention, an inward cation current such as I_{Ca} is negative, so the flux is positive.

Calcium Oscillations in the Bullfrog Sympathetic Ganglion Neuron

Bullfrog sympathetic ganglion neurons (BFSG) are excitable cells with a full complement of voltage-dependent ion channels (Yamada et al. 1998). However, the Ca^{2+} oscillations modeled here are driven by nonlinearity in the ER, with the plasma membrane playing only a passive role. As early as 1976, Kuba and Nishi observed rhythmic hyperpolarizations of the resting membrane potential when the neurons were exposed to caffeine.

The BFSG is a good model system for studying RyR-mediated Ca^{2+} oscillations in neurons. These caffeine-induced oscillations also occur when the membrane voltage is clamped at a fixed value, suggesting that voltage-gated ion channels on the plasma membrane are not involved in producing the oscillations.

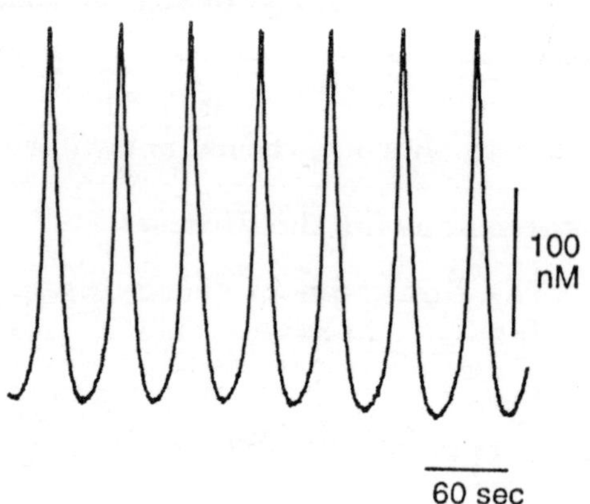

Fig. 7.2. Steady state $[Ca^{2+}]i$ oscillations recorded from a fura-2 loaded bullfrog sympathetic ganclion neuron during continuous exposure to caffeine (5 mM) under mildly depolarizing conditions (30 mM K^{+}). Data provided by Dr. David Friel, Case Western Reserve University.

Caffeine has been found to activate the RyR by shifting the Ca^{2+} dependence of channel opening to lower levels of $[Ca^{2+}]_i$. Ryanodine, a specific ligand of the RyR, blocks the response of these neurons to caffeine by locking them into a low-conductance state. The oscillations induced by caffeine have a very long period (several minutes), and the $[Ca^{2+}]_i$ spikes are characterized by a sharp upstroke followed by a plateau phase and a subsequent downstroke. The upstroke is associated with the dumping of Ca^{2+} by the ER store; the plateau phase is largely due to Ca^{2+} extrusion; and the downstroke is due to Ca^{2+} uptake by the store. Sustained Ca^{2+} entry is required during the interspike intervals, suggesting that store refilling driven by Ca^{2+} influx is necessary after each spike for the generation of the subsequent spike. Therefore, increasing the rate of Ca^{2+} entry by depolarizing the membrane leads to shortened interspike interval and increased amplitude.

Ryanodine Receptor Kinetics: The Keizer-Levine Model

Experimental investigations of the Ca^{2+}-dependent gating properties of the RyR channels have provided data for the development of a mathematical model based on channel kinetics. We begin with equations for the RyR receptor itself as studied in conditions where $[Ca^{2+}]_i$ is a parameter imposed by the experimenter. For example, in the experiments of Gyorke and Fill, cardiac RyRs in lipid bilayers were exposed to controlled Ca^{2+} elevations by flash photolysis of a caged Ca^{2+} compound, DM-nitrophen. Here we present a minimal model designed to capture the key features observed by Gyorke and Fill. Like the IP_3 receptor, the RyR exhibits fast activation and slow inactivation in response to a rise in Ca^{2+}.

A particularly interesting feature is that a second Ca^{2+} flash can "wake up" the channel and elicit more current. Another difference with the IP_3 receptor is that the steady-state open probability is monotonic with respect to $[Ca^{2+}]_i$, not bell-shaped. Keizer and Levine developed a model to account for all three features observed in those experiments. The state diagram they devised (Fig. 7.5 consists of two closed states, C_1 and C_2, and two open states, O_1 and O_2. At rest (corresponding to a background $[Ca^{2+}]_i$ level of 0.1 μM), most of the receptors are in state C_1.

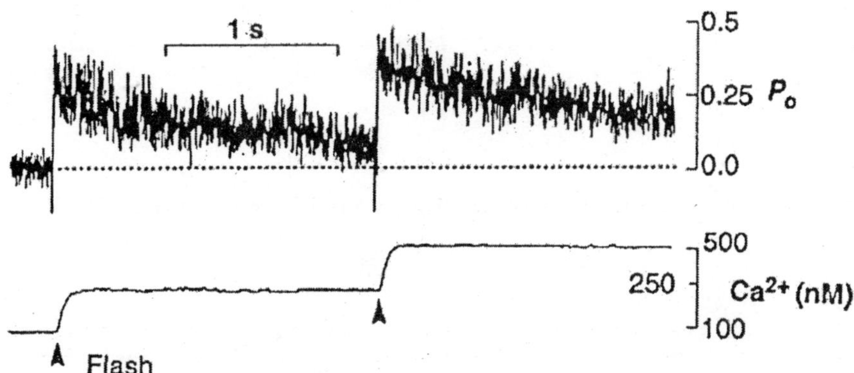

Fig. 7.3. Adaptation response of cardiac RyR in lipid bilayers to release of caged Ca^{2+}.

Both open states have the same conductance, so the net open probability is $P_0 = P_{01} + P_{02}$. The receptors respond to a step increase in Ca^{2+} with a rapid rise in P_0 as states O_1 and O_2 are populated, followed by a slow decline to a steady, plateau level as states flow into C_2, solid curve). Fig. 7.6B shows the peak and plateau values attained after a step from $[Ca^{2+}]_i$ = 0.1 μM to the indicated value. (Note that the jump to the peak depends on the initial conditions, which determine the number of receptors in C_1 available to be recruited by the Ca^{2+} step.) During the short time shown in Fig. 7.6 there is little back flow out of C_2, but a further increase in $[Ca^{2+}]_i$ draws more receptors out of C_1 into the open states Fig. 7.6A, arrow and dashed curve).

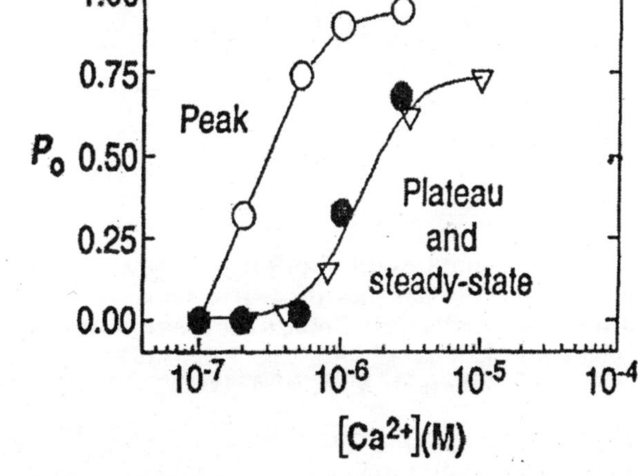

Fig. 7.4. Dependence of peak and steady-state RyR current on Ca^{2+}.

The full four-state model is described by three independent ODEs plus a conservation condition for the total number of channels: $P_{O1} + P_{O2} + P_{C1} + P_{C2} = 1$. We can apply the methods to derive a quasi-steady-state approximation assuming that P_{C2} is slow, while the other states are in rapid equilibrium:

$$Po \approx P_o^{\text{slow}} = \frac{W \cdot (1 + [Ca^{2+}]_i / K_b)^3)}{1 + (K_a / [Ca^{2+}]_i)^4 + ([Ca^{2+}]_i / K_b)^3}, \qquad \text{...(7.28)}$$

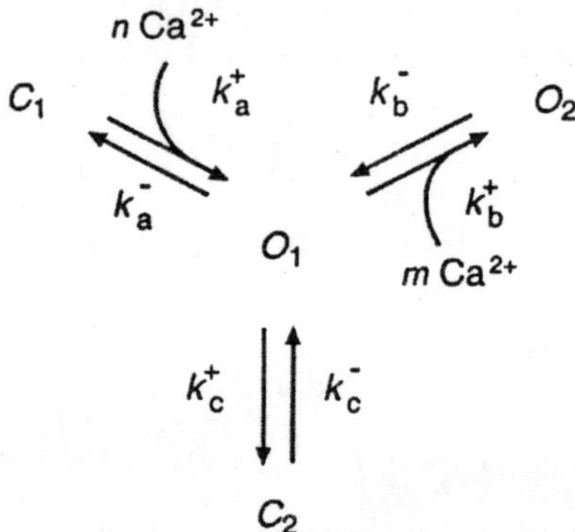

Fig. 7.5. Kinetic states of the Keizer-Levine model for the RyR.

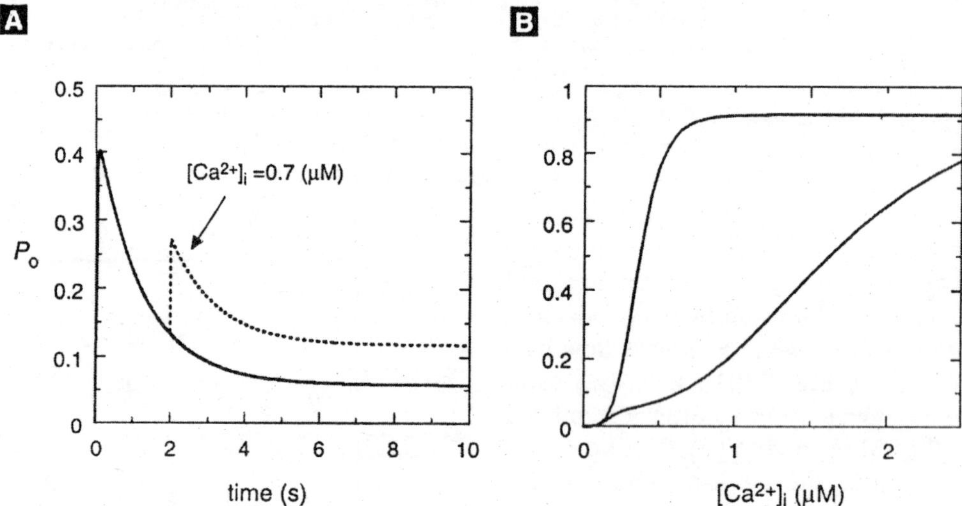

Fig. 7.6. Keizer-Levine model: (A) Time course of response of P_0 to a step increase in $[Ca^{2+}]_i$ from 0.1 to 0.35 μM (solid). Dotted curve shows "adaptation" when a second step to 0.7 μM is applied at t = 2s. (B) Peak and steady-state levels of P_0 for different values of $[Ca^{2+}]_i$. (see Keizer and Levine (1996), Fig. 7.3, and Fig. 7.4 this chapter.) Parameters: $k_a^+ = 1500/(\mu M^4)$, $k_a^+ =$ 28.8/s, $k_b^+ = 1500/(\mu M^3\ s)$, $k_b^- = 385.9/s$, $k_c^+ = 1.75/s$, $k_c^- = 0.1/s$.

where P_O^{slow} is proportional to $w = 1 - P_{C2}$, which is the fraction of noninactivated receptors. In terms of probabilities,

$$\Pr\{\text{Open}\} = \Pr\{\text{Not Inactivated}) \cdot \Pr\{\text{Open} \mid \text{Not Inactivated}\}.$$

Given $[Ca^{2+}]_i$, w is calculated from the differential equation

$$\frac{dw}{dt} = \frac{(w_\infty - w)}{\tau} \quad ...(7.29)$$

with

$$w_\infty = \frac{1 + (K_a/[Ca^{2+}]_i)^4 + ([Ca^{2+}]_i/K_b)^3}{1 + (1/K_c) + (K_a/[Ca^{2+}]_i)^4 + ([Ca^{2+}]_i/K_b)^3} \quad ...(7.30)$$

and

$$\tau = \frac{w_\infty}{k_c^-} \qquad ...(7.31)$$

As we did for the gating variables, it is natural to write a pseudo-exponential rate equation for w because changes in w are exponential for fixed $[Ca^{2+}]_i$. Note that the rate constants of the fast processes no longer appear, just the dissociation constants, defined by $K_a^4 = k_a^- / k_a^+, K_b^3 = k_b^- / k_b^+,$ and $K_c = k_c^- / k_c^+$. The slow time scale τ is inversely proportional to the rate of transition out of state C_2.

Bullfrog Sympathetic Ganglion Neuron Closed-Cell Model

We proceed to write down the equations for a closed cell using (7.15) $\left(\text{with } j_{PM}^{in} = j_{PM}^{out} = 0\right)$ together with appropriate expressions for the fluxes due to the ryanodine receptors, a leak out of the ER, and a SERCA pump:

$$\frac{d[Ca^{2+}]_i}{dt} = f_i(j_{RyR} + j_{LEAK} - j_{SERCA}) \qquad ...(7.32)$$

The RyR flux is

$$j_{RyR} = \nu_{RyR} P_0 \cdot ([Ca^{2+}]_{ER} - [Ca^{2+}]_i), \qquad ...(7.33)$$

where ν_{RyR} (also in μM/s) is proportional to the number of RyRs and P_Q is the probability that a receptor is open. (Note that here we are generalizing (7.26) for a symmetric leak channel, but symmetry is only an approximation for the RyR. See Exercise 2.) We use the quasi-steady-state approximation (7.28)-(7.31) of the previous section for P_0.

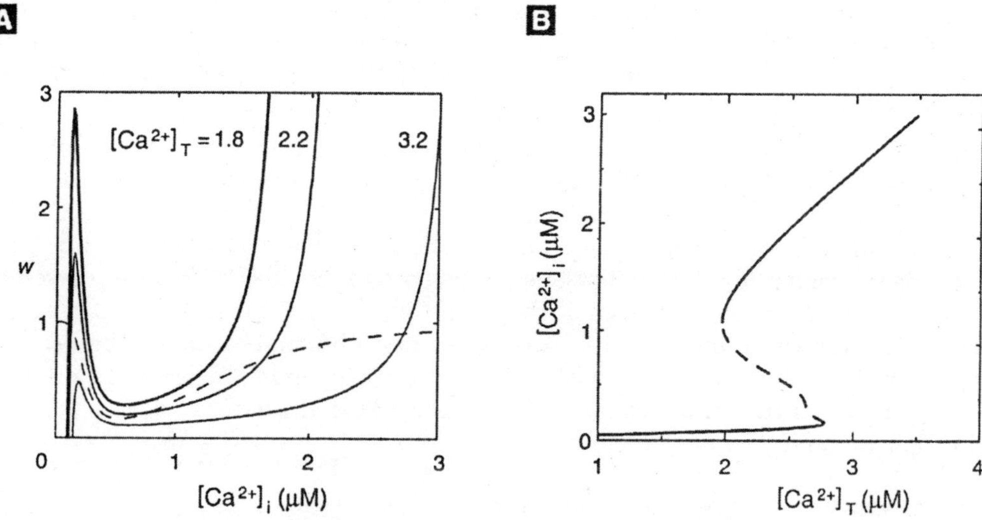

Fig. 7.7. Closed-cell Keizer-Levine model: (A) Nullclines for $[Ca^{2+}]_i$ (solid) and w (dashed). (B) Intersections of nullclines (steady-states) in (A) for various values of $[Ca^{2+}]_T$, parameters: As in Fig. 7.6 plus: $f_i = 0.01$, $\nu_{RyR} = 5/s$, $\nu_{LEAK} = 0.2/s$, $\nu_{SERCA} = 100$ μM/s, $K_{SERCA} = 0.2$ μM, $\sigma = 0.02$, $K_a = 0.4$ μM, $K_b = 0.6$ μM, $K_c = 0.1$ μM, $K_C^- = 0.1/s$, and $[Ca^{2+}]_T$ is in μM as indicated.

Because of conservation of ions, we do not need a differential equation for $[Ca^{2+}]_{ER}$, but just the algebraic relation (7.22).

The leak is given by

$$j_{LEAK} = v_{LEAK} \cdot \left([Ca^{2+}]_{ER} - [Ca^{2+}]_i\right), \qquad ...(7.34)$$

and the pump flux is

$$j_{LEAK} = v_{LEAK} \cdot \frac{[Ca^{2+}]_i^2}{[Ca^{2+}]_i^2 + K^2{}_{SERCA}} \qquad ...(7.35)$$

The closed cell model consists of the two differential equations for $[Ca^{2+}]_i$ (7.32) and w (7.29) plus the associated algebraic equations.

With the parameters given in Fig. 7.7, the model is not able to produce oscillations, but instead is bistable; it can assume either a high-$[Ca^{2+}]_i$ steady state or a low-$[Ca^{2+}]_i$ steady state. This behavior can be understood by examining the phase plane, which is possible because we reduced the kinetic equations for the RyR to one slow equation. In the closed-cell model total Ca^{2+} ($[Ca^{2+}]_T$) is constant, and this is a natural parameter to use to characterize the system. For low values of $[Ca^{2+}]_T$, the steady-state $[Ca^{2+}]_i$ is low, and w is high (few receptors are adapted). For high values of $[Ca^{2+}]_T$, steady-state $[Ca^{2+}]_i$ is high, and w is low (many receptors are adapted).

For intermediate values of $[Ca^{2+}]_T$ there are three steady states. Examination of the eigenvalues of the Jacobian in this case reveals that the low and high steady states are stable nodes (see Appendix A), while the middle steady state is a saddle. Sample nullclines are plotted in Fig. 7.7A, and the summary of how the steady states vary with $[Ca^{2+}]_T$ (i.e., the bifurcation diagram) is plotted in Fig. 7.7B. This global view reveals a large region of bistability or hysteresis bracketed by two saddle-node bifurcations (the turning points, where the determinant changes sign) that give birth to all of the steady states in the system. There is also a value of $[Ca^{2+}]_T$ near the bottom turning point where there is a Hopf bifurcation (the trace of the Jacobian changes sign.

Geometrically, the w-nullcline intersects the $[Ca^{2+}]_i$-nullcline with a more negative slope. The oscillations born at this point, however, are small in amplitude and unstable. By making a small change in the Ca^{2+}-handling parameters it is possible to obtain stable oscillations with the closed-cell model (see Exercise 4). Nonetheless, Keizer and Levine concluded that that this is not a good model for the oscillations observed by Friel and Tsien; they are not robust (i.e., the oscillations only exist for a small range of $[Ca^{2+}]_T$), and it is difficult to get oscillations with period greater than one minute. This is because the time scale of the oscillations is determined by the reciproal of k_c-which is about 10 seconds. In order to obtain robust oscillations on the minute time scale, it is necessary to add another, slower process to the system, which we do in the next section.

Bullfrog Sympathetic Ganglion Neuron Open-Cell Model

In the previous section we saw that adaptation of the RyR is not likely the mechanism behind slow oscillations in the BFSG cell, and that another, slower, process is needed. A natural extension suggested by the analysis of the closed cell is to make $[Ca^{2+}]_T$ a slow variable rather than a constant parameter. Biophysically, this allows for larger variation in the filling state of the ER. We will see that in each cycle the stores almost completely empty and refill. In other words, the net flux in and out of the cell over a cycle is 0, but there is a large efflux at some points in the cycle that is balanced by a large influx at other points.

The open-cell model expands on the closed-cell model by adding an equation to describe changes in $[Ca^{2+}]_T$ due to fluxes into the cell and PMCA pumps that remove Ca^{2+}:

$$\frac{d[Ca^{2+}]_T}{dt} = f_i\left(j_{IN} - j_{PMCA}\right). \quad ...(7.36)$$

The PMCA flux is given by

$$j_{PMCA} = \frac{v_{PMCA}[Ca^{2+}]_i^2}{K^2_{PMCA} + [Ca^{2+}]_i^2}. \quad ...(7.37)$$

Since oscillations occur under voltage clamp, we can represent influx of Ca^{2+} as a constant j_{IN} (understood as influx across the plasma membrane, in μM/s). The plasma membrane fluxes must also be included in the equation for $[Ca^{2+}]_i$:

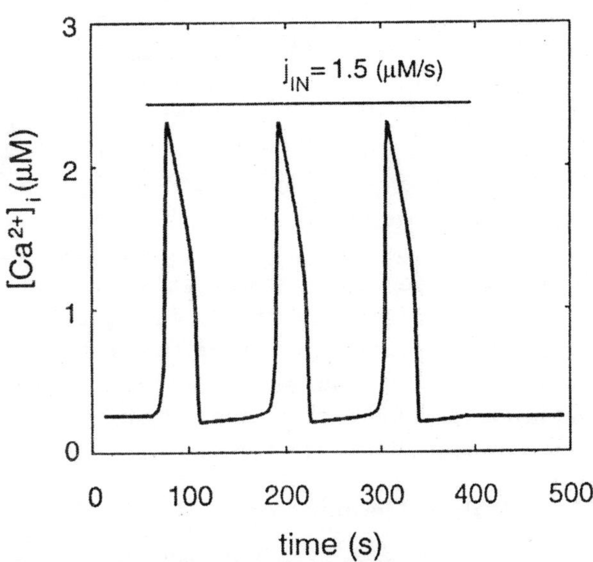

Fig. 7.8. Reduced open-cell Keizer-Levine model. At time = 50 seconds, oscillations are initiated by stepping j_{IN} from 0 to 1.5. At time = 400 seconds, oscillations are terminated by reducing j_{IN} back to 0. Parameters as in Fig. 7.7 plus: v_{PMCA} = 5.0 μM/s, K_{PMCA} = 0.6 μM, j_{IN} = 1.5 μM/s.

$$\frac{d([Ca^{2+}]_i)}{dt} = f_i\left(j_{RyR} + j_{LEAK} - j_{SERCA} + j_{IN} - j_{PMCA}\right). \quad ...(7.38)$$

The open cell model thus consists of three differential equations (7.29), (7.36), and (7.38), associated algebraic expressions for the fluxes, and the conservation relationship given by (7.20). We can collapse this system back into just two equations for $[Ca^{2+}]_i$ and $[Ca^{2+}]_T$. This is justified because variation in w is fast compared to the oscillation period. We eliminate the equation for w by writing $w = w_\infty$. The RyR channel flux is then given by

$$j_{RyR} = v_{RYR} \cdot \frac{w_\infty(1 + ([Ca^{2+}]_i / K_b)^3)}{1 + (K_a /[Ca^{2+}]_i)^4 + ([Ca^{2+}]_i / K_b)^3} ([Ca^{2+}]_{ER} - [Ca^{2+}]_i). \quad ...(7.39)$$

Solutions of the reduced open cell model ((7.36) and (7.38) with j_{RYR} replaced by the expression in (7.39)) are shown in Fig. 7.8. With adequate Ca^{2+} influx, oscillations of the right shape and duration are seen. However, if Ca^{2+} influx is suppressed, say by removal of external Ca^{2+}, the oscillations cease immediately, also in agreement with experiment (Friel and Tsien 1992).

The closed cell model can be viewed as the *fast subsystem* of the open cell. That is, the bifurcation diagram of the closed cell model with $[Ca^{2+}]_T$ as a parameter is a phase plane of the reduced open cell model with $[Ca^{2+}]_T$ as a slow variable (slow compared to $[Ca^{2+}]_i$). We illustrate this by projecting the trajectory of the open cell onto the diagram computed with the closed cell, supplemented with the nullcline for $[Ca^{2+}]_T$. Moreover, we can predict from this that if j_{IN} is increased so that the $[Ca^{2+}]_T$ nullcline intersects the S-shaped $[Ca^{2+}]_i$ nullcline above the upper limit point or below the lower limit point, there will be no oscillations. This is confirmed by the bifurcation diagram of the reduced open cell with respect to j_{IN} (Fig. 7.10). Within the oscillatory range, variation of j_{IN} produces a broad range of frequencies, but almost no change in amplitude. This follows from the relaxation oscillator character of the model.

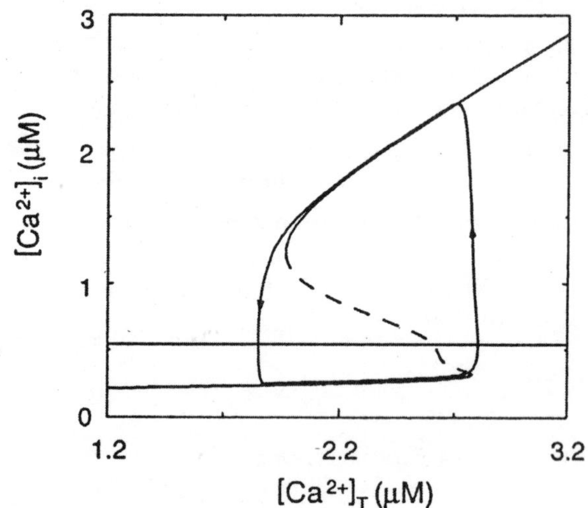

Fig. 7.9. The bifurcation diagram of the closed-cell Keizer-Levine model Fig. 7.7 reinterpreted as a phase plane of the reduced open-cell model Fig. 7.8, with superimposed trajectory and $[Ca^{2+}]_T$ nullcline (horizontal line). Direction of motion along trajectory indicated by arrows. Parameters as in Fig. 7.7 but with j_{IN} = 1.5 M/s.

The Pituitary Gonadotroph

Our next example of a whole-cell model is the coupling of electrical signaling and Ca^{2+} signaling in gonadotrophs. Gonadotrophs are hormone-secreting (endocrine) cells in the anterior part of the pituitary gland. Their physiological function is to release gonadotropins (gonad-stimulating hormones) when they themselves are stimulated by a hormone, gonadotropin-releasing hormone (GnRH), that is secreted by endocrine neurons in the hypothalamus of the brain. Ca^{2+} signaling is crucial for these cells to fulfill their secretory functions, which are part of the complex neuro-endocrine control of reproduction in mammals. Physiological levels of GnRH have been shown to trigger large-amplitude Ca^{2+} oscillations in cultured gonadotrophs. It has further been shown that these oscillatory Ca^{2+} signals are directly coupled to the exocytosis of hormone-containing vesicles in these cells. These Ca^{2+} signals can become more complex, because gonadotrophs are excitable cells capable of generating Ca^{2+} action potentials that drive Ca^{2+} from the extracellular medium into the cell in an oscillatory manner. Thus, gonadotrophs

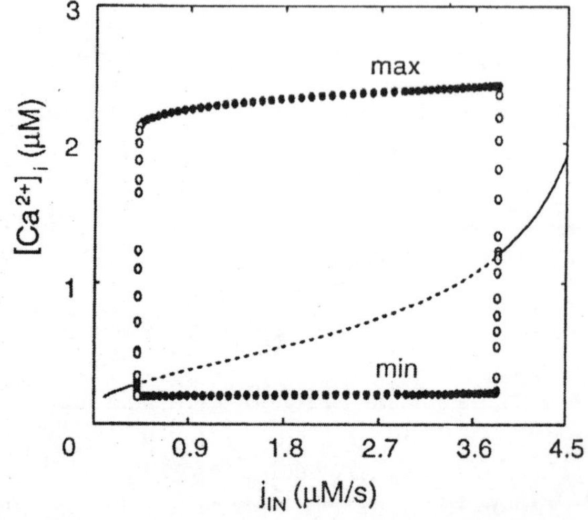

Fig. 7.10. Bifurcation diagram of the reduced open cell version of Keizer-Levine, showing the behavior of the model as in Fig. 7.7 and Fig. 7.8 for a range of values of j_{IN}.

are equipped with two distinct mechanisms of generating dynamic Ca^{2+} signals, one due to voltage gated Ca^{2+} entry (PM oscillator) and the other due to the oscillatory release of Ca^{2+} from the intracellular store (ER oscillator). These two dynamic signaling systems are coupled by the Ca^{2+} and Ca^{2+}-activated K^+ ion channels in the plasma membrane. This makes the gonadotroph an ideal system to study complex Ca^{2+} signaling mechanisms and to apply the modular approach of model development.

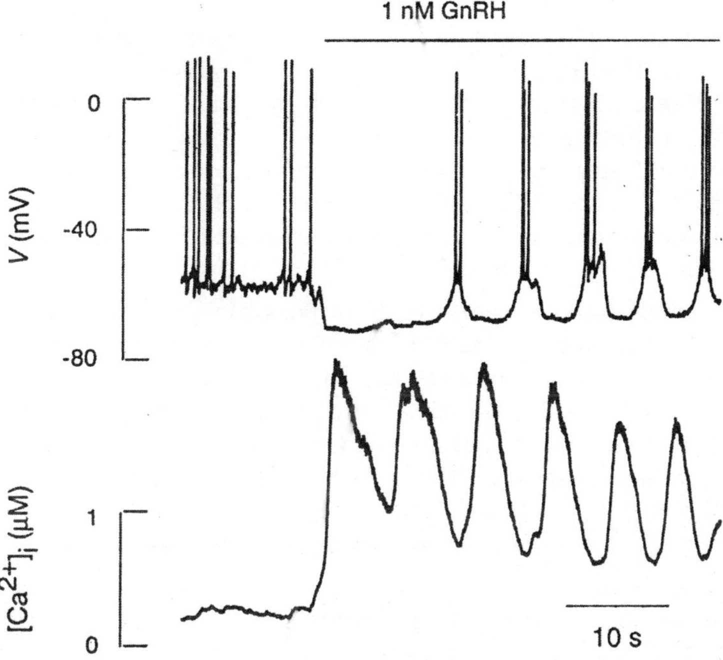

Fig. 7.11. Burst firing driven by repetitive releases of Ca^{2+} from the ER in a pituitary gonadotroph stimulated by GnRH.

One dramatic manifestation of the ER-plasma membrane interaction is shown in Fig. 7.11. The unstimulated gonadotroph exhibits low-frequency spiking and low $[Ca^{2+}]_i$. After GnRH is applied, Ca^{2+} oscillations due to repetitive releases of Ca^{2+} from the ER appear. Each Ca^{2+} rise turns on Ca^{2+}-activated K^+ channels and transiently interrupts the firing, resulting in repetitive bursting.

The ER Oscillator in a Closed Cell

As for the BFSG, it is useful to study the gonadotroph first in the closed-cell case, when the ER oscillator is decoupled from the PM oscillator. This allows us to concentrate on the intracellular mechanisms that are independent of the influence of Ca^{2+} exchange across the plasma membrane.

ER-mediated Ca^{2+} oscillations in gonadotrophs are linked to the activation of IP_3 receptor channels. In contrast to the caffeine-induced Ca^{2+} oscillations in the bullfrog sympathetic neurons, mediated by the RyR, the ER-dependent Ca^{2+} oscillations in gonadotrophs are induced by the endogenous intracellular messenger IP_3 produced when the physiological ligand GnRH binds to its receptors on the surface of the cell. Agonist-induced Ca^{2+} oscillations in a large number of cell types including gonadotrophs are generated by the dynamic interactions between IP_3 channels and SERCA pumps expressed in the membrane of the intracellular Ca^{2+} store (ER or SR). Of crucial importance are the Ca^{2+}-dependent gating properties of the IP_3 channels.

As discussed already, cytosolic free Ca^{2+} can both activate and inactivate the opening of IP_3 channels. This leads to the characteristic bell-shaped Ca^{2+}-dependence of steady-state channel opening with the peak typically located at the mid-submicromolar level of Ca^{2+}. It has also been shown that Ca^{2+}-dependent activation occurs on a faster time scale than Ca^{2+}-dependent inactivation. The De Young-Keizer model that was discussed earlier and was the first model to incorporate all these experimental data. Other models based on similar data were developed later, including important simplifications by Li andRinzel (1994), Atri et. al. (1993), and Othmer and Tang (1993). These models have in common that both positive and negative feedback reside in the IP_3R itself, in contrast with two earlier classes of models. In one class of models, oscillations depend on the interaction of two separate internal pools of Ca^{2+}, one responsive to IP_3 and one to Ca^{2+}.

In another, IP_3 itself had to oscillate. It is possible that the latter class of models corresponds to a distinct class of oscillations in cells. Yet another model that differs from all of the above in that it neither requires IP_3 nor involves calcium-induced calcium release is proposed in. Though we will not describe these other models in detail, they make good starting points for independent study. Differences in the time scales of channel gating are attributed to differences in the binding constants of IPs and Ca^{2+} to their binding sites on the IP_3 channels. This assumption, although not necessarily an accurate description of the real system, leads to a simple model that gives results that are identical to the observed ones.

Even if the time scale differences are actually due to mechanisms other than differences in the binding constants, the resulting model would be qualitatively the same mathematically. Here we use a simplified version of the model due to Li and Rinzel (1994). For a closed cell at a fixed level of IP_3, the Li-Rinzel simplification involves only two dynamic variables: $[Ca^{2+}]_i$ and h, the fraction of channels not inactivated by Ca^{2+} and hence available to open. The equations are

$$\frac{d[\mathrm{Ca}^{2+}]_i}{dt} = \frac{f_i}{V_i}\left(J_{\mathrm{ER}}^{\mathrm{out}} - J_{\mathrm{ER}}^{\mathrm{in}}\right), \quad ...(7.40)$$

$$\frac{dh}{dt} = A[K_d - ([\mathrm{Ca}^{2+}]_i + K_d)\, h], \quad ...(7.41)$$

with A a variable to control the relative time scale between the two differential equations. This simplified model gives as good a fit to the experimental data as the original model it was based on. Filling in expressions for which is the sum of a small constitutive leak and the $[Ca^{2+}]_i$ and IP3 regulated flux through the IP_3 receptors, and for $j_{\mathrm{ER}}^{\mathrm{in}}$ we have

$$J_{\mathrm{ER}}^{\mathrm{out}} = \left[L + \frac{P_{\mathrm{IP3R}}[\mathrm{IP}_3]^3[\mathrm{Ca}^{2+}]_i^3 h^3}{([\mathrm{IP}_3]+K_i)^3([\mathrm{Ca}^{2+}]_i + K_a)^3}\right]([\mathrm{Ca}^{2+}]_{\mathrm{ER}} - [\mathrm{Ca}^{2+}]_i), \quad ...(7.42)$$

$$J_{\mathrm{ER}}^{\mathrm{in}} = \frac{V_{\mathrm{SERCA}}[\mathrm{Ca}^{2+}]_i^2}{[\mathrm{Ca}^{2+}]_i^2 + K^2{}_{\mathrm{SERCA}}}, \quad ...(7.43)$$

where P_{IP3R} is the maximum total permeability of IP_3 channels (in pL/s for consistency) L is the ER leak permeability (also in pL/s) and V_{SERCA} is the nonscaled maximum SERCA pump rate (in aMol/s = 10^{-18}mol/s). The constant K_{SERCA} is the Ca^{2+} level for half maximal activation of the SERCA pump and K_i, K_a, and K_d are the dissociation constants of IP_3, Ca^{2+}-activation, and Ca^{2+}-inactivation sites on the IP_3R, respectively (all expressed in μM). As for the BSFG closed-cell model, $[Ca^{2+}]_T$ is a given constant, and we obtain $[Ca^{2+}]_{ER}$ from $([Ca^{2+}]_T - [Ca^{2+}]_i)/\sigma$. For any model, it can greatly deepen our understanding to scale the variables and parameters to achieve a dimensionless form of the equations. This process often reveals that some of the parameters do not affect the dynamical behavior independently, and they can be combined.

In the present case, the number of parameters can be reduced from 10 in the original equations (including σ) to 6 in the scaled form. There are often multiple ways of scaling the same system. One strategy is to scale the dependent variables so that their magnitudes are of order one, which facilitates comparison of magnitudes, and to achieve a maximum reduction in the number of independent parameters. The h variable in the equations above is already dimensionless. We can scale $[Ca^{2+}]_i$ by either K_a or K_d. We choose K_d, because this leaves only one independent parameter in the h equation. Once we have made this choice, $[Ca^{2+}]_{ER}$, $[Ca^{2+}]_T$, K_a, and K_{SERCA} should also be scaled by K_d:

$$\hat{c} = [Ca^{2+}]_i / K_d,$$

$$\hat{c}_E = [Ca^{2+}]_{ER} / K_d,$$

$$\hat{c}_T = [Ca^{2+}]_T / K_d,$$

$$\hat{k}_a = K_a / K_d,$$

$$\hat{k}_s = K_{SERCA} / K_d.$$

It is natural to scale the IP_3 concentration by K_i, and as a result, L and F_{SERCA} to be scaled accordingly:

$$\hat{\imath} = I / K_i$$

$$\hat{l} = L / P_{IP3R}$$

$$\hat{v}s = \left(\frac{\sigma}{K_d P_{IP3R}}\right) V_{SERCA}$$

Finally, scaling time as follows cleans up all the constants on the left-hand side of the $[Ca^{2+}]_i$ equation :

$$\hat{t} = \left(\frac{P_{IP3} f_i}{\sigma \overline{V}_i}\right) t,$$

which leaves the following to clean up the h equation :

$$\hat{a} = \left(\frac{\sigma \overline{V}_i K_d}{P_{IP3R f_i}}\right) A.$$

The final result is

$$\frac{d\hat{c}}{d\hat{t}} = \left[\hat{l} + \left(\frac{\hat{i}\hat{c}h}{(\hat{c}+1)(\hat{c}+\hat{k}_a)}\right)^3\right](\hat{c}_T - \sigma_1\hat{c}) - \frac{\hat{v}_s\hat{c}^2}{\hat{c}^2+\hat{k}_s^2}, \quad ...(7.44)$$

$$\frac{dh}{d\hat{t}} = \hat{a}[1-(\hat{c}+1)h], \quad ...(7.45)$$

where $\sigma_1 = 1 + \sigma$ and $\hat{c}_E$ can be recovered from $\hat{c}_E = (\hat{c}_T - \hat{c})/\sigma$. Note that, in order to have Ca^{2+} and time come out as recognizable quantities, the plots to follow in Section 5.3 have all been created with the dimensional forms of the models discussed, and therefore are expressed in "real world" units. Several exercises consist of comparing the dimensional and dimensionless versions of the models.

Agonist-induced and IP_3R Ca^{2+} oscillations in different cell types can have very different frequency, amplitude, and spiking profiles. However, they all share the following features: (1) oscillations occur only at intermediate levels of IP_3 concentration; (2) the oscillation amplitude is almost constant for different levels of IP_3; (3) the oscillation period decreases as the IP_3 level increases; (4) the oscillations can occur without a constant Ca^{2+} influx from the extracellular medium, even if the stores are depletable (Ca^{2+} just shuttles back and forth between the ER and the cytosol). Because the above model was based on key experimental data and is a general model of IP_3-triggered Ca^{2+} oscillations, it should reproduce all these common features. Feature (4) is automatically satisfied here, because the model is for a closed cell that is isolated from the extracellular medium. Fig. 7.12, created with the dimensional version of the model, shows that the model indeed exhibits all the other features. Properties (2) and (3) suggest that the agonist concentration is transmitted to the cell in terms of frequency rather than amplitude, termed frequency encoding. This very simple model not only reproduces well-established experimental observations but also predicts other possible ways Ca^{2+} oscillations can be induced. These can easily be shown by the phase plane analysis of (7.44)-(7.45). The equations for the two nondimensional nullclines can be solved explicitly :

$$\text{The } \hat{c}\text{-nullcline: } h = \frac{(1+\hat{i})(\hat{c}+\hat{k}_a)}{\hat{i}\hat{c}}\left[\frac{\hat{v}_S\hat{c}}{(\hat{c}^2+\hat{k}_S^2)(\hat{c}_T-\sigma_1\hat{c})} - \hat{l}\right]^{1/3} \quad ...(7.46)$$

$$\text{The } h\text{-nullcline: } h = \frac{1}{1+\hat{c}}. \quad ...(7.47)$$

If we look at plots of the dimensional versions of the nullclines shown in Figure 7.13, the $[Ca^{2+}]_i$-nullcline is "N-shaped" between two asymptotes: One is the vertical axis (i.e., $[Ca^{2+}]_i$ = 0) where h approaches negative infinity, and the other one is at $[Ca^{2+}]_i = [Ca^{2+}]_T/\sigma_1$ where h approaches positive infinity. This nullcline crosses the horizontal axis at $[Ca^{2+}]_I \approx [Ca^{2+}]i^* = K_{SERCA}\sqrt{L[Ca^{2+}]_T/V_{SERCA}}$. The value $[Ca^{2+}]_i^*$ is also the approximate value of the basal level of Ca^{2+} at equilibrium. This expression thus shows the dependence of the basal Ca^{2+} level on the key parameters K_{SERCA}, L, $[Ca^{2+}]_T$ and V_{SERCA}. The h-nullcline is a monotonically decreasing function of $[Ca^{2+}]_i$. These two nullclines can intersect each other at either one single point or at three points depending on the choice of parameter values. Since we are interested in the conditions for the occurrence of oscillations, we focus here on the case in which the two intersect at one single point. This point is the equilibrium, or steady-state, solution of the system at which the rate of Ca^{2+} release from the store exactly matches the rate of Ca^{2+} uptake. Oscillations occur when this steady state becomes unstable.

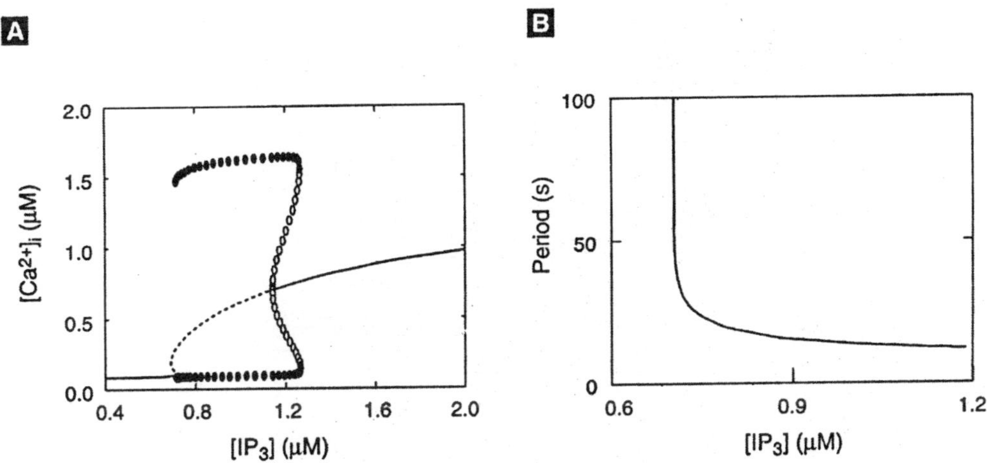

Fig. 7.12. Closed-cell gonadotroph model (dimensional equations (7.40)-(7.43)): (A) Steady-state (curves) and periodic (circles) solutions plotted against $[IP_3]$. Dotted curves and open circles denote unstable steady-state and periodic solutions. The amplitude remains almost unchanged within the range of $[IP_3]$ values that produce oscillations. (B) Oscillation period is a decreasing function of $[IP_3]$. Parameters: f_i = 0.01, $\overline{V_i}$ = 4pL, L = 0.37 pL/s, P_{IP3R} = 26640 pL/s, K_i = 1.0 μM, K_a = 0.4 μM, V_{SERCA} = 400 aMol/s, K_{SERCA} = 0.2 μM, A = 0.5/s, K_d = 0.4 μM, σ = 0.185, $[Ca^{2+}]_T$ = 2 μM, $[IP_3]$ = μM as indicated.

Phase plane analysis (Fig. 7.13) indicates that instability of the steady state can be achieved in several possible ways. In all cases, the presence of a basal level of IP_3 is a prerequisite. Notice that the h-nullcline is independent of any parameter and remains unchanged in all panels (the solid, monotonically decreasing curve). In the top panel, the $[Ca^{2+}]_i$-nullcline is plotted for three different values of $[IP_3]$. The $[Ca^{2+}]_i$-nullcline moves downward as $[IP_3]$ increases such that the intersection between the two null-clines occurs at a larger value of $[Ca^{2+}]_i$ and a lower value of h.

The following is an intuitive explanation of why this is the natural consequence when $[IP_3]$ is increased. In order to maintain balanced rates of Ca^{2+} release and uptake at the equilibrium when Ca^{2+} permeability of IP_3 channels is increased, more channels need to be inactivated at the same level of $[Ca^{2+}]_i$. Because increased inactivation happens at lower values of h, the nullcline moves downward when $[IP_3]$ increases. Instability occurs at intermediate values of $[IP_3]$, *i.e.*, when the $[Ca^{2+}]_i$-nullcline is moved downward so that it crosses the h-nullcline at locations where its slope is more negative. But the downward movement of the $[Ca^{2+}]_i$-nullcline can also be achieved at fixed basal $[IP_3]$ levels that are not themselves high enough to destabilize the equilibrium.

One alternative is to decrease the value of V_{SERCA} (middle panel). This can be achieved experimentally by applying a drug called thapsigargin that specifically blocks the SERCA pump. Another alternative is to increase the leak permeability of the ER membrane (bottom panel).

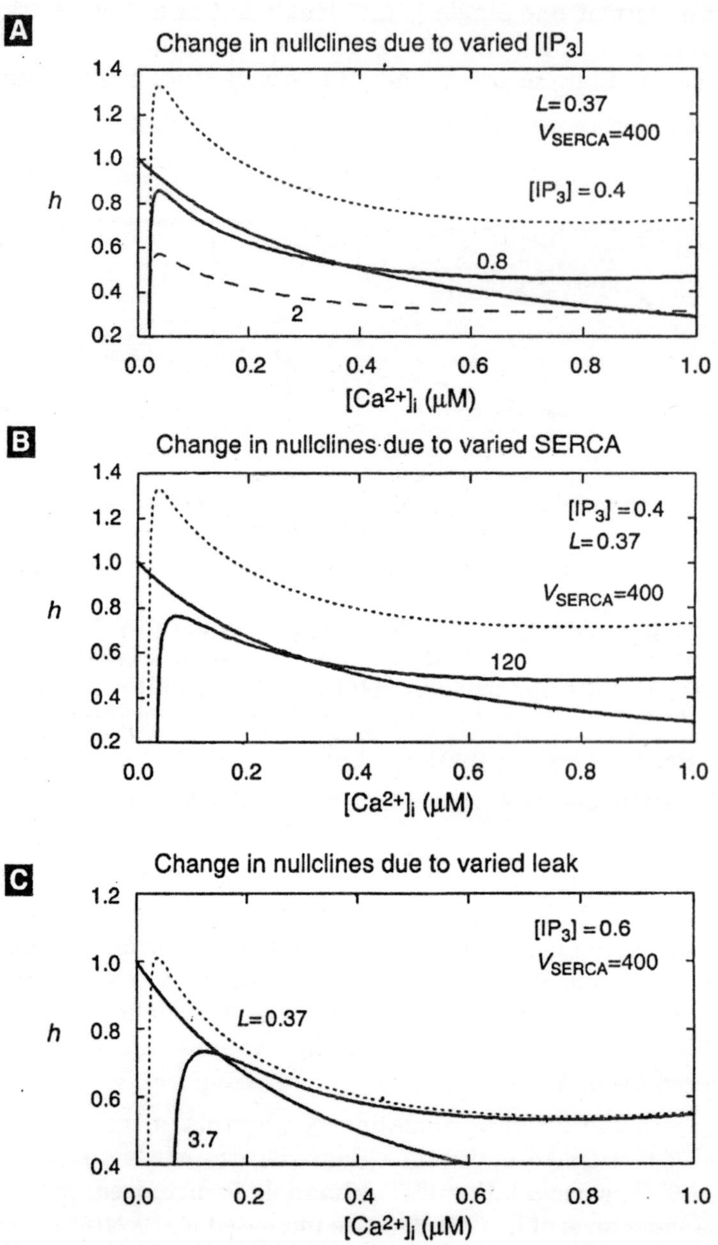

Fig. 7.13. Phase plane diagrams for the dimensional version of closed-cell gonadotroph model showing how the $[Ca^{2+}]_i$-nullcline moves in response to changes in (A) $[IP_3]$, (B) SERCA pump rate V_{SERCA}, and the leak permeability L. In each panel, the monotonically decreasing h-nullcine is always plotted in thick, solid curve while the $[Ca^{2+}]_i$-nullclines are plotted in different line types, because they correspond to different parameter valuese. The thick, solid $[Ca^{2+}]_i$-nullcline in each panel corresponds to the parameter value that destabilizes the steady state and yields oscillation.

This can be realized by applying ionomycin, a drug that makes holes in the ER membrane. There are two other ways to cause instability at basal levels of $[IP_3]$. One is to lower the value of K_a, which means increased affinity of the activation site for Ca^{2+}. The other is to increase $[Ca^{2+}]_T$. This is equivalent to overloading the ER store, which is known experimentally to result in increased excitability of the system.

Open-Cell Model with Constant Calcium Influx

Now let us study the same model in the case where the cell is no longer closed. In an open cell, Ca^{2+} can flow into the cell from the extracellular medium and vice versa. As in the BFSG, $\hat{c}_T$ is no longer a constant since unbalanced Ca^{2+} exchange across the cell surface will change its value. Returning again to the nondimensional form, such an open cell model involves three dynamical variables, $\hat{c}, \hat{c}_E$ and h, satisfying

$$\frac{d\hat{c}}{d\hat{t}} = \hat{J}_{\text{rel}}(\hat{c},h,\hat{c}_T) - \hat{J}_{\text{fil}}(\hat{c}) + \in\left[\hat{J}_{\text{in}} - \hat{J}_{\text{out}}(\hat{c})\right], \quad ...(7.48)$$

$$\frac{d\hat{c}_E}{d\hat{t}} = -\left[\hat{J}_{\text{rel}}(\hat{c},h,\hat{c}_T) - \hat{J}_{\text{fil}}(\hat{c})\right]/\sigma, \quad ...(7.49)$$

$$\frac{dh}{d\hat{t}} = \hat{a}\left[1-(\hat{c}+1)h\right], \quad ...(7.50)$$

where $\hat{j}_{\text{rel}}$ and $\hat{j}_{\text{fil}}$ are the nondimensional ER fluxes from (7.44)-(7.45):

$$\hat{J}_{\text{rel}} = \left[l + \left(\frac{\hat{i}'\hat{c}h}{(\hat{i}+1)(\hat{c}+\hat{k}_a)}\right)^3\right](\hat{c}_T - \sigma_1\hat{c}),$$

$$\hat{J}_{\text{fil}} = \frac{\hat{v}_S\,\hat{c}^2}{\hat{c}^2 + \hat{k}_S^2},$$

and

$$\hat{J}_{\text{out}} = \frac{\hat{v}_P\hat{c}^2}{\hat{c}^2 + \hat{k}_p^2} \quad ...(7.51)$$

is the nondimensional form of the plasma membrane PMCA pump (7.37).

Note that the plasma membrane flux terms in (7.48) are multiplied by a small dimensionless parameter $\in = \lambda_{ER}$ which is the ratio of plasma membrane to ER surface area (7.23) and (7.24)). The fluxes, $j_{PM}^{in}, j_{PM}^{out}, j_{ER}^{in}$, and j_{ER}^{out} are implicitly proportional to the surface area of the respective membranes, whereas the introduction of ε allows explicit comparison between the magnitudes of the respective Ca^{2+} fluxes.

Using $\hat{c}_T = \hat{c} + \sigma\hat{c}_E$, we can replace the $\hat{c}_E$ equation with an equation for $\hat{c}_T$:

$$\frac{d\hat{c}_T}{d\hat{t}} = \in\left[\hat{J}_{\text{in}} - \frac{\hat{v}_p\hat{c}^2}{\hat{c}^2 + \hat{k}_p^2}\right]. \quad ...(7.52)$$

The nondimensional formulation makes it clear that $\hat{c}_T$ is a slow variable because plasma membrane fluxes are much smaller than ER fluxes in the gonadotroph, and hence ϵ is very small.

For simplicity, we first study the case where $\hat{j}_{in}$ is a constant. This corresponds to the experimental situation in which the cells are voltage clamped. Fig. 7.14 shows a case, for the dimensional version of the model, in which the cell is clamped first at a voltage that is not too negative and J_{IN} is large enough to sustain large-amplitude Ca^{2+} oscillations and a filled store. However, when the clamping voltage is switched to a more negative value (at $t = 40$), J_{IN} is nearly 0, so that there is only Ca^{2+} extrusion but no Ca^{2+} influx at the cell surface. A similar situation can occur when the cells are placed in Ca^{2+}-deficient or zero Ca^{2+} medium. Contrary to the caffeine-induced Ca^{2+} oscillations in bullfrog sympathetic neurons of Fig. 7.8, the absence of Ca^{2+} influx does not kill the oscillation immediately. Instead, the amplitude gradually decreases as the store empties and stops only when the store is nearly depleted.

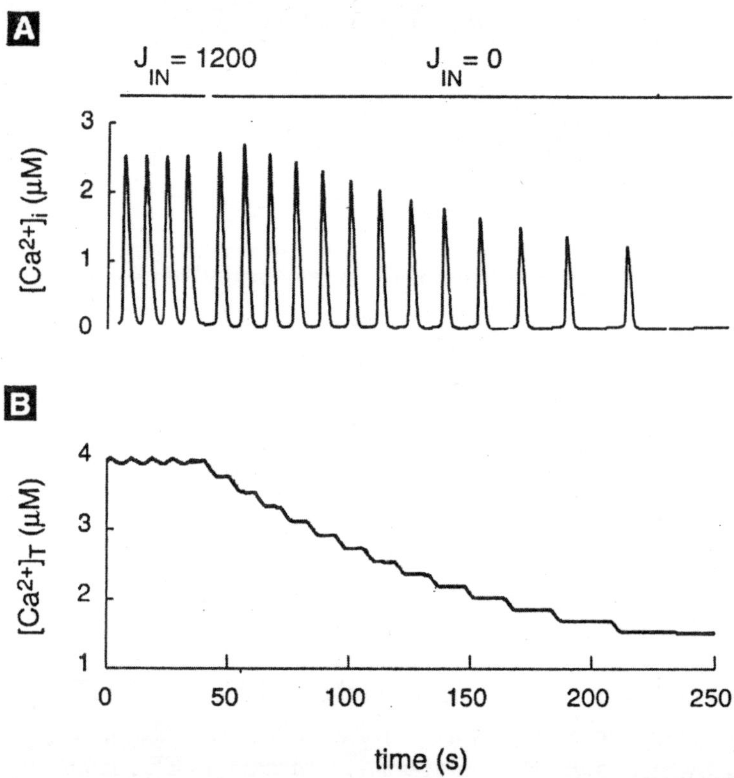

Fig. 7.14. Open-cell gonadotroph model (dimensional form of (7.48)-(7.52), see Exercise 8) with constant Ca^{2+} influx J_{IN}. Time series for $[Ca^{2+}]_i$ (A) and $[Ca^{2+}]_T$ (B) are shown for two different values of J_{IN}. When $J_{IN} = 1200$ (time between 0 and 40), sustained oscillations in $[Ca^{2+}]_i$ are obtained with high $[Ca^{2+}]_T$. After J_{IN} is switched from 1200 to 0 (time > 40), the oscillations continue, but the amplitude and the frequency decrease. Eventually, the store becomes depleted, as reflected by the low level of $[Ca^{2+}]_i$. The oscillations stop when the store becomes depleted. Parameters as in Fig. 7.12 except: $[IP_3] = 0.9\ \mu$ M, $\epsilon = 0.01$, $V_{PMCA} = 2000$ aMol/s, $K_{PMCA} = 0.3\ \mu$M J_{IN}aMol/s as indicated.

This phenomenon can be understood by the bifurcation analysis of the "fast subsystem" of the open cell model. Recall that $\hat{c}_T$ is a "slow" variable and can be treated as a parameter of the fast subsystem. Assuming that $\in = 0$, this fast subsystem is identical to the closed-cell model presented in the previous subsection. The bifurcation diagram for the dimensional version of this model with respect to $[Ca^{2+}]_T$ for a fixed value of $[IP_3]$ is shown in Fig. 7.15A. It shows that oscillations can occur only at intermediate values of $[Ca^{2+}]_T$ and that the oscillation amplitude decreases as $[Ca^{2+}]_T$ decreases. Fig. 7.15B shows that the oscillation period increases (i.e., the frequency decreases) as $[Ca^{2+}]_T$ decreases. Superimposed on the bifurcation diagram in Fig. 7.15A are the projected trajectories taken from Fig. 7.14A.

The Plasma Membrane Oscillator

Extensive studies of the plasma membrane (PM) oscillator have revealed the major types of ion channels that are expressed in gonadotrophs as well as the detailed gating properties of these channels. The oscillations in the PM potential occur spontaneously in the absence of any hormonal signal. They are generated by several voltage-gated ion currents including L-type (noninactivating) and T-type (transient) Ca^{2+} currents, a Na^+ current that is not essential, and a delayed rectifier K^+ current. There exists another K^+ channel that is of great importance to this cell: the K(Ca) (Ca^{2+}-activated K^+) channel that is sensitive to apamin.

Although the K(Ca) channels are not involved in producing the voltage spikes, they provide the key link between the ER oscillator and the PM oscillator. This is because these channels are activated when the cytosolic Ca^{2+} level reaches peak values, resulting in the disruption of tonic spiking in PM potential and the hyperpolarization of the cell. A realistic model of this oscillator based on experimental data can be found in Li et al. (1995a). Here, however, we use a much simpler model of the PM oscillator based on a modified Morris-Lecar model to demonstrate how the system works.

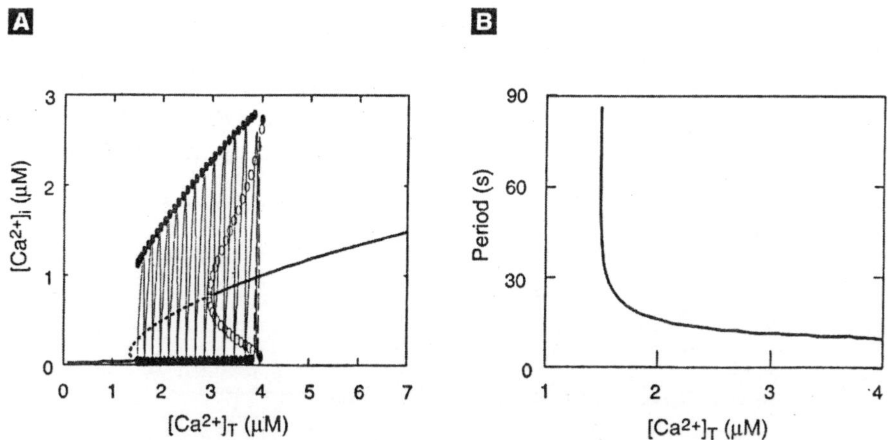

Fig. 7.15. Bifurcation in the dimensional version of the open-cell gonadotroph model: (A) Steady states (curves) and periodic solutions (circles) versus the total intracellular Ca^{2+} level $[Ca^{2+}]_T$. The oscillation amplitude decreases significantly as $[Ca^{2+}]_T$ decreases. (B) The oscillation period increases as $[Ca^{2+}]_T$ decreases. Superimposed on the bifurcation diagram in (A) are the projected trajectories for J_{IN} = 1200 (dashed) and J_{IN} = 0 (solid) from Fig. 7.14.

All we need is a model that generates Ca^{2+} action potentials spontaneously such that each AP spike is associated with the entry of a quantum of Ca^{2+} into the cell. The Morris-Lecar model is the simplest possible model of this kind, involving only an L-type Ca^{2+} current and a delayed rectifier K^+ current. The main modification involves introducing the K(Ca) current, which is absent in the original Morris-Lecar model but is crucial for gonadotrophs. Note that I_{app} and the leak currents are removed from the original model. The equations governing the gonadotroph plasma membrane oscillator are then

$$C_m \frac{dV}{dt} = -g_{Ca} m_\infty (V - V_{Ca}) - \left[gk\dot{w} + \frac{g_{K(Ca)}\left[[Ca^{2+}]_i^{\,4} \right]}{[Ca^{2+}]_i^{\,4} + K^4_{K(Ca)}} \right] (V - V_K), \quad ...(7.53)$$

$$\frac{dw}{dt} = \phi(w_\infty - w)/\tau \quad ...(7.54)$$

where $[Ca^{2+}]_i$ is again the Ca^{2+} level in the cytosol but is here a parameter for the plasma membrane model.

The voltage-dependent equilibrium open fractions for the channel states and the voltage-dependent activation time constant for the delayed rectifier are, as before:

$$m_\infty = 0.5[1 + \tanh(V - v_1)/v_2],$$
$$w_\infty = 0.5[1 + \tanh(V - v_3)/v_4],$$
$$\tau = 1/\cosh((V - v_3)/(2.v_4)).$$

At low levels of $[Ca^{2+}]_i$, the system generates continuous spiking. When $[Ca^{2+}]_i$ is larger than a certain threshold value, it hyperpolarizes the membrane potential. Also, when the plasma membrane voltage undergoes tonic spiking, Ca^{2+} enters the cell and increases $[Ca^{2+}]_i$. We take this into account in the next section.

Bursting Driven by the ER in the Full Model

A full model of Ca^{2+} signaling in agonist-stimulated gonadotrophs involves both the ER oscillator and the PM oscillator. Having studied all the ingredients of this model, we now combine the modules into a complete, realistic model described by the following five differential equations:

$$\frac{dV}{d\hat{t}} = -\hat{t}_{Ca} - \hat{t}_K - \hat{t}_{K(Ca)}, \quad ...(7.55)$$

$$\frac{dw}{d\hat{t}} = \phi(w_\infty - w)/\hat{\tau}, \quad ...(7.56)$$

$$\frac{d\hat{c}}{d\hat{t}} = \hat{j}_{rel}(\hat{c}, h, \hat{c}_T) - \hat{j}_{fil}(\hat{c}) + \in \left[-\hat{\alpha}\hat{i}_{Ca} - \frac{\hat{v}_p \hat{c}^2}{\hat{c}^2 + \hat{k}_p^2} \right], \quad ...(7.57)$$

$$\frac{dh}{d\hat{t}} = \hat{a}\left[1 - (\hat{c} + 1)h\right], \quad ...(7.78)$$

$$\frac{d\hat{c}_T}{d\hat{t}} = \in \left[-\hat{\alpha}\hat{i}_{Ca} - \frac{\hat{v}_p \hat{c}^2}{\hat{c}^2 + \hat{k}_p^2} \right], \quad ...(7.59)$$

where

$$\hat{i}_{Ca} = \hat{g}_{Ca} m_\infty (V - V_{Ca}),$$
$$\hat{l}_K = \hat{g}_K w (V - V_K),$$
$$\hat{i}_{K(Ca)} = \hat{g}_{K(Ca)} \frac{\hat{c}^4}{\hat{c}^4 + \hat{k}^4_{K(Ca)}} (V - V_K),$$

and $\hat{\alpha}$ is the non-dimensionalized factor that converts the current flux into the ion flux of Ca^{2+} (7.27). Note that $I_{K(ca)}$ is now expressed in terms of the nondimensional $\hat{c}$ (with $\hat{k}_{k_{(ca)}}/k_d$) the conductances have been scaled, eliminating C_m as a parameter. For example,

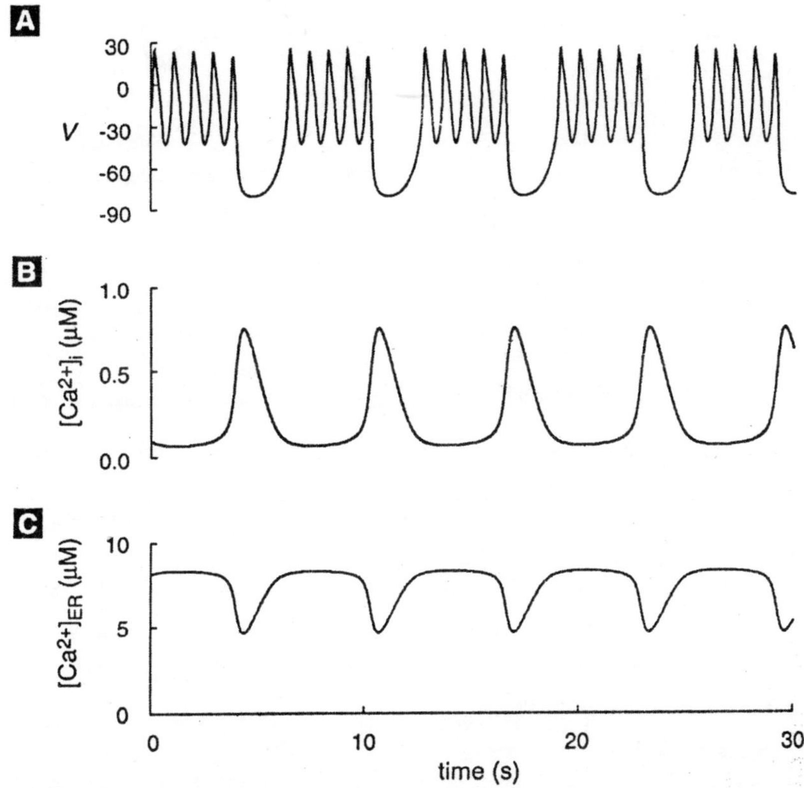

Fig. 7.16. Full gonadotroph model (Equations (7.55)-(7.59) in dimensional form): bursting as a result of interaction between the plasma membrane oscillator and the ER oscillator. Plasma membrane voltage V (A), cytosolic calcium level $[Ca^{2+}]_i$ (B), and the store calcium content $[Ca^{2+}]_{ER}$ (C) are shown. Each spike in $[Ca^{2+}]_i$ hyperpolarizes the plasma membrane by activating calcium-sensitive potassium channels. ER parameters as in Fig. 7.14 except: L = 1.48 pL/s, P = 2960 pL/s, K_i = 0 mM, A = 2/s, V_{SERCA} = 480 aMol/s, V_P = 400 aMol/s. Plasma membrane (Morris–Lecar) parameters : V_K = – 85 mV, V_{Ca} = 120 mV, $K_{K(Ca)}$ = 0.5μM, C_m = 4 μF/cm², g_K = 20 μS/cm², g_{Ca} = 20 μS/cm², $g_{K(Ca)}$ = 8 μS/cm², ϕ = 12/s, v_1 = – 3 μV, v_2 = 30 mV, v_3 = –20 mV, v_4 = 30 mV, α = 0.2 (aMol.cm²)/nC.

$$\hat{g}_{\mathrm{Ca}} = g_{\mathrm{Ca}}\left(\frac{\sigma \overline{V}_i}{Cm_{\mathrm{PIPR}} f_{\mathrm{i}}}\right).$$

The nondimensional voltage-dependent time, constant is $\tau(V)$ given by

$$\hat{\tau}(V) = \tau(V)\left(\frac{P_{\mathrm{IP3R}} f_{\mathrm{i}}}{\sigma \overline{V}_i}\right)$$

The effects of the ER oscillator on the PM oscillator are determined by $I_{\mathrm{K(Ca)}}$. Whenever $\hat{c}$ is high, it activates $I_{\mathrm{K(Ca)}}$, which hyperpolarizes the PM potential and inhibits the PM oscillator. The effects of the PM oscillator on the ER oscillator are mediated by the term which describes the Ca^{2+} influx through voltage gated Ca^{2+} channels. As shown in the previous section, this influx leads to a fuller ER store and makes the ER more excitable. Thus, the voltage spikes activate the ER oscillator. The interaction of the two oscillators gives rise to bursting of the PM potential. Compare with Fig. 7.11. Such bursting is mainly driven by the ER oscillator, which periodically hyperpolarizes the plasma membrane. However, the PM oscillations are indispensable for maintaining the bursting, because, the Ca^{2+} entry that accompanies each voltage spike compensates the loss of Ca^{2+} to the extracellular medium and is thus key in preventing the store from depleting for details of such interactions).

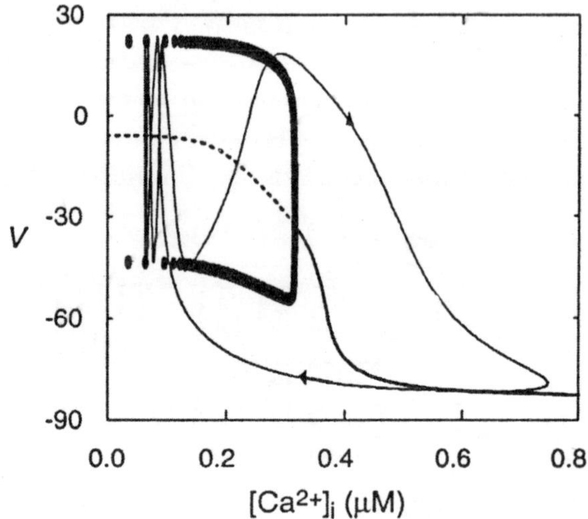

Fig. 7.17. Full gonadotroph model: bifurcation diagram for the dimensional form of the fast subsystem (plasma membrane oscillator) with superimposed trajectory from Fig. 7.16.

In Fig. 7.17 the dimensional bifurcation diagram of the fast subsystem (the PM oscillator) with respect to $[Ca^{2+}]_i$ is shown with the trajectory of the bursting oscillations in the top panel of Fig. 7.16 superimposed. Notice that the PM voltage V follows the diagram well during the spiking phase, when $[Ca^{2+}]_i$ changes slowly. But during the $[Ca^{2+}]_i$ spike, which is fast, the trajectory does not follow the diagram closely.

The Pancreatic Beta Cell

Pancreatic β-cells secrete insulin, the hormone that maintains blood plasma glucose within narrow limits in the face of variable food ingestion and physical activity patterns. Insulin allows muscle to take up glucose for immediate energy, causes the liver to store glucose as glycogen for medium-term energy storage, and signals fat cells to use glucose for fat deposition for long-term energy storage. Diabetes is a disease in which glucose is chronically elevated, leading to blindness, kidney failure, limb amputation, cardiovascular disease, and death. Type I ("juvenile") diabetes is the result of an absolute lack of insulin following auto-immune destruction of the β-cells. Type II ("adult onset") diabetes, the more common variety, involves a relative lack of insulin, usually as a result of two defects: insulin resistance (higher than normal concentrations of insulin are required for glucose processing) and failure of the β-cells to produce enough insulin to compensate. The rising tide of diabetes in the industrialized nations suggests that Type II diabetes is a maladaptive response to a toxically energy-rich food environment, with genes governing the susceptibility of individuals. A number of aspects of this system have been modeled, including the kinetics of insulin secretion, the etiology of diabetes as β-cells fail to compensate, insulin action and whole-body plasma insulin oscillations. Here we limit our attention to the electrical activity and Ca^{2+} oscillations used by the β-cell to regulate insulin secretion.

In terms of cell physiology, the β-cell closely resembles the pituitary gonadotroph. Both are endocrine cells in which secretion is controlled by $[Ca^{2+}]_i$, which is in turn regulated to a large degree by bursting, and they share many mechanistic elements such as K(Ca) channels and IP_3 receptors. However, there are important and interesting differences from that point on. Bursting in β-cells is primarily driven by the plasma membrane oscillator, though the ER probably plays a significant role at least in modulating the burst mechanism. These characteristics are shared with many neurons, and β-cells provided an early paradigm for modeling of bursting.

The models discussed in this section are complementary to the Keizer-Maki model. The latter considers insulin oscillations but does not explicitly treat the Ca^{2+} influx that leads to insulin secretion. Here we treat Ca^{2+} influx in detail, but assume that secretion increases with mean $[Ca^{2+}]_i$. We will focus on membrane-potential-driven $[Ca^{2+}]_i$ oscillations, though β-cells have occasionally been observed to exhibit oscillations driven by the ER, similar to those in gonadotrophs.

Ironically, the Keizer-DeYoung model for the IP_3 receptor was originally developed to explain this marginal phenomenon in β-cells, and was then ready at hand for application to gonadotrophs. One indication of the differences between β-cell and gonadotroph bursting is apparent in Fig. 7.18. This simultaneous recording of membrane potential and $[Ca^{2+}]_i$ oscillations shows that $[Ca^{2+}]_i$ is high during the depolarized spiking phase of the bursts, rather than during the hyperpolarized silent phase. An important aspect of β-cell function that we will also have to neglect in this chapter is the organization of the cells into electrically coupled populations, called the islets of Langerhans.

Here we take advantage of the observation that the cells in an islet are synchronized. That is why the voltage of the single peripheral cell shown in Fig. 7.18B is in register with the $[Ca^{2+}]_i$ signal (Fig. 7.18A), which comes from many cells in the islet. This allows us to study a simple single-cell model, which can be taken as representative of the whole islet. More subtle effects of electrical coupling will be discussed elsewhere. See also Sherman (1997).

Chay-Keizer Model

The model we use to illustrate β-cell bursting is based loosely on the model of Sherman et al. (1988), recast in Morris-Lecar form. Bursting occurs in response to glucose, so no applied current is needed. Morris-Lecar by itself can account for the spiking during the active phase of a burst. In order to obtain cycling between bursting and silent phases, we need to add a slow negative-feedback current. The first hypothesis, proposed by Atwater, Rojas, and colleagues and made into a mathematical model by Chay and Keizer (1983), was that bursting was mediated by a K(Ca) current. The idea was that $[Ca^{2+}]_i$ would build up slowly during the spiking phase until the inhibitory effect of the increased K^+ current reached a sufficiently high level to terminate the spiking. The K(Ca) current is represented as

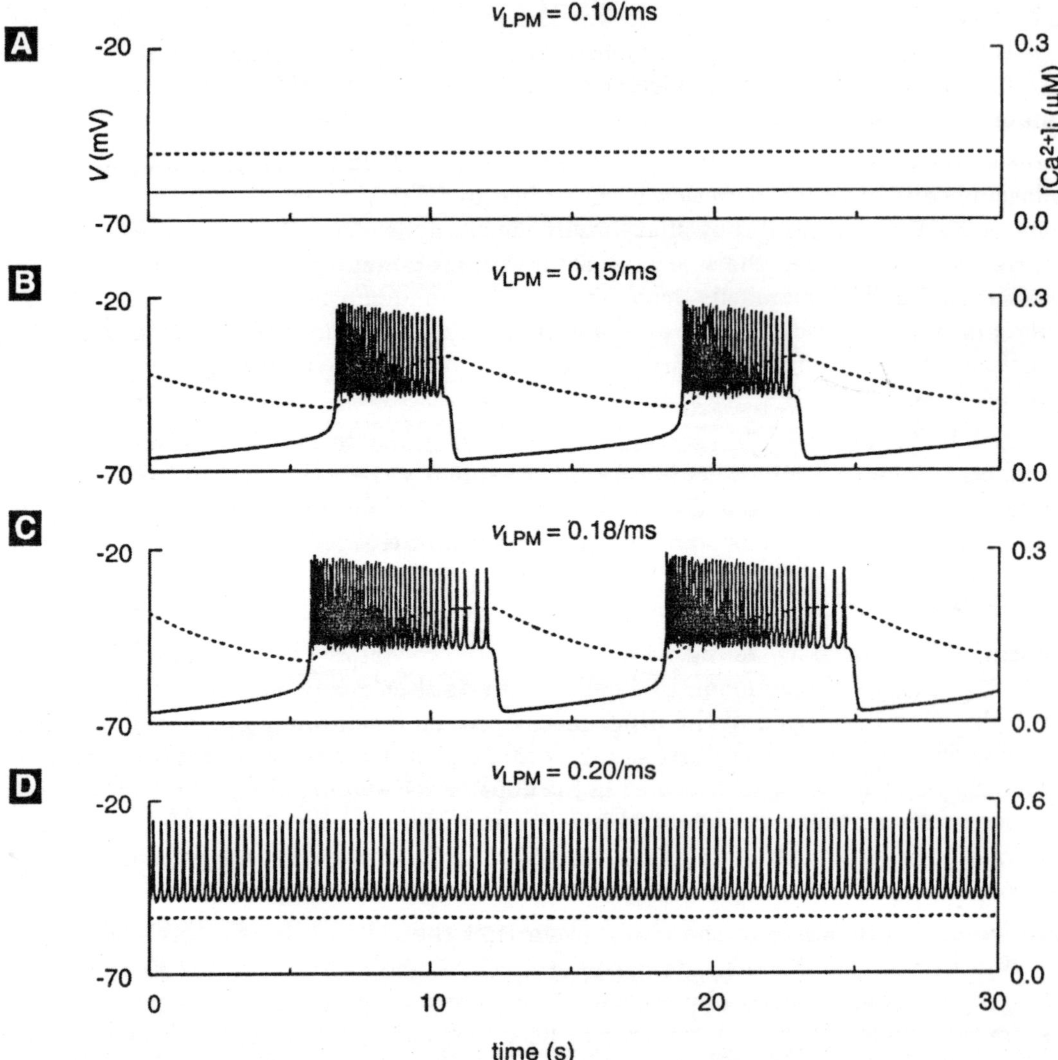

Fig. 7.18. Bursting and glucose sensing of the Chay-Keizer model, (7.61)-(7.63). Parameters: C_m = 5300 fF, g_{Ca} = 1000 pS, V_{Ca} = 25 mV, g_K = 2700 pS, V_K = –75 mV, I_{app} = 0, v_1 = –20 mV, v_2 = 24, v_3 = –16 mV, v_4 = 11.2 mV, ϕ = 0.035/ms, g

$$I_{K(Ca)} = g_{K(Ca)} \frac{[Ca^{2+}]_i}{K_{K(Ca)} + [Ca^{2+}]_i}(V - V_K), \qquad ...(7.60)$$

and the modified Morris-Lecar equations, supplemented by a slow equation for $[Ca^{2+}]_i$, are

$$C_m \frac{dV}{dt} = -I_{Ca} - I_K - I_L - I_{K(Ca)}, \qquad ...(7.61)$$

$$\frac{dw}{dt} = \phi \frac{w_\infty - w}{\tau}, \qquad ...(7.62)$$

$$\frac{d\left[Ca^{2+}]_i\right]}{dt} = f_i(-\alpha I_{Ca} - v_{LPM}[Ca^{2+}]_i). \qquad ...(7.63)$$

$[Ca^{2+}]_i$ as a parameter that controls the behaviour of the fast subsystem, consisting of V and n. In analogy to Fig. 7.9, we construct the bifurcation diagram of the fast subsystem with respect to $[Ca^{2+}]_i$. Fig. 7.20A shows the resulting three-branched Z-shaped curve (Z-curve) of steady states. The Z-curve is actually the projection into the V-$[Ca^{2+}]_i$ plane of the three-dimensional curve on which V and n are at steady state. When $[Ca^{2+}]_i$ is high, $gK_{(Ca)}$ is large, and the system goes to the lower branch of low-voltage steady states. When $[Ca^{2+}]_i$ is low, $gK_{(Ca)}$ is small, and V rises above the threshold for activation of the voltage-dependent Ca^{2+} and K^+ currents, generating the depolarized upper branch. For intermediate values of $[Ca^{2+}]_i$ there are three steady states.

The new steady states are born via saddle-node bifurcations, with the saddle points making up the middle branch of the Z-curve. The fast subsystem consists of two variables, not just one as was the case in Fig. 7.9, so it can itself undergo oscillations, even with $[Ca^{2+}]_i$ fixed. These oscillations arise via a Hopf bifurcation on the upper branch and correspond to the fast spikes during a burst. For intermediate values of $[Ca^{2+}]_i$, the system is bistable: It can either be at rest on the lower branch of the Z-curve or oscillate on the upper branch. Bursting is a repetitive alternation between these two states. In order for this to occur, $[Ca^{2+}]_i$ must have appropriate slow dynamics, such 'that it increases when V is high and decreases when V is low. We can interpret this geometrically by ignoring n and viewing the V-$[Ca^{2+}]_i$ bifurcation diagram as a V-$[Ca^{2+}]_i$ phase plane.

The Z-curve serves as a nullcline for V, and we add a nullcline for $[Ca^{2+}]_i$, which increases when the phase point is above the $[Ca^{2+}]_i$ nullcline and decreases when the point is below the nullcline. Thus, in order for $[Ca^{2+}]_i$ to rise during the active phase of each burst and fall during the silent phase, the $[Ca^{2+}]_i$-nullcline must intersect the middle branch of the Z-curve. If instead the $[Ca^{2+}]_i$-nullcline intersects the lower branch, the cell will remain silent. If the $[Ca^{2+}]_i$-nullcline intersects the spiking branch, the cell can fire continuously; the influx of $[Ca^{2+}]_i$ during each spike exactly balances the removal. To understand this analysis in depth, it is helpful to construct the phase planes corresponding to different values of $[Ca^{2+}]_i$. Of particular interest is the key role of the homoclinic orbit (labeled HC in Fig. 7.20A) in terminating the active phase of each burst. See also Sherman (1997) or Rinzel and Ermentrout (1998). Returning to Fig. 7.19, the sequence of panels illustrates how β-cells modify their electrical activity according to the level of glucose.

As glucose concentration increases, the rate of the PMCA is here hypothesized to increase as well. For low glucose (low pump rate), the cell is electrically silent (top panel). For glucose above a threshold concentration, bursting appears (second panel). Further increases in glucose result in longer active spiking phases, or plateaus, and shorter silent phases (third panel). Finally, for very high glucose, the cell remains permanently in the active phase, spiking continuously. Comparing these statements to the bifurcation analysis, we find that those about glucose and pump rate are equivalent to statements about how the $[Ca^{2+}]_i$-nullcline intersects the Z-curve.

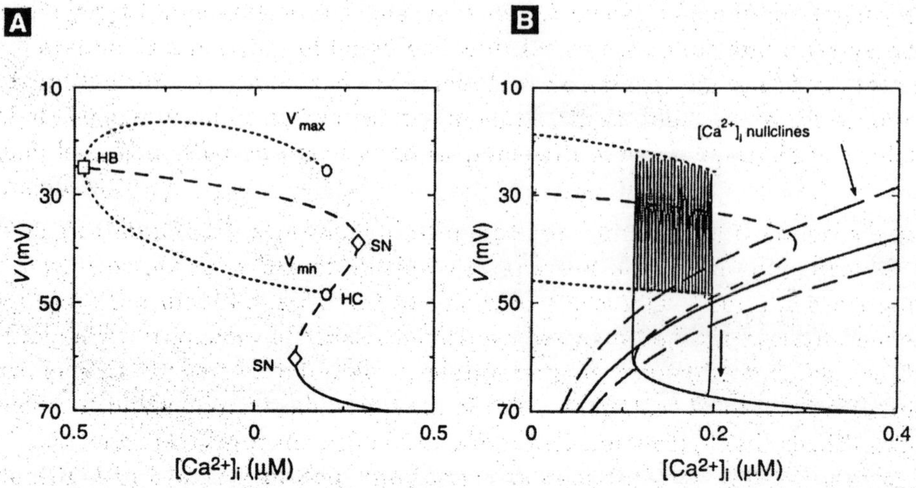

Fig. 7.19(A). Bifurcation diagram of Chay-Keizer fast equations (7.61)-(7.62) using $[Ca^{2+}]_i$ as a parameter. Hopf bifurcation (HB-square); saddle-nodes (SN-diamonds); homoclinic orbit (HC - circles). (B) Projection of V-$[Ca^{2+}]_i$ trajectory for V_{LPM} = 0.13 ms^{-1} onto the bifurcation diagram. Vertical arrow next to trajectory indicates direction of increasing time. Nullclines for $[Ca^{2+}]_i$ are shown for three cases: V_{LPM} = 0.10 (bottom), V_{LPM} = 0.13 (middle), and V_{LPM} = 0.20 (top). The corresponding solutions are silent, bursting, and continuously spiking, respectively. Compare with Fig. 7.19.

Chay-Keizer with an ER

The ER is significant in β-cells for two reasons. One is that an important potentiator of insulin secretion, acetylcholine (ACh), works by promoting production of IP_3 and dumping the stores, similar to GnRH in gonadotrophs. This has been modeled in Bertram et al. (1995) and Chay (1997).

The second reason is that the ER strongly influences the kinetics of $[Ca^{2+}]_i$, and hence all $[Ca^{2+}]_i$-dependent processes. Here we consider only the latter aspect, showing that even a passive ER, which does not actively dump Ca^{2+}, can have profound effects on bursting. To do this we append an equation for $[Ca^{2+}]_{ER}$ to the Chay-Keizer model and add appropriate flux terms, a passive conductance and a linearized version of the SERCA pump (with leading constant V_{LSP}), to the $[Ca^{2+}]_i$ equation:

$$Cm\frac{dV}{dt} = -I_{Ca} - I_K - I_{K(ATP)} - I_{K(Ca)}, \quad ...(7.64)$$

$$\frac{dw}{dt} = \phi\frac{w_\infty - w}{\tau}, \quad ...(7.65)$$

$$\frac{d\left[Ca^{2+}\right]_i}{dt} = f_i(-\alpha I_{Ca} - V_{LPM}[Ca^{2+}]_i])$$

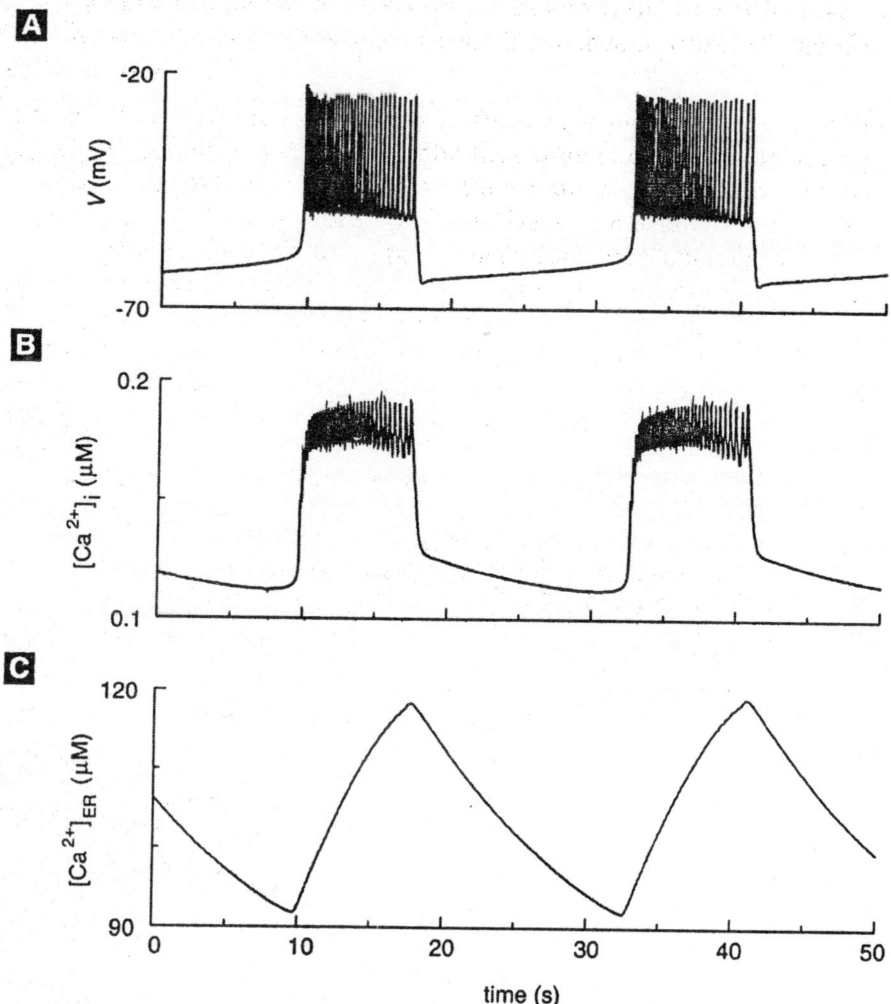

Fig. 7.20. Bursting in the Chay-Keizer model with ER, (7.64)-(7.67). Parameters as in Fig. 7.19, except f_i = 0.01, with the following parameters added for the ER: P_{IP3R} = 0.0008/ms, λ = 2, σ = 0.032, V_{LSP} = 0.6/ms

$$+\frac{f_i}{\lambda_{ER}}\left(P_{IP3R}\left(\left[Ca^{2+}\right]_{ER}-\left[Ca^{2+}\right]_i\right)-V_{LSP}\left[Ca^{2+}\right]_i\right), \qquad ...(7.66)$$

$$\frac{d\left[Ca^{2+}\right]_{ER}}{dt} = \frac{f_i}{\sigma\lambda_{ER}}\left(-P_{IP3R}\left(\left[Ca^{2+}\right]_{ER}-\left[Ca^{2+}\right]_i\right)+V_{LSP}\left[Ca^{2+}\right]_i\right). \qquad ...(7.68)$$

Compare these equations with (7.23) and (7.24). Although $[Ca^{2+}]_{ER}$ does not directly affect the plasma membrane, it has indirect effects through $I_{K(Ca)}$. This model is essentially equivalent to Theresa Chay's last β-cell model.

Bursting with this model is shown in Fig. 7.21. Here f_i has been increased to 0.01, which means that in the absence of the ER only very fast bursting can occur (see Exercise 12). In this case, bursting with periods of 10-60 seconds, as typically observed in islets, depends on slow kinetics supplied by the ER. Note that $[Ca^{2+}]_{ER}$ rises and falls slowly the way $[Ca^{2+}]_i$ does in the original Chay-Keizer model. In contrast, $[Ca^{2+}]_i$ shows two distinct time scales, fast jumps at the beginning and end of each burst, a slowly rising plateau during the active phase, and a slow tail during the silent phase.

The fast jumps reflect the intrinsic kinetics of $[Ca^{2+}]_i$, while the slow portions reflect slow uptake and release of Ca^{2+} by the ER. This complex time course matches the characteristics of $[Ca^{2+}]_i$ seen in experiments (see Fig. 7.18 and also Valdeolmillos et al. (1989)) better than Chay-Keizer. It also demonstrates that it is possible for slow negative feedback to operate through the K(Ca) channel, or some other Ca^{2+}-sensitive channel, even though $[Ca^{2+}]_i$ does not itself appear to be slow. The examples discussed here represent only the tip of the iceberg of the β-cell field. The precise contributions of the mechanisms we have treated, K(Ca) channels, $g_{K(ATP)}$ channels, and the ER, as well as some that we have not treated, such as inactivation of Ca^{2+} channels, are not settled. However, it seems likely that complex interactions of all of these will be necessary to explain the diverse phenomena observed. The mechanisms of other important regulators of cell electrical activity and $[Ca^{2+}]_i$, such as cAMP and epinephrine, remain to be elucidated. Nonetheless, the basic mechanisms and concepts presented here should prepare the reader sufficiently to explore the exercises and the literature on his or her own.

8 DIMENSIONAL FORMULATION

All of the models considered in previous chapters have relied on the implicit assumption that chemical concentrations are uniform in space. This assumption is reasonable when the region of space in which the reaction takes place is confined and quite small. However, there are many situations in which chemical concentrations are not uniform in space. A well-known example in which nonuniform distributions are crucial is the propagation of an action potential along the axon of a nerve fiber. When a nerve cell "fires," a wave of membrane depolarization is initiated at the base of the axon (where it connects to the cell body and propagates along the axon out to its terminus. During propagation, large spatial gradients in membrane potential and local currents are created.

The interaction between these spatial gradients and voltage-sensitive ion channels in the axonal membrane drives the wave along the axon. In order to understand the propagation of a nerve impulse, we must first master the basic principles of molecular diffusion and the interactions between chemical reaction and diffusion. Many other questions arise in molecular cell biology that demand at least an elementary understanding of molecular diffusion. For instance, how long does it take for a chemical signal generated at the cell membrane to diffuse to the nucleus? Why are expensive transport systems required to move some materials in cells, for example between a nerve cell body and synapses in axons and dendrites? How fast can molecules or ions pass through protein channels in membranes? Nonuniform Ca^{2+} concentration in the vicinity of Ca^{2+} channels has been reported. Therefore, a simplifying assumption is made that the Ca^{2+} concentration is high in a small region adjacent to the channel (domain Ca^{2+}).

To improve on the domain Ca^{2+} approximation and other simplified approaches to spatial nonuniformity, we will need the spatial modeling principles described here. A more sophisticated example of spatiotemporal organization in living cells is the phenomenon of Ca^{2+} waves that propagate through eggs after fertilization. Similar to Ca^{2+} waves in eggs are waves of cyclic AMP that propagate through fields of slime mold amoebae. By directing the motion of the amoebae, these chemical waves organize the complex behaviours of this primitive multicellular organism: aggregation of simple amoebae into a multicellular slug, motility of the slug, and formation of the fruiting body. Other interesting examples of spatial organization include gap-gene expression in early fruit fly embryos, seashell patterns, and medial ring placement at cell division. Although we will not attempt to model any of these phenomena in this book, a starting point for such investigations is this chapter. The chapter is organized

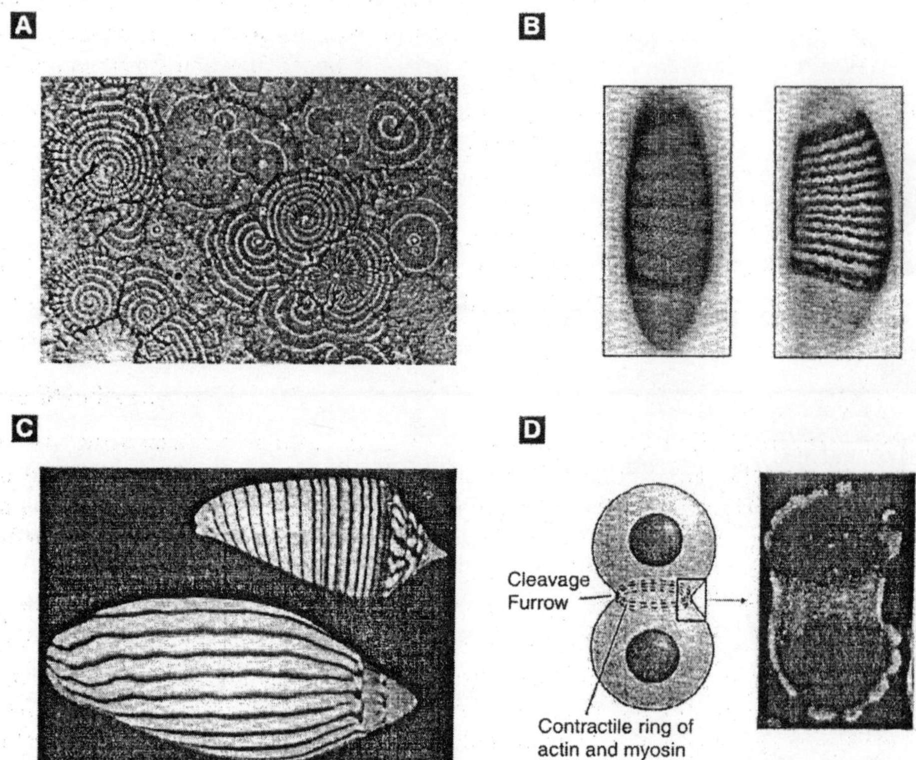

Fig. 8.1. Spatial phenomena in cell biology. (A) Waves of aggregation in fields of slime mold amoebae. The light and dark bands correspond to regions where the amoebae are actively moving or not. The amoebae collect at the center of each pattern, where they form a multicellular slug. The collective motion of the cells is organized by waves of cyclic AMP that propagate through the extracellular medium. (B) Stripes of gene expression (*ftz* and *eve*) in a fruit fly embryo at (left) 3 hr. after fertilization and (right) $3\frac{1}{2}$ hr. after fertilization. For an introductory discussion of the segmentation genes in Drosophila. (C) Pigmentation patterns on sea shells. The cleavage furrow in a dividing cell. A dividing slime mold amoeba is stained for actin and myosin. The actomyosin ring in the center of the cell contracts like a purse-string to divide the cell in half.

along the following lines: First, we consider diffusion in one dimension, such as we might find in a long thin tube like a nerve axon.

We distinguish between a *conservation law* (how the law of conservation of matter relates molecular flux to local changes in concentration) and a *constitutive, relation* (how molecular flux is determined by concentration gradients, fluid transport, and electrophoresis). These principles are expressed in the precise mathematical terms of *partial differential equations* (PDEs). We show the exact solution to these equations for a number of important illustrative cases. Because PDEs cannot be solved exactly in most realistic situations, we next describe a numerical procedure, called the method of lines, that is easily implemented. Also, because very few spatial nonuniformities are effectively one-dimensional, we show how to formulate the conservation law and constitutive relations in two and three dimensions. We then couple molecular diffusion to nonlinear chemical reactions in order to study wave propagation in one spatial dimension. The theory is applied to the FitzHugh-Nagumo equations of nerve impulse propagation introduced in.

ONE-DIMENSIONAL FORMULATION

Conservation in One Dimension

Many equations in biology are consequences of *conservation laws.* A conservation law is simply a mathematical statement describing how some quantity is created or destroyed or moves about.

Consider a chemical species C whose concentration $c(x, t)$ varies in time and space, where the spatial variation is restricted to one spatial variable x. This situation is illustrated in Fig. 8.2, where the chemical species C is contained in a long, thin tube with constant cross-sectional area A. In any fixed region R along the tube, the conservation of C can be expressed in words as

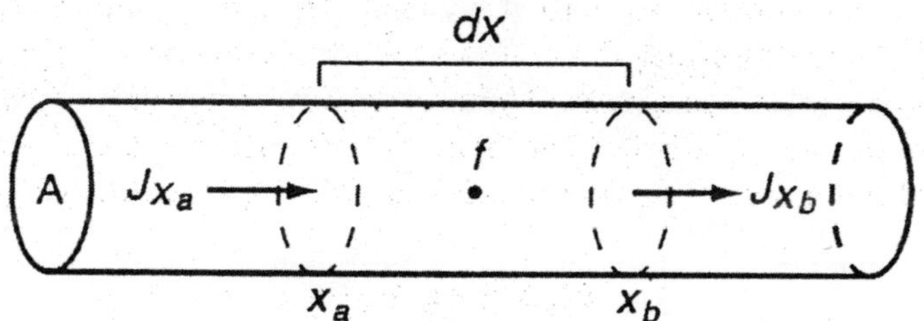

Fig. 8.2. Conservation in one dimension.

time rate of change of the total amount of C within R =
rate at which C flows in to R
– rate at which C flows out of R
+ rate at which C is produced within R
– rate at which C is destroyed within R. ...(8.1)

The total amount of chemical C contained in a small slice of tube between x and $x + dx$ is $c(x,t)Adx$. At any time t, the total amount of C in some arbitrary interval $x_a < x < x_b$ can be computed by integrating $c(x, t)\,A$ over that interval:

$$\text{Total amount of C in the interval } [x_a, x_b] = \int_{x_a}^{x_b} c(x,t)A\,dx. \quad ...(8.2)$$

It is important to distinguish between concentration (amount/ volume) and the "total amount." If c has units of micromolar (micromol/liter), then the total amount has units of micromoles.

Now suppose that C is free to move about inside the tube, so that C moves in and out of the interval by crossing the boundaries of the interval at $x = x_a$ and $x = x_b$. If we denote by $J(x, t)$ the rate at which C moves across the boundary at position x from left to right at time t, then the net movement, or flux, of C into the interval is

$$\text{net rate of entry of C} = AJ(x_a, t) - AJ(x_b, t). \quad ...(8.3)$$

Since the net rate of entry has units of amount/time and A has units of area, the flux rate $J(x, t)$ has units of amount/area/time. It is also important to remember that $J(x, t)$ is positive when C moves to the right, and negative when C moves to the left.

The total amount of C in the interval can also change because of the production or destruction of C within the interval. If we let $f(x, t, c)$ denote the net rate of increase of C (production—destruction) per unit volume at location x and time t, then the total amount of C produced in the interval at time t is

$$\text{net rate of production of C} = \int_{x_a}^{x_b} f(x, t, c(x, t))A\, dx. \quad ...(8.4)$$

Note that the presence of c in the definition of f allows for the possibility that the rate of production of C depends on c itself. Since the units of the net rate of production of C are amount/time, the units off must be amount/time/volume. When f is positive, the region is a source (leading to an increase in the total amount of C), and when f is negative, it is a sink. The function f is often called a source function.

The conservation law (8.1) can now be written in mathematical symbols as

$$\frac{d}{dt}\int_a^b c(x,t)\,dx = J(x_a,t) - J(x_b,t) + \int_{x_a}^{x_b} f(x,t,x(x,t))\,dx, \quad (8.5)$$

where the constant A has been factored out. The flux terms can be replaced by

$$J(x_b,t) - J(x_a,t) = \int_{x_a}^{x_b} \frac{\partial}{\partial x} J(x, t)\,dx, \quad ...(8.6)$$

allowing all the terms in (8.5) to be written as integrals:

$$\frac{d}{dt}\int_{x_a}^{x_b} c(x,t)\,dx = \int_{x_a}^{x_b} \frac{\partial}{\partial x} J(x,t)\,dx + \int_{x_a}^{x_b} f(x,t,c(x,t))\,dx. \quad ...(8.7)$$

If the function $c(x, t)$ is smooth enough, then the differentiation and integration can be interchanged, and (8.7) can be rewritten as

$$\int_{x_a}^{x_b}\left[\frac{\partial}{\partial x} C(x,t)\,dx - J(x, t)\cdot\frac{\partial}{\partial x} - f(x,t,c(x,t))\,d\right]dx = 0. \quad ...(8.8)$$

Since the interval is arbitrary, the only way this equality can hold is if the integrand is zero. Therefore, we replace (8.8) by the equivalent conservation law in differential form:

$$\frac{\partial c}{\partial t} - \frac{\partial J}{\partial x} = f(x,t,c). \quad ...(8.9)$$

Notice that in this equation there are two independent variables (x and t), and that the equation contains partial derivatives with respect to both of these. Such equations are called partial differential equations. Since time is one of the independent variables, and this equation describes the evolution of $c(x, t)$ in time, (8.9) is called an evolution equation because it describes how the concentration of C evolves (changes) as time proceeds.

Pick's Law of Diffusion

Equation (8.9) is underdetermined because it is a single equation relating two unknowns: the concentration c and the flux J. To resolve this problem, an additional equation relating c and J is needed.

In contrast to the conservation law (8.9), which follows indubitably from the general principle of material conservation, the relation between c and J must be determined empirically and is not universally valid. To make this distinction, the secondary relation between c and J is usually called a constitutive equation.

One such constitutive relation is called Fick's law, and states that C moves from regions of high concentration to regions of low concentration, at a rate proportional to the concentration gradient. In mathematical symbols, this diffusive flux is

$$J(x, t) = -D\frac{\partial}{\partial x}c(x, t), \qquad ...(8.10)$$

where the proportionality constant D is called the diffusion constant. The negative sign signifies that C moves spontaneously from regions of high concentrations to regions of low concentrations. The value of D depends on the size of C, as well as properties of the medium in which it is diffusing. The constant D has units of length²/time. Diffusion coefficients of some typical biochemicals are given in Table 8.1.

Table 8.1. Molecular Weight and Diffusion Coefficients of Some Biochemical Substances in Dilute Aqueous Solution.

Substance	*Molecular Weight*	*$D/10^7\ cm^2/s$*
glucose	192	660
insulin	5734	210
cytochrome c	13,370	11.4
myoglobin	16,900	11.3
β-lacroglobulin	37,100	7.5
serum albumin	68,500	6.1
hemoglobin	64,500	6.9*
catalase	247,500	4.1
urease	482,700	3.46
fibrinogen	339,700	1.98
myosin	524,800	1.10
tobacco mosaic virus	40,590,000	0.46

Using Fick's law, (8.9) becomes the reaction-diffusion equation

$$\frac{\partial c}{\partial t} - \frac{\partial}{\partial x}\left(D\frac{\partial c}{\partial x}\right) = f(x, t, c). \qquad ...(8.11)$$

In this equation, the term $\frac{\partial c}{\partial t}\left(D\frac{\partial c}{\partial x}\right)$ is the diffusion term, and f is the reaction term. When f is zero, that is, when there are no sources or sinks, (8.11) becomes the diffusion equation

$$\frac{\partial c}{\partial t} = \frac{\partial}{\partial x}\left(D\frac{\partial c}{\partial x}\right). \qquad ...(8.12)$$

Advection

Suppose that there is a uniform macroscopic flow of the solvent, with speed v along the x-axis, which carries solutes along with it. Then, during a small time Δt, all of the C between

$x = x_a$ and $x = x_a - v\Delta t$ will cross the point $x = x_a$. The total amount of C crossing x_a during this time is found by multiplying the concentration $c(x, t)$ by the fluid volume $Av\,\Delta t$. The corresponding flux is therefore (after dividing by Δt to get amount per unit time)

$$J(x,t) = vc(x,t). \qquad ...(8.13)$$

This flux is called the advective flux. Note that whereas the diffusive flux was proportional to the concentration gradient, the advective flux is proportional to the concentration itself.

If there is both diffusive flux and advective flux, then the total flux is the sum of the two:

$$J(x,t) = vc(x,t) - D\frac{\partial}{\partial x}c(x,t). \qquad ...(8.14)$$

Using this constitutive relation, (8.9) becomes a reaction-advection-diffusion equation,

$$\frac{\partial c}{\partial t} + \frac{\partial}{\partial x}\left(vc - D\frac{\partial c}{\partial x}\right) = f(x,t,c). \qquad ...(8.15)$$

Flux of Ions in a Field

If the substance C is an ion and there is an electrical potential gradient, then there will also be a flux of C because of the influence of the potential on the ion. In this case the flux of ions is given by the Nernst-Planck equation

$$J = -D\left(\frac{\partial c}{\partial x} + \frac{zF}{RT}c\frac{\partial \phi}{\partial x}\right), \qquad ...(8.16)$$

where ϕ is the electric potential, z is the number of positive charges on the ion (a negative integer if the ion is negatively charged), F is Faraday's constant, R is the universal gas constant, and T is absolute temperature. Notice that according to this equation, there is movement because of both the concentration gradient and the potential gradient.

The Cable Equation

Suppose that our long one-dimensional tube is bounded by a membrane, as in a nerve axon. In this case, we wish to keep track of the electrical potential across the membrane, rather than some chemical species within the tube. Nonetheless, the rules of conservation are the same, so the derivation of the governing equation is similar.

Suppose the total current along the interior of the axon is I, positive from left to right, and the transmembrane current per unit membrane area is I_T, positive outward. Then, conservation of current implies that

$$I(x_a,t) - I(x_b,t) = \int_{x_a}^{x_b} SI_T\,dx, \qquad ...(8.17)$$

where S is the circumference of the tube. This conservation law can be expressed using integrals as

$$-\int_{x_a}^{x_b} \frac{\partial I}{\partial x}dx = \int_{x_a}^{x_b} SI_T\,dx, \qquad ...(8.18)$$

and since the interval is arbitrary, the integrands must be equal, so that

$$-\frac{\partial I}{\partial x}=SI_T. \qquad ...(8.19)$$

Recall from Chapter 1 that the total transmembrane current consists of two components, a capacitive current and the ionic currents

$$-\frac{\partial I}{\partial x}=S\left(Cm\frac{\partial V}{\partial t}+I_{ion}\right), \qquad ...(8.20)$$

where V is the transmembrane potential. Finally, the relationship between current and potential is given by the constitutive relationship known as Ohm's law (also called the core conductor assumption),

$$I=\frac{A}{R_c}\frac{\partial \phi_i}{\partial x} \qquad ...(8.21)$$

where R_c is the cytoplasmic resistance (with units Ohms length), and ϕ_i is the intracellular potential. With this constitutive relationship, our equation becomes

$$\frac{\partial}{\partial x}\left(\frac{A}{R_c}\frac{\partial \phi_i}{\partial x}\right)=S\left(Cm\frac{\partial V}{\partial t}+I_{ion}\right). \qquad ...(8.22)$$

Finally, we close the model by assuming that the membrane is in a highly conductive bath, so that the extracellular potential ϕ_e is a constant. Since $V=\phi_i-\phi_e$, we arrive at the cable equation

$$\frac{\partial}{\partial x}\left(\frac{A}{R_c}\frac{\partial V}{\partial x}\right)=S\left(Cm\frac{\partial V}{\partial t}+I_{ion}\right). \qquad ...(8.23)$$

For a tube of uniform circular cross section and diameter d, $A/s = d/4$. Typical parameter values for a variety of cells are shown in Table 8.2.

Boundary and Initial Conditions

In the study of ordinary differential equations, it is necessary to specify initial data before one can find a solution trajectory. With partial differential equations, one must specify both initial data and boundary data before a solution can be found. Roughly speaking, there must

Table 8.2. Typical Cable Parameter Values for a Variety of Excitable Cells.

Parameter *Units*	*d* *10^{-4} cm*	*Rc* *Ω cm*	*Rm* *10^3 Ω cm^2*	*Cm* *μF/cm^2*	*λm* *cm*
squid giant axon	500	30	1	1	0.65
lobster giant axon	75	60	2	1	0.25
crab giant axon	30	90	7	1	0.24
earthworm giant axon	105	200	12	0.3	0.4
marine worm giant axon	560	57	1.2	0.75	0.54
mammalian cardiac cell	20	150	7	1.2	0.15
barnacle muscle fiber	400	30	23	20	0.28

be one condition for each degree of freedom. Thus, since reaction-diffusion equations are of first order in time, there must be one initial condition for each unknown function. Since they are of second order in space, there must be two boundary conditions (conditions at some points in space) for each unknown function.

Initial conditions usually specify the values of the dependent variables at some initial time (usually $t = 0$) at which the solution is known or specified by experimental conditions. Boundary conditions reflect certain physical conditions of the experiment. For example, if the concentration c is specified to be some function $f(t)$ at some boundary point, say $x = x_a$, then the condition $c(x_a, t) = f(t)$ is applied, called a Dirichlet boundary condition. If, on the other hand, the flux at a point is specified, then the condition $-D\frac{\partial c}{\partial x}(x_a, t) = g(t)$, called a Neumann boundary condition, is applied. If the flux is related to the value of c at the boundary, then the Robin condition, $(x_a, t) = h(t) - \alpha c(x_a, t)$, is applied.

It is often convenient to assume that a domain is infinite, even though there is no such thing as an infinitely long tube. Even with infinite domains, however, boundary conditions must be specified as constraints on the behaviour of the dependent variable in the limit that $x \to \pm\infty$.

IMPORTANT EXAMPLES WITH ANALYTIC SOLUTIONS

Diffusion Through a Membrane

Consider a membrane separating two large regions of space that contain some chemical C. The concentration on the left is c_1, and the concentration on the right is c_2 (Fig. 8.3). There is a small pore in the membrane (a one-dimensional channel of length L) through which the chemical C can freely pass. Suppose that the two regions of space are so large that their concentrations are not changing, even if chemical is flowing from one region to the other.

Let us assume that the transport of C across the membrane has been going on for some time, so that the process is at steady state, i.e., the concentration $c(x, t)$ is independent of time ($\frac{\partial c}{\partial t} = 0$). In this case, $c(x)$ must satisfy the "boundary value problem"

$$\frac{\partial^2 c}{\partial x^2} = 0, \quad c(0) = c_1, c(L) = c_2. \qquad \text{...(8.24)}$$

The solution of this problem is quite easy to find, being

$$x(0) = c_1\left(1 - \frac{x}{L}\right) + c_2\frac{x}{L}. \qquad \text{...(8.25)}$$

Consequently, the steady flux through the channel is proportional to the concentration difference across the membrane,

$$J = -Dc_x = \frac{D}{L}(c_1 - c_2). \qquad \text{...(8.26)}$$

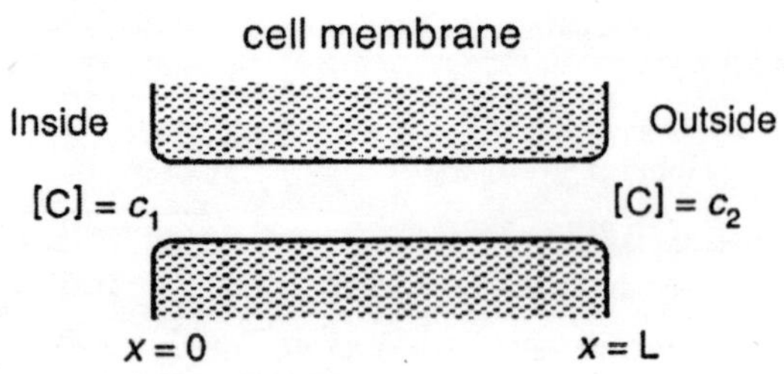

Fig. 8.3. Simple diagram of a pore through a membrane.

Ion Flux Through a Channel

Suppose that the chemical moving through the channel is an ion, and that there is a potential difference across the channel, with $\phi(0) = \phi_1$ and $\phi(L) = \phi_2$. We make the simplifying approximation that the potential gradient through the channel is constant:

$$\frac{d\phi}{dx} = \frac{\Delta\phi}{L} = \frac{V}{L}, \text{where } V = \phi_1 - \phi_2 \qquad ...(8.27)$$

If the process is in steady state so that the ion flux everywhere in the channel is the same constant, then, from (8.16),

$$J = -D\left(c_x + \alpha c\frac{V}{L}\right), \qquad ...(8.28)$$

where $\alpha = zF/RT$. We solve this differential equation for $c(x)$ subject to the boundary condition that $c(0) = c_1$ and find that

$$c(x) = c_1 e^{-\alpha V\frac{x}{L}} - \frac{JL}{D\alpha V}\left(1 - e^{-\alpha\Delta\phi x/L}\right). \qquad ...(8.29)$$

Now we can determine the flux J by requiring that $c(L) = c_2$, so that

$$J = -\frac{D}{L}\alpha V\left(\frac{c_2 - c_1 e^{-\alpha V}}{1 - e^{-\alpha V}}\right). \qquad ...(8.30)$$

This expression for flux can be converted to an ionic current I_c by multiplying by zF, in which case we obtain

$$I_c = -\frac{D}{F} zF\alpha V\left(\frac{c_2 - c_1 e^{-\alpha V}}{1 - e^{-\alpha V}}\right). \qquad ...(8.31)$$

This expression is the famous Goldman-Hodgkin-Katz current equation, and it has the important property that $I_C = 0$ when

$$\Delta V = \frac{RT}{zF}\ln\frac{c_1}{c_2}, \qquad ...(8.32)$$

which is called the reversal potential or Nernst potential for the channel. The Nernst potential is the transmembrane potential when the ion is at equilibrium across the membrane for a given transmembrane concentration ratio c_1/c_2.

Voltage Clamping

A typical experiment in electrophysiology is to hold fixed the transmembrane potential at some place on the membrane. For example, with a long axon, one might clamp the voltage at one end and determine the resulting voltage profile along the axon.

For a passive membrane (*e.g.*, a dendritic membrane), the voltage profile should satisfy (8.23) with $I_{ion} = V/R_m$, where R_m is the passive membrane resistance (ohm.cm^2). The steady-state voltage profile $V(x)$ must satisfy

$$\frac{A}{R_c}\frac{\partial^2 V}{\partial^2 x} = \frac{SV}{R_m}, \qquad ...(8.33)$$

subject to the boundary conditions $V(0) = V_{\text{fixed}}$ and $\frac{\partial V}{\partial x}(L) = 0$ if the for end is sealed. The solution of this problem is

$$V(x) = V_{\text{fixed}} \frac{e^{(L-x)/\lambda_m} - e^{(x-L)\lambda_m}}{e^{L/\lambda_m} - e^{-L/\lambda_{m,}}}, \qquad ...(8.34)$$

where $\lambda_m = \sqrt{\frac{AR_m}{SR_c}}$ is the length constant for the axon. For a long axon (L is many length constants), this solution reduces to

$$V(x) = V_{\text{fixed}} e^{-\frac{x}{\lambda m,}} \qquad ...(8.35)$$

a simple exponential decay away from the voltage-clamped end. Some examples of space constants for a variety of excitable tissues are included in Table 8.2.

Diffusion in a Long Dendrite

All of the above examples examined steady behaviour, after initial transients have decayed. However, reaction-diffusion equations also contain information about the temporal evolution of the process to steady state.

Consider calcium diffusing in a long dendrite. Suppose caged calcium is photore-leased from a small region around $x = 0$. If we denote by $c(x, t)$ the concentration of calcium along the length of the dendrite at each time t, then the model becomes

$$\frac{\partial c}{\partial t} = D\frac{\partial^2 c}{\partial x^2}, \quad -\infty < x < \infty, t > 0 \qquad ...(8.36)$$

$$c(x,0) = C_0 \delta(x), \qquad ...(8.37)$$

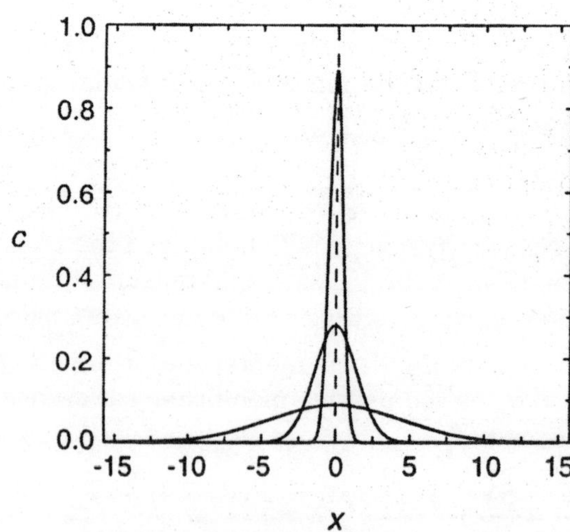

Fig. 8.4. Gaussian at 0.1, 1, and 10s

where C_0 is the total amount of released calcium, and $\delta(x)$ is the Dirac delta function. Because the dendrite is long, we view the domain as infinite. Since we do not expect the concentration of calcium to become appreciable at $x = \pm\infty$ in any finite time, we require $\lim_{x \to \pm\infty} c(x, t) = 0$.

It can be shown that the solution of this model is

$$c(x,t) = \frac{C_0}{\sqrt{4\pi Dt}} \exp\left(-\frac{x^2}{4Dt}\right), \qquad \text{...(8.38)}$$

which is illustrated in Fig. 8.4. For each fixed t, this solution is a Gaussian function, and over time, the function becomes wider and the maximal value (at $x = 0$) declines,

$$c(0,t) = \frac{C_0}{\sqrt{4\pi Dt}} \qquad \text{...(8.39)}$$

At any other point $x \neq 0$, the solution is biphasic, initially increasing to a maximum value and then decreasing back to zero. The maximum is attained when $Dt/x^2 = \frac{1}{2}$. This time behavior is illustrated in Fig. 8.5.

We can readily calculate that

$$\langle x^2 \rangle = \int_{-\infty}^{\infty} x^2 c(x,t)\, dx = 2Dt, \qquad \text{...(8.40)}$$

so that the "root mean square" (rms) distance moved in time t is

$$\sqrt{\langle x^2 \rangle} = x_{\text{rms}} = \sqrt{2Dt}. \qquad \text{...(8.41)}$$

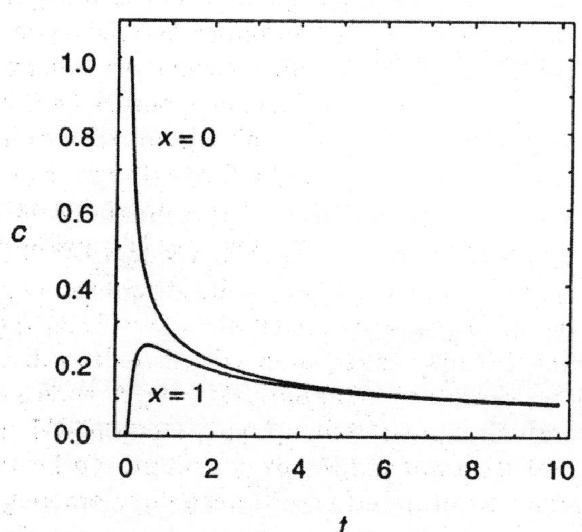

Fig. 8.5. Plot of $c(0, t)$ and $c(x \neq 0, t)$ and from (8.38).

Diffusion into a Capillary

Suppose that a long capillary (open at one end) filled with water is inserted into a solution of known chemical concentration c_0, and the chemical species diffuses into the capillary through the open end. The concentration of the chemical species should depend only on the distance down the tube and so is governed by the diffusion equation

$$\frac{\partial c}{\partial t} = D\frac{\partial^2 c}{\partial x^2}, \quad 0 < x < \infty, \ t > 0 \qquad ...(8.42)$$

where for convenience we assume that the capillary is infinitely long. Because the solute bath in which the capillary sits is large, it is reasonable to assume that the chemical concentration at the tip is fixed at $C(0, t) = C_0$, and because the tube is initially filled with pure water, $C(x, 0) = 0$.

The solution of this problem is given by

$$C(x,t) = 2C_0\left(1 - \frac{1}{\sqrt{2\pi}}\int_{-\infty}^{z} \exp\left(-\frac{s^2}{2}ds\right)\right), \ z = \frac{x}{\sqrt{2Dt}}. \qquad ...(8.43)$$

If the cross-sectional area of the capillary is A, then the total number of molecules that enter the capillary in a fixed time T is

$$N = A\int_0^{\infty} C(x,T)\,dx = 2C_0 A\sqrt{\frac{TD}{\pi}} \qquad ...(8.44)$$

From this equation it is possible to determine the diffusion coefficient by solving (8.44) for D, yielding

$$D = \frac{\pi N^2}{4C_0^2 A^2 L}. \qquad ...(8.45)$$

Segel, Chet, and Henis used this formula to estimate the diffusion coefficient for bacteria. With C_0 at 7×10^7/ml, and times T = 2, 5, 10, 12.5, 15, and 20 minutes, they counted N of 1800, 3700, 4800, 5500, 6700, and 8000 bacteria in a capillary of length 32 mm with 1 μl total capacity. In addition, with concentrations C_0 of 2.5, 4.6, 5.0, and 12.0 $\times 10^7$ bacteria per milliliter, counts of 1350, 2300, 3400, and 6200 were found at T = 10 minutes. Using (8.45) a value of D in the range of 0.1-0.3 cm^2/hour was found.

A second useful piece of information is found from (8.43) by observing that $C(x, t)/C_0$ is constant on any curve for which z is constant. Thus, the curve $t = x^2 lD$ is a level curve for the concentration, and gives a measure of how fast the substance is moving into the capillary. The time $t = x^2/D$ is called the diffusion time for the process. To give some idea of the effectiveness of diffusion in various cellular contexts, in Table 8.3 are shown typical diffusion times for a variety of cellular structures. Clearly, diffusion is quite effective when distances are short, but totally inadequate for longer distances, such as along a nerve axon. Obviously, biological systems must employ other transport mechanisms in these situations in order to survive.

Table 8.3. Estimates of Diffusion Times for Cellular Structures of Typical Dimensions, Computed from the Relation $t = x^2/D$ using $D = 10^{-5}$ cm²/s.

×	*t*	*Example*
10 nm	100 ns	thickness of cell membrane
1 μM	1 ms	size of mitochondrion
10 μM	100 ms	radius of small mammalian cell
100 μM	10s	diameter of a large muscle fiber
250 μM	60s	radius of squid giant axon
1 mm	16.7 min	half-thickness of frog sartorius muscle
2 mm	1.1 h	half-thickness of lens in the eye
5 mm	6.9 h	radius of mature ovarian follicle
2 cm	2.6 d	thickness of ventricular myocardium
1 m	31.7 yrs	length of a nerve axon

Numerical Solution of the Diffusion Equation

In order to attack more complex situations of reaction and diffusion, it is usually necessary to resort to numerical solutions of the partial differential equation. With the advent of cheap powerful computers, this approach has become increasingly useful. Here we describe the simplest numerical method to solve reaction-diffusion equations. While other more sophisticated numerical methods are available, this method is adequate for our purposes, and can be readily implemented using your favourite numerical integrator.

Consider the problem of determining calcium concentration following the photorelease of caged calcium in a sealed dendrite 40 microns long. We define the spatial variable x to extend from 0 to 40 microns, and the starting time $t = 0$ to be the time at which the caged calcium is released. The equations we wish to solve are

$$\frac{\partial c}{\partial t} = D\frac{\partial^2 c}{\partial x^2}, \qquad \text{...(8.46)}$$

where

$$c(x,0) = \begin{cases} C_0 & 20\,\mu m < x < 30\,\mu\text{m} \\ 0 & \text{elsewhere} \end{cases} \qquad \text{...(8.47)}$$

and ...(8.48)

$$c_x(0, t) = c_x(40, t) = 0, \qquad \text{...(8.49)}$$

and where C_0 is the concentration of calcium released. Because the dendrite is closed to calcium flux at its ends, no-flux boundary conditions are specified at both ends. The caged calcium is initially confined to the region between 20 and 30 microns.

To solve this problem numerically, we subdivide the spatial domain ($0 < x < 40$) into N equal intervals, with $\Delta x = 40/N$ denoting the length of each interval. If the $N+1$ endpoints of these intervals are denoted by x_n, where $n = 0,1,2,..., N$, then we define an approximation to $c(x, t)$ at these points by $c(x_n, t) = c_n(t)$.

Recall from calculus that the definition of the partial derivative of c is

$$\frac{\partial c(x,t)}{\partial x} = \lim_{\Delta x \to 0} \frac{c(x+\Delta x, t) - c(x,t)}{\Delta x} \quad \text{...(8.50)}$$

It follows that if Ax is small, but not zero, we have an approximation to the partial derivative:

$$\frac{\partial c(x,t)}{\partial x} \approx \frac{c(x+\Delta x, t) - c(x,t)}{\Delta x} \quad \text{...(8.51)}$$

In a similar way an approximation to the second partial derivative is found with Ax small, but not zero:

$$\frac{\partial^2 c(x,t)}{\partial x^2} \approx \frac{c(x+\Delta x, t) - 2c(x,t) + c(x-\Delta x, t)}{\Delta x^2}. \quad \text{...(8.52)}$$

Using this approximation at each spatial grid point $x = x_n$, we derive a set of ordinary differential equations

$$\frac{\partial c_n(t)}{\partial t} = \frac{D}{\Delta x^2}\left(c_{n+1}(t) - 2c_n(t) + c_{n-1}(t)\right). \quad \text{...(8.53)}$$

Notice that this approximation is valid only at interior grid points with $n = 1, 2, \ldots, N-1$, since for $n = 0$ or $n = N$ equation (8.53) references points c_{-1} and c_{N+1} that are outside the domain, and therefore are not known. However, if we invoke the no-flux boundary conditions, and use the approximation (8.51), we learn that

$$c_{-1}(t) = c_0(t),\ c_{N+1}(t) = c_N(t) \quad \text{...(8.54)}$$

These we apply to (8.53) for $n = 0$ and $n = N$ and obtain

$$\frac{\partial c_0(t)}{\partial t} = \frac{D}{\Delta x^2}\left(c_1(t) - c_0(t)\right), \quad \text{...(8.55)}$$

and

$$\frac{\partial c_N(t)}{\partial t} = \frac{D}{\Delta x^2}\left(c_{N-1}(t) - c_N(t)\right), \quad \text{...(8.56)}$$

The system of ordinary differential equations (8.53), (8.55), and (8.56) is a closed system of $N + 1$ equations in $N + 1$ unknowns that can be simulated with any standard differential equation solver. This conversion of a partial differential equation to a system of ordinary differential equations using difference approximations for the spatial derivatives is called the method of lines.

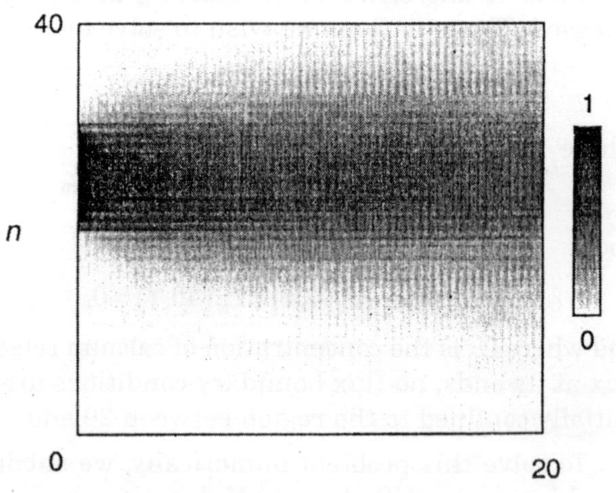

Fig. 8.6. Numerical solution of equations (8.46)-(8.49).

The initial conditions are found directly from the initial condition for the partial differential equation, with one minor adjustment. Since $c_n(t) = c(x_n, t)$, we set $c_n(0) = c(x_n, 0)$, wherever that is well-defined. However, the initial profile has a jump discontinuity at $x = 20$ and $x = 30$, so it is preferable to define the value of c at these points to be the average of the limiting values from the left and right. The solution of this problem is shown in Fig. 8.6.

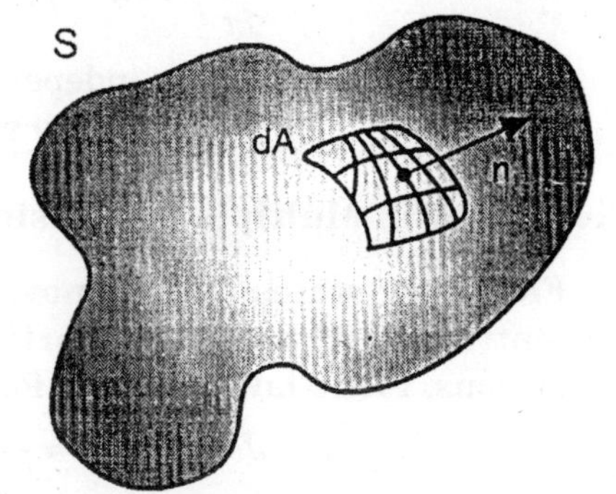

Fig. 8.7. Schematic diagram of a multi-dimensional region.

Multidimensional Problems

The multidimensional formulation of a reaction-diffusion equation is an easy generalization from one dimension. The primary difference is that in multiple dimensions, flux is a vector rather than a scalar. As a vector, flux indicates not only the rate, but also the direction, of transport, and the derivation of the conservation law is an exercise in multidimensional calculus.

Conservation Law in Multiple Dimensions

Consider a chemical species C whose concentration $c(x, y, z, t)$ varies in both time and in some three-dimensional region with volume V. The verbal expression of conservation (8.1) remains valid. At any time t, the total amount of C in the volume can be computed by integrating $c(x,y, z, t)$ over the volume:

$$\text{total amount of C} = \int_V c(x, y, z, t)\,dV. \qquad \text{...(8.57)}$$

Now suppose that C is free to move about randomly, so that C moves in and out of the volume by passing through the volume's, surface S. The flux $J(x, y, z, t)$ is a vector, since C can move in any direction. If we denote by $\boldsymbol{n}(x, y, z)$ the outward unit normal vector on S (see Fig. 8.7), then the net flux of C into the region is given by

$$\text{net rate of entry of } C = -\int_S \cdot J(x, y, z, t)\cdot \boldsymbol{n}(x, y, z)\,dA, \qquad \text{...(8.58)}$$

where dA is the surface integration element. Because **n** is the outward normal, $J.\,\boldsymbol{n}$ is positive when the motion is from inside to outside, which accounts for the negative sign in this equation. The rate of production of C in the volume can be written as $f(x, y, z, t,c)$, where as before, this rate is allowed to depend on c itself. Thus, the total rate of production of C in the region is given by

$$\text{net rate of production of } C = f(x, y, z, t, c(x, y, z, t))\,dV, \qquad \text{...(8.59)}$$

The conservation equation can now be expressed mathematically as

$$\frac{d}{dt}\int_V c\,dV = -\int_S J\cdot \boldsymbol{n}\,dA + \int_V f\,dV. \qquad \text{...(8.60)}$$

The surface integral can be replaced by a volume integral using the divergence theorem, which yields the multidimensional integral form

$$\frac{d}{dt}\int_V c\,dV = -\int_V \nabla\cdot J\,dV + \int_V f\,dV. \qquad \text{...(8.61)}$$

where ∇. is the divergence operator. As before, if the function $c(x, y, z, t)$ is smooth enough, and since the volume V is arbitrary, we can rewrite (8.8) in differential form:

$$\frac{\partial c}{\partial t} + \nabla \cdot J = f. \quad ...(8.62)$$

Note that there are four independent variables $(x, y, z,$ and $t)$ and that the equation contains partial derivatives with respect to all four variables.

Fick's Law in Multiple Dimensions

Fick's law states that C moves from regions of high concentration to regions of low concentration, at a rate proportional to the concentration gradient. Thus, in multiple dimensions, Fick's law takes the form

$$J(x, y, z, t) = -D\nabla c(x, y, z, t), \quad ...(8.63)$$

where the diffusion constant D is the proportionality constant, and the negative sign ensures that C moves down the concentration gradient. Even in multiple dimensions, the units of D are length^2/time.

Using Fick's law, (8.62) can be rewritten as a reaction-diffusion equation:

$$\frac{\partial c}{\partial t} - \nabla \cdot (D\nabla c) = f. \quad ...(8.64)$$

Advection in Multiple Dimensions

Multidimensional advective flux has the same appearance as in the one-dimensional case:

$$J(x, y, z, t) = vc(x, y, z, t). \quad ...(8.65)$$

Notice, however, that the velocity v is a vector, so the flux vector points in the direction of the velocity vector.

If the random and biased directional motions coexist, the total flux is the vector sum of the diffusive and drift fluxes:

$$J(x, y, z, t) = Vc(x, y, z, t) - D\nabla c(x, y, z, t). \quad ...(8.66)$$

Using this constitutive relation in (8.62), the multidimensional reaction-advection-diffusion equation is

$$\frac{\partial c}{\partial t} + V\nabla c - \nabla \cdot (D\nabla c) = f. \quad ...(8.67)$$

Boundary and Initial Conditions for Multiple Dimensions

As in one dimension, we must specify both initial and boundary conditions to pose the problem completely. The only difference here is that the functions involved are multidimensional, and so, when the spatial domain is complex, can be quite complicated.

The Dirichlet boundary condition in multiple dimensions specifies the values of the dependent variable c on the boundary, via $c(x, y, z, t) = f(x, y, z, t)$ with x,y, z restricted to the boundary. Similarly, the Neumann boundary condition specifies the flux of c on the boundary via $\boldsymbol{n}.\nabla c = g$. Finally, the Robin condition specifies some relationship between the flux of c and the value of c on the boundary via $-\boldsymbol{n}.\nabla Dc = h + \alpha c$.

Diffusion in Multiple Dimensions: Symmetry

If the diffusion constant D does not vary in space or time, then the diffusive term can be written

$$\nabla \cdot (D\nabla c) = D\nabla \cdot (\nabla c) = D\nabla^2 c. \qquad ...(8.68)$$

In this expression, ∇^2 is the "Laplacian operator," which in Cartesian coordinates is

$$\nabla^2 c = \frac{\partial^2 c}{\partial x^2} + \frac{\partial^2 c}{\partial y^2} + \frac{\partial^2 c}{\partial z^2}. \qquad ...(8.69)$$

If the spatial domain is more naturally described by other coordinate systems, then the representation of the Laplacian changes accordingly. For example, if the domain is a long cylindrical tube, and the concentration is not expected to be uniform in tubular cross-sections, then cylindrical coordinates (r, θ, z), where $x = r \cos \theta$, $y = r \sin \theta$, are most appropriate. In these coordinates

$$\nabla^2 c = \frac{1}{r}\frac{\partial}{\partial r}\left(r\frac{\partial c}{\partial r}\right) + \frac{1}{r^2}\frac{\partial^2 c}{\partial \theta^2} + \frac{\partial c^2}{\partial z^2}. \qquad ...(8.70)$$

If the domain is a sphere, then spherical coordinates (r, θ, ϕ), where $x = r \sin \phi \cos \theta$, $y = r \sin \phi \sin \theta$, $z = r \cos \phi$, are most appropriate, in which case the Laplacian operator is

$$\nabla^2 c = \frac{1}{r^2}\frac{\partial}{\partial r}\left(r^2\frac{\partial c}{\partial r}\right) + \frac{1}{r^2 \sin\theta}\frac{\partial}{\partial \theta}\left(\sin\theta\frac{\partial c}{\partial \theta}\right) + \frac{1}{r^2 \sin^2\theta}\frac{\partial^2 c}{\partial \phi^2}. \qquad ...(8.71)$$

An important reason for using other coordinate systems is that there may be symmetries that allow the problem to be reduced. For example, suppose that a spherical cell of radius R is suddenly immersed into a large bath containing a high concentration of glucose, and that the glucose can move across the membrane and then diffuse throughout the cell. If the concentration of glucose in the cell is initially uniform ($c = c_0$), then the solution should be independent of ϕ and θ for all time. This implies that

$$\frac{\partial c}{\partial \theta} = \frac{\partial c}{\partial \phi} = \frac{\partial^2 c}{\partial \phi^2} = 0. \qquad ...(8.72)$$

Thus, a reasonable model for this problem is

$$\frac{\partial c}{\partial t} = \frac{D}{r^2}\frac{\partial}{\partial r}\left(r^2\frac{\partial c}{\partial r}\right), \qquad ...(8.73)$$

$$c(r,0) = c_0, \qquad ...(8.74)$$

$$D\frac{dc}{dr} = j \text{ at } r = R, \qquad ...(8.75)$$

where c_0 is the initial cytosolic glucose concentration, and j is the rate of entry of glucose through the plasma membrane.

Traveling Waves in Nonlinear Reaction-Diffusion Equations

Consider a reaction-diffusion equation with a nonlinear source term:

$$\frac{\partial c}{\partial t} = D\frac{\partial^2 c}{\partial x^2} + f(c), \qquad ...(8.76)$$

where $f(c)$ is the cubic polynomial $f(c) = Ac(1 - c)(c - \alpha)$, with $0 < a < \frac{1}{2}$. While real chemical reactions are not modeled exactly by a cubic polynomial, the reaction term has features that resemble those of several more realistic reactions, and so is worthy of our attention. This equation can be (and has been) used to understand features of action potential propagation in nerve axons, calcium fertilization waves in frog eggs, and cyclic AMP waves in slime molds.

A key feature of this reaction term is that it has three zeros (0, α, and 1), two of which (0 and 1) are stable. Linear stability is determined by the sign of $f'(c)$ at the rest point, and if $f'(c_0) < 0$, the rest point c_0 is stable. In this problem, however, there is a stronger type of stability in that the solution of the ordinary differential equation $\frac{dc}{dt} = f(c)$ approaches either $c = 0$ or $c = 1$ starting from any initial position except $c = \alpha$.

The function $f(c)$ can be thought of as a switch. If c is somehow pushed slightly away from 0, it returns quickly to 0. However, if c is pushed away from 0 and exceeds α, then it goes to 1. Thus, the level a is α threshold for c. Because it has two stable rest points, equation (8.76) is often called the bistable equation.

Traveling Wave Solutions

An interesting and important problem is to determine the behavior of the bistable equation when a portion of the region is initially above the threshold α and the remainder is initially at zero. To get some idea of what to expect it is useful to perform a numerical simulation. For this numerical simulation we use the method of lines to solve the differential equations

$$\frac{dc_0}{dt} = \frac{D}{\Delta x^2}\left(c_1(t) - c_0(t)\right) + f(c_0), \qquad ...(8.77)$$

$$\frac{dc_n}{dt} = \frac{D}{\Delta x^2}\left(c_{n+1}(t) - 2c_n(t) + c_{n-1}(t)\right) + f(c_n),\ n = 1,2,\ldots,N-1, \qquad ...(8.78)$$

$$\frac{dc_N}{dt} = \frac{D}{\Delta x^2}\left(c_{N-1}(t) - c_N(t)\right) + f(c_N). \qquad ...(8.79)$$

The simulation shows that the variable c quickly changes into a profile that is a transition between $c = 0$ on the bottom and $c = 1$ on the top (Fig. 8.8). After this transitional profile is formed, it moves without change of shape from top to bottom at (what appears to be) a constant velocity.

This numerical solution suggests that we should try to find a translationally invariant solution. A translationally invariant solution is one that does not change its value along any straight line $x + st = x_0$, for an appropriately chosen value of s, the wave speed. Thus, we look for special solutions of the bistable equation of the form

$$c(x,t) = U(x + st), \qquad ...(8.80)$$

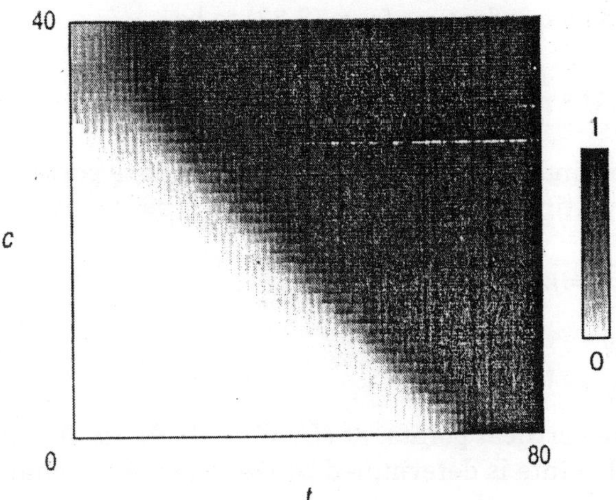

Fig. 8.8. Numerically computed solution of the bistable equation, with A = 1, α = 0.1 and *D* = 1.

with the additional property that $\lim_{\xi\to\infty} U(\xi) = 0, \lim_{\xi\to\infty} U(\xi) = 1$. Notice that since

$$\frac{\partial c(x,t)}{\partial x} = \frac{d}{d\xi}U(\xi)\frac{\partial \xi}{\partial x} = \frac{d}{d\xi}U(\xi) \text{ and } \frac{\partial c(x,t)}{\partial t} = \frac{d}{d\xi}U(\xi)\frac{\partial \xi}{\partial t} = s\frac{d}{d\xi}U(\xi), \qquad ...(8.81)$$

in this translating coordinate system, the bistable equation becomes the ordinary differential equation

$$s\frac{dU}{d\xi} = D\frac{d^2U}{d\xi^2} + f(U). \qquad ...(8.82)$$

There are two ways to try to solve (8.82). An exact solution can be found in the special case that *f* is a cubic polynomial. There are several other examples of functions *f* for which exact solutions can be found, but this method does not work in most cases. A more general method is to examine (8.82) in the phase plane, which we will do below.

The exact solution can be found for the cubic polynomial *f* as follows. Since we want we guess a relationship between $\lim_{\xi\to-\infty} U(\xi) = 0, \lim_{\xi\to\infty} U(\xi) = 1$, $dU/d\xi$ and *U* of the form

$$\frac{dU}{d\xi} = aU(1-U), \qquad ...(8.83)$$

for some positive number *a*. It follows that

$$\frac{d^2U}{d\xi^2} = a\,(1-2U)\frac{dU}{d\xi}. \qquad ...(8.84)$$

Substituting this into (8.82) and factoring out $U(1 - U)$, we find that

$$as = a^2D(1-2U) + A(U-\alpha). \qquad ...(8.85)$$

This identity holds for all *U* only if

$$a^2 = \frac{A}{2D}, s = \sqrt{AD/2}\cdot(1-2\alpha). \qquad ...(8.86)$$

The solution is found by quadrature from (8.83) to be

$$U(\xi) = \frac{1}{2} + \frac{1}{2}\tanh\left(\frac{1}{2}\sqrt{\frac{A}{2D}}\xi\right). \quad ...(8.87)$$

The analysis used for finding traveling waves using phase portraits works for any bistable function *f*. We begin by writing the traveling wave (8.82) as the two-dimensional system

$$\frac{dU}{d\xi} = W, \quad ...(8.88)$$

$$\frac{dW}{d\xi} = sW - f(U). \quad ...(8.89)$$

This system has three critical points, at $(U, W) = (0, 0), (\alpha, 0)$, and $(1, 0)$. The lin-earized stability of these critical points is determined by the roots of the characteristic equation

$$\lambda^2 - s\lambda + f'(U_0) = 0 \quad ...(8.90)$$

where U_0 is any one of the three steady rest values of U.

If $f'(U_0)$ is negative, then the critical point $(U_0, 0)$ is a saddle point. To find a traveling wave solution, we seek a trajectory that leaves the saddle point at $(U, W) = (0, 0)$ and ends up at the saddle point at $(U, W) = (1, 0)$. We can implement this (almost) numerically. If we start with an initial point close to the origin along the straight line $W = \lambda U$ in the positive quadrant, with λ the positive root of the characteristic equation $\lambda^2 - s\lambda + f'(0) = 0$, and integrate for a while, one of two things will occur. If s is relatively small, the trajectory will cross the U-axis before reaching $U = 1$, while if s is relatively large, the trajectory will increase beyond $U = 1$ and become quite large. By adjusting the parameter s one can find trajectories that barely miss hitting the point $(U, W) = (1, 0)$ by crossing the U-axis or by exceeding $U = 1$ and becoming large (Fig. 8.9). A trajectory that comes close to the saddle point at $(U, W) = (1, 0)$ is a numerical approximation to the traveling wave solution, and the value of s for which this nearly connecting trajectory is attained is a good approximation for the wave speed.

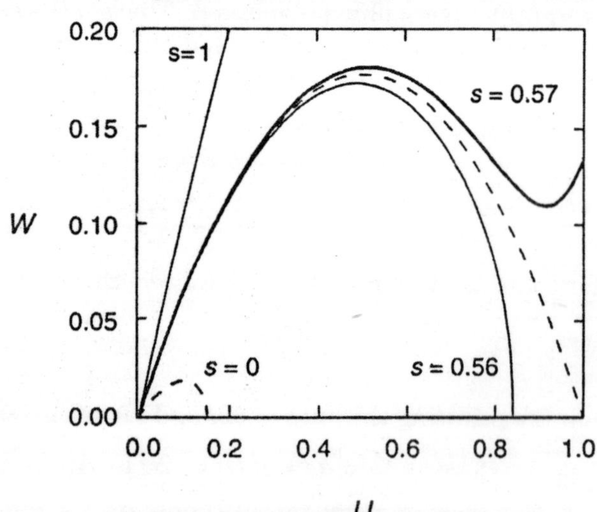

Fig. 8.9. Phase plane portrait of possible traveling wave trajectories.

Traveling Wave in the Fitzhugh-Nagumo Equations

As we have seen in earlier chapters, chemical reaction schemes in cell biology can be quite complicated, involving many species. Furthermore, some species may be free to move, while others are not. Models of nerve axons, for example, include both diffusing species (transmembrane potential) and nondiffusing variables (the ion-gating variables, because ion channels are embedded in the membrane and do not move on the millisecond time scale of an action potential). Similarly, the Ca^{2+} wave induced by fertilization of a frog egg involves both cytosolic calcium, which is a diffusing variable, and ER calcium which, (to a first approximation) is not.

Perhaps the best known example of a reaction-diffusion system is the Hodgkin-Huxley equations discussed already, which describe action potential propagation in a nerve axon. In this model there are four dependent variables: transmembrane potential V and three gating variables, m, n, and h. The equation for transmembrane potential $V(x, t)$ is the cable equation

$$C_{\mathrm{m}}\frac{\partial V}{\partial t}=\frac{R}{2\rho}\frac{\partial^2 V}{\partial x^2}+I_{\mathrm{ion}}(V, m, n, h), \qquad \text{...(8.91)}$$

where C_m is the membrane capacitance, R is the axonal radius, ρ is the axoplasmic resistivity, and I_{ion} is the current carried into the axon by ions crossing through voltage-sensitive channels, and $\frac{R}{2\rho}\frac{\partial^2 V}{\partial x^2}$ is the net current along the axon carried by ions in response to spatial gradients of intracellular potential. For the giant axon of the squid, Hodgkin and Huxley report that $R = 240\ \mu$m, $\rho = 0.35\ \Omega$.m and $C_m = 0.01$ F/m^2.

In the Hodgkin-Huxley equations given in Section 2.5, I_{ion} is a complicated function of transmembrane potential and the gating variables. To simplify the function, FitzHugh lumped the three gating variables into one (called w). The resulting equations in spatial form are

$$\frac{\partial V}{\partial t}=D\frac{\partial^2 V}{\partial x^2}+\frac{B}{V_1 V_2}V(V-V_1)(V_2-V)-C\sqrt{V_1 V_2}w, \qquad \text{...(8.92)}$$

$$\frac{\partial w}{\partial t}=\frac{\in}{\sqrt{V_1 V_2}}(V-V_3 w), \qquad \text{...(8.93)}$$

where $D = R/(2\rho C_m) \approx 0.03$ m^2/s (for squid giant axon), V_1, V_2, and V_3 are positive "voltage" constants, and B, C, and $\in$ are rate constants with units 1/s. It is also assumed that $\in \ll B, C$.

By defining $v = V/\sqrt{V_1 V_2}$, we transform (8.92) and (8.93) into

$$\frac{\partial v}{\partial t}=D\frac{\partial^2 v}{\partial x^2}+Bv(v-\beta)(\delta-v)-Cw \qquad \text{....(8.94)}$$

$$\frac{\partial w}{\partial t}=\in(v-\gamma w), \qquad \text{....(8.95)}$$

where $\beta = V_1/\sqrt{V_1 V_2}$, $\delta = V_2/\sqrt{V_1 V_2}$ and $\gamma = V_3/\sqrt{V_1 V_2}$. Because $\in \gg B, C$, we can use reduction of scale arguments (see Chapter 4) to justify the assumption that $w(x, t) = w_0 =$ constant. In this case, (8.94) and (8.95) reduce to a single reaction-diffusion equation:

$$\frac{\partial v}{\partial t}=D\frac{\partial^2 v}{\partial x^2}+Bv(v-\beta)(\delta-v)-Cw_0 \qquad \text{....(8.96)}$$

Let us assume that the "reaction" part of (8.96), $G(v) = Bv(v - \beta)(\delta - v) - Cw_0$, has three real steady states

$$G(v_i) \equiv Bv_i(v_i-\beta)(\delta-v_i)-Cw_0=0, \text{ for } i = 1, 2, 3;\ v_1 < v_2 < v_3.$$

By defining $c = (v - v_1)/(v_3 - v_1)$, (8.96) becomes identical to (8.76), with A now some nonlinear function of B, β, δ, and Cw_0. Hence, from the results leading to (8.86), we know that for appropriate choices of β, δ and Cw_0, (8.96) supports traveling wave solutions of velocity

$$\delta s = \sqrt{AD/2} \cdot (1 - 2\alpha).$$

To estimate the velocity of propagation of an action potential wave front, we must have, in addition to $D \approx 0.03$ m²/s, estimates of the rate constant A and the threshold α in the $f(c)$ term of (8.76). Given that the amplitude of an action potential is $\approx$ 100 mV, and that the threshold for initiation is $\approx$ 20 mV (from rest), we set $\alpha \approx 0.2$. During the rise of an action potential, V increases with a doubling time of a fraction of a millisecond (say, 0.2 ms). Trajectories of the reaction equation $dc/dt = f(c)$ depart from the unstable steady state according to

$$c(t) - \alpha = (c_0 - \alpha)\exp\big(A\alpha(1-\alpha)t\big). \qquad \text{...(8.97)}$$

To verify this, set $c = y + \alpha$ and linearize $dc/dt = f(\alpha)$ to get $dy/dt - f'(\alpha)y$ and then solve to obtain (8.97). The doubling time for departure from the unstable steady state is

$$\frac{l_n 2}{A\alpha(1-\alpha)} \approx 0.2 \text{ ms.}$$

or $A \approx 2.\ 10^4$/s. Hence, if the upstroke of the action potential can be approximated by the FitzHugh-Nagumo equations, it should propagate at velocity

$$s \approx \sqrt{\frac{(2 \cdot 10^4/\text{s})(3 \cdot 10^2 \text{m}^2/\text{s})}{2}} .(1 - 0.4),$$

or approximately 10 m/s. This compares favourably with the observed velocity of 20 m/s.

MOLECULAR DYNAMICS

Molecular dynamics simulation, which provides the methodology for detailed microscopical modeling on the atomic scale, is a powerful and widely used tool in chemistry, physics, and materials science. This technique is a scheme for the study of the natural time evolution of the system that allows prediction of the static and dynamic properties of substances directly from the underlying interactions between the molecules. Dynamical simulations monitor time-dependent processes in molecular systems in order to study their structural, dynamic, and thermodynamic properties by numerically solving an equation of motion, which is the formulation of the rules that govern the motion executed by the molecule.

Molecular dynamics (MD) provides information about the time dependence and magnitude of fluctuations in both positions and velocities, whereas the Monte Carlo approach provides mainly positional information and gives only little information on time dependence. Depending on the desired level of accuracy, the equation of motion to be numerically solved may be the classical equation of motion (Newton's), a stochastic equation of motion (Langevin's), a Brownian equation of motion, or even a combination of quantum and classical mechanics (QM/MM).

TYPES OF MOTIONS

Macromolecules in general, and proteins in particular, display a broad range of characteris-tic motions. These range from the fast and localized motions characteristic of atomic fluctuations to the slow large-scale motions involved in the folding transition. Many of these motions, on each and every time scale, have an important role in the biological function of proteins. For example, localized side-chain motion controls the diffusion of oxygen into and out of myoglobin and hemoglobin. A more extensive "medium-scale" structural transition is involved, for example, in the hemoglobin R to T allosteric transition, which makes it such an efficient transport agent. Finally, prion proteins exhibit a global structural transition of biological importance.

These proteins undergo a global transition from an α-helical structure to a predominantly (β-sheet structure, which is implicated in the onset of Creutzfeldt – Jacob disease (CJD) in humans and the "mad cow" disease in cattle (bovine spongiform encephalopathy; BSE). Table 9.1 gives a crude overview of the different types of motion executed by a protein and their characteristic time scales and amplitudes. These should be regarded as rough guidelines, because individual motions in specific systems may differ significantly from these estimates. Note that the motions executed by a protein span almost 20 orders of magnitude in characteristic time scales, from femtoseconds (10^{-15} s) to hours (10^4-10^5 s). They also cover a wide range of amplitudes (0.01-100 Å) and energies (0.1-100 kcal/ mol). An important characteristic of biomolecular motion is that the different types of motion are interdependent and coupled to one another. For example, a large-scale dynamic transition cannot occur without involving several medium-scale motions, such as helix rearrangements.

Medium-scale motions cannot occur without involving small-scale motions, such as side-chain movement. Finally, even side-chain motions cannot occur without the presence of the very fast atomic fluctuations, which can be viewed as the "lubricant" that enables the whole molecular construction to move. From the point of view of dynamic simulations, this has serious implications. It indicates that even in the study of slow large-scale motions (of biological importance) it is not possible to ignore the fast small-scale motions, which eventually are the ones that impose limitations on the simulation time step and length.

Table 9.1. An Overview of Characteristic Motions in Proteins

Type of motion	*Functionality, examples*	*Time and amplitude scales*
Local motions		
Atomic fluctuation Side chain motion	Ligand docking flexibility Temporal diffusion pathways	Femtoseconds (fs) to picosesonds (ps) (10^{-15}–10^{-12} s); less than 1 Å
Medium-scale motions		
Loop motion Terminal-arm motion Rigid-body motion	Active site conformation adaptation Binding specificity	Nanoseconds (ns) to micro-seconds (μs) (10^{-9}–10^{-6}s); 1 – 5 Å
Large-scale motions		
Domain motion Subunit motion	Hinge-bending motion Allosteric transitions	Microseconds (μs) to milli-seconds (ms) (10^{-6}-10^{-3} s); 5-10 Å
Global motion		
Helix-coil transition Folding/unfolding Subunit association	Hormone activation Protein functionality	Milleseconds (ms) to hours (10^{-3}-10^4 s); more than 10 Å

THE STATISTICAL MECHANICS BASIS OF MOLECULAR DYNAMICS

A classical system is described by a classical Hamiltonian, H, which is a function of both coordinates $\mathbf{r}$ and momenta $\mathbf{p}$. For regular molecular systems, where the potential energy function is independent of time and velocity, the Hamiltonian is equal to the total energy,

$$H = H(\mathbf{r}, \mathbf{p}) \cdot K(\mathbf{p}) + U(\mathbf{r}) = \sum_i \frac{p_i}{2m_i} + U(\mathbf{r}) \qquad \text{...(9.1)}$$

where $K(\mathbf{p})$ is the kinetic energy, $U(\mathbf{r})$ is the potential energy, p_i is the momentum of particle i, and m_i the mass of particle i. A microscopic state of the system is therefore characterized by the set of values $\{\mathbf{r}, \mathbf{p}\}$, which corresponds to a point in the space defined by both coordinates $\mathbf{r}$ and momenta $\mathbf{p}$ (known as "phase space").

To obtain thermodynamic averages over a "canonical" ensemble, which is characterized by the macroscopic variables (N, V, T), it is necessary to know the probability of finding the system at each and every point (= state) in phase space. This probability distribution, $\rho(\mathbf{r}, \mathbf{p})$, is given by the Boltzmann distribution function,

$$\rho(\mathbf{r}, \mathbf{p}) = \frac{\exp\left[-H(\mathbf{r}, \mathbf{p})/k_B T\right]}{Z} \quad ...(9.2)$$

where the canonical partition function, Z, is an integral over all phase space of the Boltz-mann factors exp $[-H(\mathbf{r}, \mathbf{p})/k_B T]$, and k_B is the Boltzmann factor. Once this distribution function is known it can be used to calculate phase space averages of any dynamic variable $A(\mathbf{r}, \mathbf{p})$ of interest. Examples for dynamic variables are the position, the total energy, the kinetic energy, fluctuations, and any other function of **r** and/or **p**. These averages,

$$\langle A(\mathbf{r}, \mathbf{p})\rangle_Z = \int_V d\mathbf{r} \int_{-\infty}^{\infty} d\mathbf{p}\, \rho(\mathbf{r}, \mathbf{p})\, A(\mathbf{r}, \mathbf{p}) \quad ...(9.3)$$

are called "thermodynamic averages" or "ensemble averages" because they take into account every possible state of the system. However, in order to calculate these thermody-namic averages, it is necessary to simultaneously know the Boltzmann probability [Eq. (2)] for each and every state {**r**, **p**}, which is an extremely difficult computational task. An alternative strategy for calculating systemwide averages is to follow the motion of a single point through phase space instead of averaging over the whole phase space all at once. That is, in this approach the motion of a single point (a single molecular state) through phase space is followed as a function of time, and the averages are calculated only over those points that were visited during the excursion.

Averages calculated in this way are called "dynamic averages." The motion of a single point through phase space is obtained by integrating the system's equation of motion. Starting from a point {**r**(0), **p**(0)}, the integration procedure yields a trajectory that is the set of points {**r**(t), **p**(t)} describing the state of the system at any successive time t. Dynamic averages of any dynamical variable $A(\mathbf{r}, \mathbf{p})$ can now be calculated along this trajectory as follows:

$$\langle A(\mathbf{r}, \mathbf{p})\rangle_\tau = \frac{1}{\tau}\int_0^\tau A(\mathbf{r}(t), \mathbf{p}(t))\,dt \quad ...(9.4)$$

where τ is the duration of the simulation. Compared to the previous approach, dynamic averaging is easier to perform. The two averaging strategies can be summarized as follows:

Thermodynamic average. An average over *all* points in phase space at a *single* time.

Dynamic average. An average over a *single* point in phase space at *all* times.

It is hoped that the point that is being dynamically followed will eventually cover all of phase space and that the dynamic average will converge to the desired thermodynamic average. A key concept that ties the two averaging strategies together is the *ergodic hypothesis.* This hypothesis states that for an infinitely long trajectory the thermodynamic ensemble average and the dynamic average become equivalent to each other,

$$\lim_{\tau\to\infty}\langle A(\mathbf{r}, \mathbf{p})\rangle\tau = \langle A(\mathbf{r}, \mathbf{p})\rangle_Z \quad ...(9.5)$$

In other words, the ergodic hypothesis claims that when the trajectory becomes long enough, the point that generates it will eventually cover all of phase space, so the two averages become identical. For this hypothesis to hold, the system has to be at equilibrium (technically, at a stationary state). Also, there must not be any obstacle, such as a frag-mented topology, that will prevent an infinitely long trajectory from covering all of phase space. A system that obeys these two conditions is said to be *ergodic,* and its hypothesis is the theoretical justification for using molecular dynamic simulations as a means for calculating thermodynamic averages of molecular systems. It is tacitly assumed that finite molecular dynamics trajectories are "long enough" in the ergodic sense.

NEWTONIAN MOLECULAR DYNAMICS

A. Newton's Equation of Motion

The temporal behavior of molecules, which are quantum mechanical entities, is best described by the quantum mechanical equation of motion, *i.e.,* the time-dependent Schrödinger equation. However, because this equation is extremely difficult to solve for large systems, a simpler classical mechanical description is often used to approximate the motion executed by the molecule's heavy atoms. Thus, in most computational studies of biomolecules, it is the classical mechanics Newtonian equation of motion that is being solved rather than the quantum mechanical equation.

In its most simplistic form, Newton's equation of motion (also known as Newton's second law of motion) states that

$$F_i = m_i a_i = m_i \ddot{r}_i \qquad ...(9.6)$$

where F_i is the force acting on particle *i,* m_i is the mass of particle *i,* a_i is its acceleration, and $\ddot{r}_i$ is the second derivative of the particle position **r** with respect to time. The force F_i is determined by the gradient of the potential energy function, $U(\mathbf{r})$, which is a function of all the atomic coordinates **r**,

$$F_i = -\nabla_i U(\mathbf{r}) \qquad ...(9.7)$$

Equation (9.7) is a second-order differential equation. A more general formulation of Newton's equation of motion is given in terms of the system's Hamiltonian, H [Eq. (9.1)]. Put in these terms, the classical equation of motion is written as a pair of coupled first-order differential equations:

$$\ddot{r}_k = \frac{\partial H(\mathbf{r}, \mathbf{p})}{\partial p_k};\ \dot{p}_k = \frac{\partial H(\mathbf{r}, \mathbf{p})}{\partial r_k}; \qquad ...(9.8)$$

By substituting the definition of H [Eq. (9.1)] into Eq. (8), we regain Eq. (9.6). The first first-order differential equation in Eq. (9.8) becomes the standard definition of momentum, *i.e.,* $p_i = w_i r_i = m_i v_i$, while the second turns into Eq. (9.6). A set of two first-order differential equations is often easier to solve than a single second-order differential equation.

B. Properties of Newton's Equation of Motion

Newton's equation of motion has several characteristic properties, which will later serve as "handles" to ensure that the numerical solution is correct (Section V.C). These properties are

Conservation of energy. Assuming that U and H do not depend explicitly on time or velocity (so that $\partial H/\partial t = 0$), it is easy to show from Eq. (9.8) that the total derivative dH/dt is zero;

i.e., the Hamiltonian is a constant of motion for Newton's equation. In other words, there is conservation of total energy under Newton's equation of motion.

Conservation of linear and angular momentum. If the potential function U depends only on particle separation (as is usual) and there is no external field applied, then Newton's equation of motion conserves the total linear momentum of the system, ***P***,

$$\boldsymbol{P} = \sum_i p_i \qquad ...(9.9)$$

and the total angular momentum, ***L***,

$$\boldsymbol{L} = \sum_i r_i \times p_i = \sum_i m_i r_i \times \dot{r}_i \qquad ...(9.10)$$

Time reversibility. The third property of Newton's equation of motion is that it is reversible in time. Changing the signs of all velocities (or momenta) will cause the molecule to retrace its trajectory. If the equations of motion are solved correctly, then the numerical trajectory should also have this property. Note, however, that in practice this time reversibility can be reproduced by numerical trajectories only over very short periods of time because of the chaotic nature of large molecular systems.

C. Molecular Dynamics: Computational Algorithms

Solving Newton's equation of motion requires a numerical procedure for integrating the differential equation. A standard method for solving ordinary differential equations, such as Newton's equation of motion, is the finite-difference approach. In this approach, the molecular coordinates and velocities at a time $t + \Delta t$ are obtained (to a sufficient degree of accuracy) from the molecular coordinates and velocities at an earlier time t. The equations are solved on a step-by-step basis. The choice of time interval Δt depends on the properties of the molecular system simulated, and Δt must be significantly smaller than the characteristic time of the motion studied.

A good starting point for understanding finite-difference methods is the Taylor expansion about time t of the position at time $t + \Delta t$,

$$\boldsymbol{r}(t+\Delta t) = \mathbf{r}(t) + \dot{\mathbf{r}}(t)\Delta t + \frac{1}{2}\ddot{\mathbf{r}}(t)\Delta t^2 + ... \qquad ...(9.11)$$

Alternatively, this can be written as

$$\mathbf{r}(t+\Delta t) = \mathbf{r}(t) + \mathbf{v}(t)\Delta t + \frac{1}{2}\mathbf{a}(t)\Delta t^2 + ... \qquad ...(9.12)$$

where $\mathbf{v}(t)$ is the velocity vector and $\mathbf{a}(t)$ is the acceleration. Because the integration proceeds in a stepwise fashion, and recalling Eq. (9.6), it is convenient to rewrite the above expansion in a discrete form. Using $\mathbf{r}_n$ to indicate the position at step n (at time t) and $\mathbf{r}_{n+1}$ to indicate the position at the next step, $n + 1$ (at time $t + \Delta t$), Eq. (9.12) can be written as

$$\mathrm{r}_{n+1} = \mathbf{r}_n + \mathbf{v}_n\Delta t + \frac{1}{2}\left(\frac{\mathbf{F}_n}{\mathbf{m}}\right)\Delta t^2 + O\left(\Delta t^3\right) \qquad ...(9.13)$$

where $O(\Delta t^n)$ is the terms of order Δt^n or smaller. With this information the velocity $\mathbf{v}_n + 1$ at time $n + 1$ can be crudely estimated, for example, as

$$\mathbf{v}_{n+1} = (\mathbf{r}_{n+1} - \mathbf{r}_n)/2$$

Together, Eqs. (9.13) and (9.14) form an integration algorithm. Given the position $\mathbf{r}_n$, the velocity $\mathbf{v}_n$, and the force $\mathbf{F}_n$ at step n, these equations allow one to calculate (actually, estimate) the position $\mathbf{r}_{n+1}$ and velocity $\mathbf{v}_{n+1}$ at step $n + 1$. The formulation is highly trivial and results in a low quality integration algorithm (large errors). Other, more accurate, algorithms have been developed using the same kind of reasoning. In the following sub-sections we survey some of the more commonly used finite-difference integration algorithms, highlighting their advantages and disadvantages.

1. Verlet Integrator

The most common integration algorithm used in the study of biomolecules is due to Verlet. The Verlet integrator is based on two Taylor expansions, a forward expansion $(t + \Delta t)$ and a backward expansion $(t - \Delta t)$,

$$\mathbf{r}_{n+1} = \boldsymbol{r}_n + \boldsymbol{v}_n\Delta t + \frac{1}{2}\left(\frac{\boldsymbol{Fn}}{\boldsymbol{m}}\right)\Delta t^2 + O\left(\Delta t^3\right) \qquad \text{...(9.15a)}$$

$$r_{n+1} = \boldsymbol{r}_n + \boldsymbol{r}_n\Delta t + \frac{1}{2}\left(\frac{\boldsymbol{Fn}}{\boldsymbol{m}}\right)\Delta t^2 - O\left(\Delta t^3\right) \qquad \text{...(9.15b)}$$

The sum of the two expansions yields an algorithm for propagating the position,

$$\boldsymbol{r}_{n+1} = 2\boldsymbol{r}_n - \boldsymbol{r}_{n-1} + \frac{\boldsymbol{Fn}}{\boldsymbol{m}}\Delta t^2 - O\left(\Delta t^4\right) \qquad \text{...(9.16)}$$

Translated into a stream of commands, this algorithm is executed in two steps:

1. Use the current position $\boldsymbol{r}_n$ to calculate the current force $\boldsymbol{F}_n$.
2. Use the current and previous positions $\boldsymbol{r}_n$ and $\boldsymbol{r}_{n-1}$, together with the current force $\boldsymbol{F}_n$ (calculated in step 1) to calculate the position in the next step, $\boldsymbol{r}_{n+1}$, according to Eq. (9.16).

These two steps are repeated for every time step for each atom in the molecule. Sub-tracting Eq. (9.15b) from Eq (9.15a) yields a complementary algorithm for propagating the velocities,

$$\boldsymbol{v}_n = \frac{\mathbf{r}_{n+1} - \mathbf{r}_{n+1}}{2\Delta t} + O\left(\Delta t^2\right) \qquad \text{...(9.17)}$$

Fig. 9.1a gives a graphical representation of the steps involved in a Verlet propagation. The algorithm embodied in Eqs. (9.16) and (9.17) provides a stable numerical method for solving Newton's equation of motion for systems ranging from simple fluids to biopolymers. Like any algorithm, the Verlet algorithm has advantages as well as disadvantages.

Advantages of the Verlet Algorithm are

1. The position integration is quite accurate [errors on the order of $O(\Delta t^4)$] and is independent of the velocity propagation, a fact that simplifies the position integration and reduces memory requirements.
2. The algorithm requires only a single force evaluation per integration cycle (computationally, force evaluations are the most "expensive" part of the simulation).

3. This formulation, which is based on forward and backward expansions, guarantees time reversibility (a property of the equation of motion).

Disadvantages of the Verlet Algorithm are

1. The velocity propagation is subject to relatively large errors, on the order $O(\Delta t^2)$. Recall that an accurate estimate of the velocity is required for the kinetic energy evaluations. An added inconvenience is that $\boldsymbol{v}_n$ can be computed only if $\boldsymbol{r}_{n+1}$ is already known.

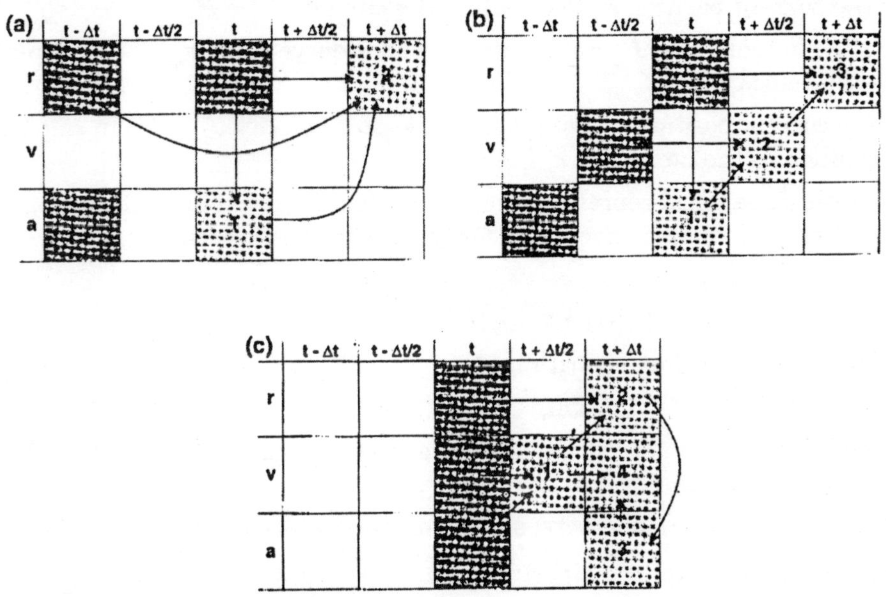

Fig. 9.1. A stepwise view of the Verlet integration algorithm and its variants, (*a*). The basic Verlet method, (*b*) Leap-frog integration, (*c*) Velocity Verlet integration. At each algorithm dark and light gray cells indicate the initial and calculating variables, respectively. The numbers in the cells represent the orders in the calculation procedures. The arrows point from the data that are used in the calculation of the variable that is being calculated at each step.

2. Further numerical errors are introduced when an $O(\Delta t^2)$ term is added to an $O(\Delta t^0)$ term.
3. The Verlet algorithm is not "self-starting." A lower order Taylor expansion [e.g., Eq. (9.13)] is often used to initiate the propagation.
4. It must be modified to incorporate velocity-dependent forces or temperature scaling.

2. Leap-Frog Integrator

Modifications to the basic Verlet scheme have been proposed to tackle the above deficiencies, particularly to improve the velocity evaluation. One of these modifications is the leap-frog algorithm, so called for its half-step scheme: Velocities are evaluated at the mid-point of the position evaluation and vice versa. The algorithm can be written as

$$\boldsymbol{r}_{n+1} = \boldsymbol{r}_n + \boldsymbol{v}_{n+1/2}\Delta t \quad \text{...(9.18a)}$$

$$\boldsymbol{v}_{n+1/2} = \boldsymbol{v}_{n-1/2} + \frac{\boldsymbol{F}_n}{\boldsymbol{m}}\Delta t \quad \text{...(9.18b)}$$

where $\boldsymbol{v}_{n \pm 1/2}$ stands for the velocity at the mid-step time $[t \pm (1/2)\Delta t]$. Elimination of the velocities from these equations shows that the method is algebraically equivalent to the Verlet algorithm. Cast in the form of execution instructions, the leap-frog algorithm involves three steps:

1. Use the current position $\boldsymbol{r}_n$ to calculate the current force $\boldsymbol{F}_n$.
2. Use the current force $\boldsymbol{F}_n$ and previous mid-step velocity $\boldsymbol{v}_{n-1/2}$ to calculate the next mid-step velocity $\boldsymbol{v}_{n+1/2}$.
3. Use the current position $\boldsymbol{r}_n$ and the next mid-step velocity $\boldsymbol{v}_{n+1/2}$ (from step 2) to calculate the position in the next step, $\boldsymbol{r}_{n+1}$.

Fig. 9.1*b* gives a graphical representation of the steps involved in the leap-frog propagation. The current velocity $\boldsymbol{v}_n$, which is necessary for calculating the kinetic energy, can be calculated as

$$\boldsymbol{v}_n = (\boldsymbol{v}_{n+1/2} + \boldsymbol{v}_{n-1/2})/2 \quad \text{...(9.19)}$$

Advantages of the leap-frog algorithm are

1. It improves velocity evaluation.
2. The direct evaluation of velocities gives a useful handle for controlling the simulation temperature (via velocity scaling).
3. It reduces the numerical errors, since here $O(\Delta t^1)$ terms are added to $O(\Delta t^0)$ terms.

Disadvantages of the leap-frog algorithm are

1. It still does not handle the velocities in a completely satisfactory manner, because the velocities at time t are only approximated by Eq. (9.19).
2. This algorithm is computationally a little more expensive than the Verlet algorithm.

3. Velocity Verlet Integrator

An even better handling of the velocities is obtained by another variant of the basic Verlet integrator, known as the "velocity Verlet" algorithm. This is a Verlet-type algorithm that stores positions, velocities, and accelerations all at the same time t and minimizes round-off errors. The velocity Verlet algorithm is written

$$\boldsymbol{r}_{n+1} = \boldsymbol{r}_n + \boldsymbol{v}_n\Delta t + \frac{1}{2}\left(\frac{\boldsymbol{F}_n}{\boldsymbol{m}}\right)\Delta t^2 \quad \text{...(9.20a)}$$

$$\boldsymbol{v}_{n+1} = \boldsymbol{v}_n + \frac{1}{2}\left[\frac{\boldsymbol{F}_n}{\boldsymbol{m}} + \frac{\boldsymbol{F}_{n+1}}{\boldsymbol{m}}\right]\Delta t \quad \text{...(9.20b)}$$

Again, elimination of the velocities from these equations recovers the Verlet algo-rithm. In practice, the velocity Verlet algorithm consists of the following steps:

1. Calculate the position $\boldsymbol{r}_{n+1}$ at time $t + \Delta t$ from Eq. (9*a*).
2. Calculate the velocity at mid-step $\boldsymbol{v}_{n+1/2}$ using the equation

$$\boldsymbol{v}_{n+1/2} = \boldsymbol{v}_n + \frac{1}{2}\left(\frac{\boldsymbol{F}_n}{\boldsymbol{m}}\right)\Delta t \qquad ...(9.21)$$

3. Calculate the force $\boldsymbol{F}_{n+1}$ at time $t + \Delta t$.
4. Finally, complete the velocity move to $\boldsymbol{v}_n$ by using

$$\boldsymbol{v}_{n+1} = \boldsymbol{v}_{n+1/2} + \frac{1}{2}\left(\frac{\boldsymbol{F}_{n+1}}{\boldsymbol{m}}\right)\Delta t \qquad ...(9.22)$$

At this point, the kinetic energy at time $t + \Delta t$ is available. Fig. 9.1*c* gives a graphical representation of the steps involved in the velocity Verlet propagation. Advantages of the velocity Verlet algorithm are

1. It is numerically very stable.
2. It is convenient and simple [the code of this method is a straightforward transcription of Eqs. (9.20-(9.22)].
3. It provides an accurate evaluation of velocities and hence of the kinetic energy.

The main disadvantage of this algorithm is that it is computationally a little more expensive than the simpler Verlet or leap-frog algorithms (though the added accuracy often outweighs this slight overhead).

MOLECULAR DYNAMICS: SIMULATION PRACTICE

A. Assigning Initial Values

Newton's equation of motion is a second-order differential equation that requires two initial values for each degree of freedom in order to initiate the integration. These two initial values are typically a set of initial coordinates $\{\mathbf{r}(0)\}$ and a set of initial velocities $\{\boldsymbol{v}(0)\}$.

1. Initial Coordinates

The initial coordinates $(\boldsymbol{r}(0)\}$ are usually obtained from experimentally determined molecular structures, mainly from X-ray crystallography and NMR experiments. Alternatively, the initial coordinates can be based on computer models generated by a variety of modeling techniques. Note, however, that even the experimentally determined structures must often undergo some preparation steps before they can be used as initial structures in a dynamic simulation.

First, it is not possible to determine hydrogen atom positions by X-ray crystallography. Thus the coordinates for the many hydrogen atoms in the molecule are missing from X-ray coordinate files. These coordinates must be added to the initial structure before the simulation is started. Several algorithms are available for ensuring reasonable placement of hydrogens.

In some cases, whole parts of the protein are missing from the experimentally determined structure. At times, these omissions reflect flexible parts of the molecule that do not have a well-defined structure (such as loops). At other times, they reflect parts of the molecule (*e.g.*, terminal sequences) that were intentionally removed to facilitate the crystallization process. In both cases, structural models may be used to fill in the gaps.

After all the coordinates are accounted for, it is good practice to refine the initial structure by submitting it to energy minimization. The role of this minimization is to relieve local stresses due to nonbonded overlaps, as well as to relax bond length and bond angle distortions in the experimental structure. The origin of these stresses is due both to the empirical nature of the energy function and to the average nature of the experimentally determined structures.

2. Initial Velocities

Unlike the initial coordinates, which can be obtained experimentally, the only relevant information available about atomic velocities is the system's temperature *T*, which determines the velocity distribution. In the absence of a better guideline, initial velocities (v_x, v_y, v_z) are usually randomly assigned from the standard Maxwellian velocity distribution at a temperature *T*,

$$P(v)dv = \left(\frac{m}{2\pi k_B T}\right)^{1/2} \exp\left[\frac{-mv^2}{2k_B T}\right]dv \qquad ...(9.23)$$

This initial assignment is, of course, not at equilibrium. In particular, the expected velocity correlation between neighbouring atoms is not guaranteed, and most likely it is nonexistent (*i.e.,* in general, neighboring atoms, such as bonded pairs, are expected to move at similar velocities). Furthermore, the random assignment process may accidentally assign high velocities to a localized cluster of atoms, creating a "hot spot" that makes the simulation unstable. To overcome this problem, it is common practice to start a simulation with a "heat-up" phase.

Velocities are initially assigned at a low temperature, which is then increased gradually allowing for dynamic relaxation. This slow heating continues until the simulation reaches the desired temperature. In practice, heating is performed by increasing atomic velocities, either by reas-signing new velocities from a Maxwellian distribution [Eq. (9.23)] at an elevated temperature or by scaling the velocities by a uniform factor. This heating process, as well as the dynamic simulation that follows, requires a measurable definition of the system's temperature *T* at time *t*. According to the equipartition theorem, the temperature, *T(t),* at any given time *t* is defined in terms of the mean kinetic energy by

$$T(t) = \frac{1}{k_B N_{dof}} \sum_{i=1}^{N_{dof}} m_i \mid v_i \mid^2 \qquad ...(9.24)$$

where N_{dof} is the number of unconstrained degrees of freedom in the system ($N_{dof} = 3N - n$, where *N* is the number of atoms and *n* is the number of constraints). It is clear from this expression that scaling the velocities by a factor of $[T_0/T(t)]^{1/2}$ will result in a mean kinetic energy corresponding to a desired temperature T_0. Another problem related to the initial velocity assignment is the large total linear momentum ***P*** and total angular momentum ***L*** formed [Eqs. (9.9) and (9.10)]. These momenta cause a drift of the molecule's center of mass relative to the reference frame.

Numerically, this drift hampers the computational efficiency of the simulation, overshadowing the smaller internal motions by the physically irrelevant translational drift and global rotation. Because Newton's equation of motion conserves linear and angular momenta, these momenta will not go away unless they are actively taken out. The ideal zero-drift situation is reached by periodically zeroing these momenta during the equilibration phase of the simulation.

B. Selecting the Integration Time Step

The size of the time step Δt is an important parameter that determines the magnitude of the error associated with each of the foregoing integration algorithms. On the one hand, a small time step means better integration quality. But on the other hand it also means that more integration steps are required for the same length of simulation. Thus, every simulation involves a trade-off between economy and accuracy. The time step in molecular dynamics simulations is one of the most important factors that balance this trade-off. In general, one would like to choose the largest possible time step that will still ensure an accurate simulation.

An appropriate time step should be small by comparison to the period of the fastest motion (highest frequency motion) in the system being simulated. If i is the period of the fastest motion, a good rule of thumb for selecting Δt is

$$\tau / \Delta t \approx \cdot 20 \qquad \text{...(9.25)}$$

For biomolecules, such as proteins, the fastest motions are the stretching vibrations of the bonds connecting hydrogen atoms to heavy atoms (X — H stretching). The frequency of these motions is in the vicinity of 3000 cm^{-1}, which means periods of about 10 fs (1×10^{-14} s). Thus, an appropriate time step for simulating biomolecules would be $\Delta t \approx 0.5$ fs. This extremely small time step is the main reason that, with only a few exceptions, molecular dynamics simulations are limited today to the nanosecond (10^{-9} s) time scale. Naturally, much effort is invested in developing advanced algorithms that allow for larger time steps that enable longer simulations.

A basic rationale, common to many such approaches, is to remove the fast (high frequency) motions from the numerical integration and account for them in some other way. Because of the characteristic coupling between types of motion this task is far from simple. A first step in this direction is to take the $X - H$ stretching motions out of the numerical integration. If these stretching motions were accounted for in some other way, then the time step would be determined by the next fastest molecular motion, *i.e.,* the $X - X$ stretching modes with frequencies around 1500 cm^{-1}. According to relation (9.25), this elimination will increase the time step by a factor of 2 (to $\Delta t \approx 1.0$ fs), extending the length of the simulation by a similar factor at only a slight additional computational cost. The algorithm that is usually employed to account for the hydrogen positions is SHAKE (and its variant RATTLE [17]).

Stated in a simplistic way, the SHAKE algorithm assumes that the length of the $X - H$ bond can be considered constant. Because in a numerical simulation there are always fluctuations, this means that the deviation of the current length $d_k(t)$ of the kth bond from its ideal (constant) bond length d_k^0 must be smaller than some tolerance value ε,

$$s_k = \left[d_k(t)^2 - d_k^{o2}\right] / d_k^{o2} < \varepsilon \qquad ...(9.26)$$

SHAKE is an iterative procedure that adjusts the atomic positions (of the hydrogen atoms in this case) after each integration step (of the heavy atoms) in order to simultaneously satisfy all the constraints. It iterates until s_k is smaller than $\in$ for all values of k, SHAKE can be applied not only to $X - H$ type bonds but also to all bond-stretching motions in the system, allowing for time steps as large as 2 or 3 fs (depending on the details of the system). Although, SHAKE can in principle be applied to bending motions too, it was found that such constraints result in low quality simulations. This is due to the fact that the important coupling between bond angle motion and torsional motion is neglected [18].

C. Stability of Integration

An important issue in any numerical study is that of the accuracy and stability of the simulation. A simulation become unstable when it has picked up errors along the way and has become essentially meaningless. In general, it is unrealistic to expect that any approximate method of solution will follow the exact classical trajectory indefinitely. It is our goal, however, to maintain a stable simulation for at least the duration of interest for the specific study. Thus, the stability of the simulation must be gauged at all times. If one is lucky, an unstable simulation will also crash and not reach its designated termination.

It may happen, however, that even though the unstable simulation reaches its designated termination, its content carries little merit. The best gauges for the stability of any simulation are the constants of motion of the physical equation that is numerically solved, *i.e.*, quantities that are expected to be conserved during the simulation. Since numerical fluctuations cannot be avoided, a namic variable $A(\boldsymbol{r}, \boldsymbol{p})$ is considered numerically "conserved" if the ratio of its fluctuations to its value is below an acceptable tolerance ε, $\Delta A/A < \varepsilon$.

Conservation of Energy

Newton's equation of motion conserves the total energy of the system, E (the Hamiltonian), which is the sum of potential and kinetic energies [Eq. (9.1)]. A fluctuation ratio that is considered adequate for a numerical solution of Newton's equation of motion is

$$\frac{\Delta E}{E} < 10^{-4} \text{ or } \log_{10}\left(\frac{\Delta E}{E}\right) < -4 \qquad ...(9.29)$$

Conservation of Linear and Angular Momenta

After equilibrium is reached, the total linear momentum P [Eq. (9.9)] and total angular momentum L [Eq. (9.10)] also become constants of motion for Newton's equation and should be conserved. In advanced simulation schemes, where velocities are constantly manipulated, momentum conservation can no longer be used for gauging the stability of the simulation.

Time Reversibility

Newton's equation is reversible in time. For a numerical simulation to retain this property it should be able to retrace its path back to the initial configuration (when the sign of the time step Δt is changed to $-\Delta t$). However, because of chaos (which is part of most complex systems), even modest numerical errors make this backtracking possible only for short periods of time. Any two classical trajectories that are initially very close will eventually exponentially diverge from one another. In the same way, any small perturbation, even the tiny error associated with finite precision on the computer, will cause the computer trajectories to diverge from each other and from the exact classical trajectory. Nonetheless, for short periods of time a stable integration should exhibit temporal reversibility.

D. Simulation Protocol and Some Tricks for Better Simulations

Every molecular dynamics simulation consists of several steps. Put together, these steps are called a "simulation protocol." The main steps common to most dynamic simulation protocols are the following.

1. *Preparation of the data.* Preparation of the initial coordinates (adding hydrogen atoms, minimization) and assignment of initial velocities.
2. *Heating up.* Gradual scaling of the velocities to the desired temperature, accompanied by short equilibration dynamics.
3. *Equilibration.* A relatively long dynamic simulation, whose goal is to ensure that the simulation is stable and free of erratic fluctuations. This step may take from tens of picoseconds to several hundred picoseconds.
4. *Production.* When the simulation is "equilibrated," the dynamic simulation is considered reliable. From this point on, the trajectory generated is stored for further analysis. Typical "production runs" take from several hundred picosec-onds up to tens of nanoseconds (depending on the size of the system and the available computer power).
5. *Analysis.* The resulting trajectory is submitted to careful analysis.

We have presented a simple protocol to run MD simulations for systems of interest. There are, however, some "tricks" to improve the efficiency and accuracy of molecular dynamics simulations. Some of these techniques, which are discussed later in the book, are today considered standard practice. These methods address diverse issues ranging from efficient force field evaluation to simplified solvent representations. One widely used trick is to apply bookkeeping to atom-atom interactions, commonly referred to as the neighbour list, which is illustrated in Fig. 9.2. If we simulate a large N-particle system and use a cutoff that is smaller than the simulation box, then many particles do not contribute significantly to the energy of a particle i. It is advantageous, therefore, to exclude from the expensive energy calculation particle pairs that do not interact. This technique increases the efficiency of the simulations.

Other techniques can be found elsewhere in this volume. The following list includes pointers to several of these techniques :

1. Constant-temperature and constant-pressure simulations.
2. Constraint and multiple time step methods.
3. Periodic boundary conditions.
4. Long-range interactions and extended electrostatics.
5. Solvation models.

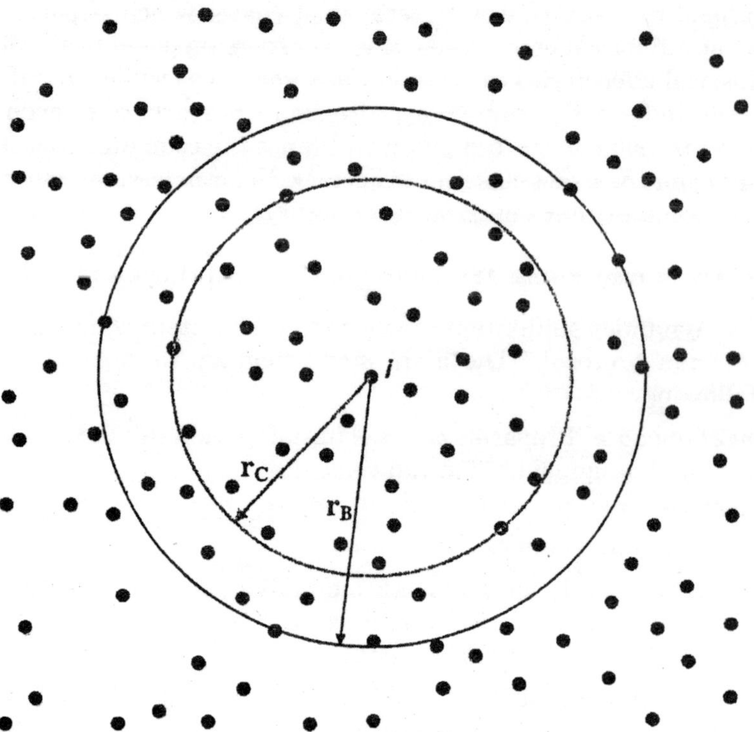

Fig. 9.2. A particle i interacts mainly with particles that are within the cutoff radius r_C. The "neighbor list" contains only those particles that are within a sphere of radius $r_B > r_C$. Particles outside this sphere will not contribute to the force or energy affecting particle *i*. The use of a neighbor list that is periodically updated during the simulation reduces the computer time required in calculat-ing pairwise interactions.

ANALYSIS OF DYNAMIC TRAJECTORIES

An important issue, the significance of which is sometime underestimated, is the analysis of the resulting molecular dynamics trajectories. Clearly, the value of any computer simulation lies in the quality of the information extracted from it. In fact, it is good practice to plan the analysis procedure before starting the simulation, as the goals of the analysis will often determine the character of the simulation to be performed.

A. Time Series

The direct output of a dynamic simulation is a numerical trajectory, *i.e.,* a series of system "snapshots" (coordinates and/or velocities) taken at equal time intervals from the full trajectory (the sampling interval $\Delta\tau$ is typically much larger than Δt). The size of the trajectory sampling interval, $\Delta\tau$, should be determined according to the time scale of the phenomenon that is being studied. For example, a 1 ps sampling interval may be a good choice for studying phenomena that take many tens of picoseconds but is clearly inappropriate for studying fast processes on the subpicosecond time scale. Calculating any dynamic variable, *A(t),* along the trajectory results in a "time series."

Dynamic variables can be any function of coordinates and/or velocities. They may be relatively straightforward, such as total energy, specific bond lengths, or torsion angles of interest, or quite complex. Examples of the latter are the end-to-end distance in a protein (a quantity useful for studying protein folding), distances between a hydrogen bond donor and an acceptor, an angle formed between two helices, and so forth. The most straightforward analytical procedure is to plot the time series as a function of time. Such plots give a quick and easy overview of the simulation and are especially useful for picking up trends (*e.g.*, drifts) or sudden transitions from one state to another. Because a time series consists of instantaneous values taken at the trajectory sampling points, they tend to be very noisy. The level of noise can be reduced by simple smoothing procedures.

A common smoothing procedure is to slide an "*N*-point window" along the data points and plot the "window averages" $\{A(t)\}_N$ as a function of time [instead of plotting $A(t)$ itself]. The width of the window, N, and the suitability of the smoothing approach depend on the property that is being studied and on the ratio between the sampling interval $\Delta\tau$ and the characteristic time of the noise. For example, noise due to bond-stretching motions (time scale of 10-50 fs) can be smoothed by 0.5-1 ps windows. Alternatively, simply increasing the size of the sampling interval $\Delta\tau$ beyond the characteristic time scale of the "noise," to 0.5-1 ps in this case, will also reduce the effect of the noise.

B. Averages and Fluctuations

Time series plots give a useful overview of the processes studied. However, in order to compare different simulations to one another or to compare the simulation to experimental results it is necessary to calculate average values and measure fluctuations. The most common average is the root-mean-square (rms) average, which is given by the second moment of the distribution,

$$\langle A\rangle_{\text{rms}} = \left(\langle A^2\rangle\right)^{1/2} = \left[\frac{1}{N_S}\sum_{i=1}^{N_S}(A_i)^2\right]^{1/2} \qquad ...(9.28)$$

where A is any dynamic variable and N_s is the number of "snapshots" in the trajectory. Root-mean-square fluctuations are calculated in the same way, with the fluctuation ΔA, which is described as a difference with respect to an average A, replacing the values A in Eq. (9.28).

Higher moments of the distribution are often of interest too, especially when nonisotropic or anharmonic processes are being studied. The third moment of the distribution reflects the skewness α_3 defined as

$$\alpha_3 = \langle A^3\rangle / \langle A^2\rangle^{3/2} \qquad ...(9.29)$$

while the fourth moment reflects the excess kurtosis α_2 defined as

$$\alpha_3 = \langle A^4\rangle / \langle A^2\rangle^{2} - 3 \qquad ...(9.30)$$

Both α_3 and α_4 are zero for a Gaussian distribution.

C. Correlation Functions

A powerful analytical tool is the time correlation function. For any dynamic variable *A (t)*, such as bond lengths or dihedral angles, the time autocorrelation function $C_A(t)$ is defined as

$$C_A(t) = \langle a(t)\ A(0)\rangle \quad ...(9.31)$$

This function measures the correlation of the property A(t) to itself at two different times separated by the time interval t, averaged over the whole trajectory. The auto-correlation function is *reversible* in time [*i.e.,* $C_A(t) = C_A(-t)$], and it is *stationary* (*i.e.,* $\langle A(t + t)A(t)\rangle = \langle A(t)A(0)\rangle$). In practical terms, the autocorrelation function is obtained by averaging the terms $(A(s + t)\ A(s))$ while sliding s along the trajectory.

A time "cross-correlation" function between dynamic variables $A(t)$ and $B(t)$ is defined in a similar way:

$$C_{AB}(t) = \{A(t)B(0)\} \quad ...(9.32)$$

An important property of the time autocorrelation function $C_A(t)$ is that by taking its Fourier transform, $F\{C_A(t)\}_\omega$, one gets a spectral decomposition of all the frequencies that contribute to the motion. For example, consider the motion of a single particle in a harmonic potential (harmonic, oscillator). The "time series" describing the position of the particle as a function of time is given by the cos ($\omega_0 t$) function (Fig. 9.3a). The autocorrelation function is given by a cosine function with a period $2\pi/\omega_0$(Fig. 9.3b), and its Fourier transform gives a spectrum with a single sharp peak at ω_0 (Fig. 9.3c). The resulting frequency can be used to extract the (real or effective) local force constant $K_0 = m\omega_0^2$, where m is the mass of the system.

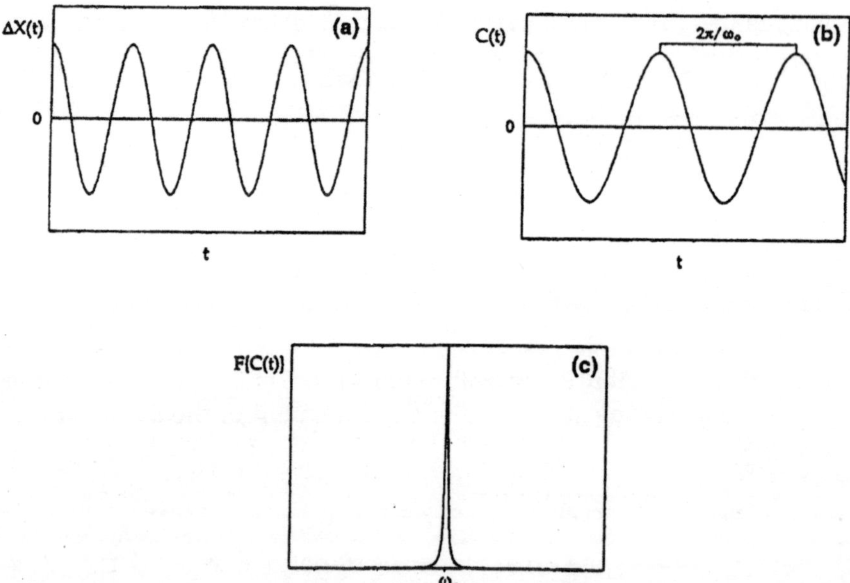

Fig. 9.3(*a*). A "time series" describing the position as a function of time of a particle moving in a one-dimensional harmonic well (harmonic oscillator), (*b*) The autocorrelation function of that motion; (*c*) its Fourier transform spectrum.

D. Potential of Mean Force

The potential of mean force is a useful analytical tool that results in an effective potential that reflects the average effect of all the other degrees of freedom on the dynamic variable of interest. Equation (9.2) indicates that given a potential function it is possible to calculate the probability for all states of the system (the Boltzmann relationship). The potential of mean force procedure works in the reverse direction. Given an observed distribution of values (from the trajectory), the corresponding effective potential function can be derived. The first step, in this procedure is to organize the observed values of the dynamic variable, *A,* into a distribution function $\rho(A)$. From this distribution the "effective potential" or "potential of mean force," $W(A)$, is calculated from the Boltzmann relation:

$$W(A) = -RT \ln [\rho(A)] \qquad ...(9.33)$$

The advantage of a potential of mean force is that it reflects the effect of environmental factors on the behavior of the dynamic variable of interest. Such an effect may be the damping of the solvent or a correlated fluctuation, which reduces effective transition barriers.

E. Estimating Standard Errors

Computer simulation is an experimental science to the extent that calculated dynamic properties are subject to systematic and statistical errors. Sources of systematic error consist of size dependence, poor equilibration, non-bond interaction cutoff, etc. These should, of course, be estimated and eliminated where possible. It is also essential to obtain an estimate of the statistical significance of the results. Simulation averages are taken over runs of finite length, and this is the main cause of statistical imprecision in the mean values so obtained. Statistical errors of dynamic properties could be expressed by breaking a simulation up into multiple blocks, taking the average from each block, and using those values for statistical analysis.

In principle, a block analysis of dynamic properties could be carried out in much the same way as that applied to a static average. However, the block lengths would have to be substantial to make a reasonably accurate estimate of the errors. This approach is based on the assumption that each block is an independent sample. Another approach is to run multiple MD simulations with different initial conditions. The recent work by Karplus and coworkers [19] observes that multiple trajectories with different initial conditions improve conformational sampling of proteins. They suggest that multiple trajectories should be used rather than a single long trajectory.

OTHER MD SIMULATION APPROACHES

A. Stochastic Dynamics

There are cases in which one is interested in the motion of a biomolecule but wishes also to study the effect of different solvent environments on this motion. In other cases, one may be interested in studying the motion of one part of the protein (*e.g.,* a side chain or a loop) as moving in a "solvent bath" provided by the remainder of the protein.

One way to deal with these issues is, of course, to explicitly include all the additional components in the simulation (explicit water molecules, the whole protein, etc.). This solution is computationally very expensive, because much work is done on parts of the system that are of no direct interest to the study. Another way is to reduce the magnitude of the problem by eliminating the explicit solvent degrees of freedom from the calculation and representing them in another way. Methods of this nature, which retain the framework of molecular dynamics but replace the solvent by a variety of simplified models, are discussed in. An alternative approach is to move away from Newtonian molecular dynamics toward stochastic dynamics.

The basic equation of motion for stochastic dynamics is the Langevin equation,

$$m_i\ddot{r}_i = -\nabla_i U(\mathrm{r}) - m_i\beta_i v_i(t) + R_i(t) \qquad \text{...(9.34)}$$

which has three terms on the right-hand side. The first term accounts for molecular interactions and is the same as that used in Newtonian molecular dynamics [Eqs. (9.6) and (9.7). The other two terms are unique to stochastic dynamics and represent solvent effects. The second term is a dissipative drag force reflecting friction due to the solvent. This term is proportional to the velocity v_i, and the friction coefficient β_i, (which is related to the diffusion constant D_i). The third term is a random force R_i,(t) that represents stochastic collisions between solvent molecules and the solute.

The stochastic force introduces energy into the system, and the friction force removes energy from the system. The Langevin random force $R_i(t)$ on atom i is obtained from a random Gaussian distribution of zero mean and a variance related to the friction coefficient,

$$\langle R_i(t)\rangle = 0 \qquad \text{...(9.35}a\text{)}$$

$$\langle R_i(t)R_i(0)\rangle = 6m_i k_B T\delta(t) = 2D_i\,\delta(t) \qquad \text{....(9.35}b\text{)}$$

Sometimes an additional term, F_i^{mean}, is added to the right-hand side of Eq. (34). This "mean force" represents an average effect of degrees of freedom not explicitly treated in the simulation.

Because of the additional velocity-dependent forces (the dissipative forces), the straightforward finite-difference algorithms of the Verlet type cannot be used to integrate Langevin's equation. An algorithm may be derived that reduces to the Verlet algorithm in the limit of vanishing friction $(\beta_i \to 0)$. This algorithm is obtained by adding the Langevin terms to the Verlet equation described in Eqs. (9.16) and (9.17). The resulting algorithm is of the order $O\left(\Delta t^3\right)$ and is valid for $\beta_i\Delta t < 1$.

B. Brownian Dynamics

Biomolecular motions that involve substantial displacements of the molecular surface, such as the motion of heavy particles in aqueous solution, are usually damped owing to the high viscosity of the surrounding solvent. In many such cases the damping effects are sufficiently great that internal forces are negligible and the motion has the character of a random walk. In other cases, such as the diffusion pairing of a substrate with its enzyme, the specific details of the internal dynamics are of little interest to the intermolecular motion. In such cases a further level of simplification can be introduced into the equation of motion.

The relevant approximation applicable to such cases is the Brownian equation of motion, which is a diffusional analog of the molecular dynamics method. The Brownian equation can be easily derived from the stochastic Langevin equation presented in Eq. (9.34). If the inertial term on the left-hand side of the equation is small compared to the force terms on the right-hand side, it can be neglected, resulting in the diffusional Brownian equation of motion,

$$\mathrm{v}_i(t) = \dot{r}_i = \frac{-\nabla V_i(\mathrm{r}) + F_i^{\mathrm{mean}} + R_i(t)}{m_i \beta_i} \qquad \text{...(9.36)}$$

where the properties of the stochastic force $R_i(t)$ and its dependence on the friction coefficient D_i are given by Eqs. (9.35*a*) and (9.35*b*).

As with Newtonian molecular dynamics, a number of different algorithms have been developed to calculate the diffusional trajectories. An efficient algorithm for solving the Brownian equation of motion was introduced by Ermak and McCammon [21]. A detailed survey of this and other algorithms as well as their application can be found in Ref. 2.

C. Molecular Dynamics in Alternative Ensembles

The original molecular dynamics (MD) technique was used only to study the natural time evolution of a classical system of N particles in a volume V. In such simulations, the total energy, E, is a constant of motion, so that the time averages obtained by such a conventional MD simulation are equivalent to microcanonical ensemble averages. It is often scientifically more relevant to perform simulations in other ensembles, such as the canonical *(NVT)* that is associated with a Helmholtz free energy or the isothermal-isobaric (NPT) ensemble that is associated with a Gibbs free energy. Two rather different types of solutions to the problem of simulating alternative ensembles with the MD method have been proposed.

The first approach is based on introducing simple velocity or position rescaling into the standard Newtonian MD. The second approach has a dynamic origin and is based on a reformulation of the Lagrangian equations of motion for the system (so-called extended Lagrangian formulation). In this section, we discuss several of the most widely used constant-temperature or constant-pressure schemes.

1. Constant-Temperature MD

The simplest method that keeps the temperature of a system constant during an MD simulation is to rescale the velocities at each time step by a factor of $(T_0/T)^{1/2}$, where T is the current instantaneous temperature [defined in Eq. (9.24)] and T_0 is the desired temperature. This method is commonly used in the equilibration phase of many MD simulations and has also been suggested as a means of performing "constant temperature molecular dynamics". A further refinement of the velocity-rescaling approach was proposed by Berendsen et al., who used velocity rescaling to couple the system to a heat bath at a temperature T_0. Since heat coupling has a characteristic relaxation time, each velocity v is scaled by a factor λ, defined as

$$\lambda = \left[1 + \frac{\Delta t}{2\tau_T}\left(\frac{T_0}{T} - 1\right)\right]^{1/2} \qquad \text{...(9.37)}$$

In this expression, Δt is the size of the integration time step, τ_T is a characteristic relaxation time, and T is the instantaneous temperature. In the simulation of water, they found a relaxation time of $\tau r = 0.4$ ps to be appropriate. However, this method does not correspond exactly to the canonical ensemble.

An alternative method, proposed by Andersen, shows that the coupling to the heat bath is represented by stochastic impulsive forces that act occasionally on randomly selected particles. Between stochastic collisions, the system evolves at constant energy according to the normal Newtonian laws of motion. The stochastic collisions ensure that all accessible constant-energy shells are visited according to their Boltzmann weight and therefore yield a canonical ensemble.

To carry out this method, values are chosen for T_0, the desired temperature, and v, the mean frequency with which each particle experiences a stochastic collision. If successive collisions are uncorrected, then the distribution of time intervals between two successive stochastic collisions, $P(v, t)$, is of the Poisson form,

$$P(v, t) = v \exp(-vt) \qquad ...(9.38)$$

A constant-temperature simulation now consists of the following steps:

1. Start with an initial set of positions and momenta and integrate the equation of motion.
2. The probability that any particular particle suffers a stochastic collision in a time interval of Δt is v Δt.
3. If particle i has been selected to undergo a collision, obtain its new velocity from a Maxwellian velocity distribution, defined in Eq. (9.23), corresponding to the desired temperature T_0. All other particles are unaffected by this collision.

Another popular approach to the isothermal (canonical) MD method was shown by Nose. This method for treating the dynamics of a system in contact with a thermal reservoir is to include a degree of freedom that represents that reservoir, so that one can perform deterministic MD at constant temperature by reformulating the Lagrangian equations of motion for this extended system. We can describe the Nose approach as an illustration of an extended Lagrangian method.

Energy is allowed to flow dynamically from the reservoir to the system and back; the reservoir has a certain thermal inertia associated with it. However, it is now more common to use the Nose scheme in the implementation of Hoover. To construct Nose-Hoover constant-temperature molecular dynamics, an additional coordinate, s, and its conjugate momentum p_s are introduced. The Hamiltonian of the extended system of the N particles plus extended degrees of freedom can be expressed as

$$H_{\text{Nose}} = \sum_{i=1}^{N} \frac{P_1^2}{2m_i s^2} + U(q) + \frac{P_s^2}{2Q} + \frac{g}{\beta} \ln s \qquad ...(9.39)$$

where β is $1/k_BT$, Q is an effective "mass" associated with s, and g is a parameter related to the degrees of freedom in the system. The microcanonical distribution in the augmented set of variables in Eq. (9.39) is equivalent to a canonical distribution of the variables r_i and ϕ_i/s. One of the disadvantages of the original Nose approach, however, is that s can be interpreted as a scaling factor of the time step. This implies that the real time-step fluctuations occur during a simulation. In a simulation it is not convenient to work with fluctuating time intervals. The real-variable formulation is therefore recommended. However, showed that the equations derived by Nose can be further simplified. He derived a slightly different set of equations that dispense with the time-scaling parameter s. To simplify the equations, we can introduce the thermodynamic friction coefficient, $\xi = p_s/Q$. The equations of motion then become

$$\dot{r} = \frac{p_i}{m_i} \qquad \text{...(9.40}a\text{)}$$

$$\dot{p}_i = -\frac{\partial U}{\partial r_i} - \xi p_i \qquad \text{...(9.40}b\text{)}$$

$$\dot{\xi} = \left(\sum_i \left(\frac{p_i^2}{m_i}\right) - \frac{g}{\beta}\right).Q^{-1} \qquad \text{...(9.40}c\text{)}$$

$$\frac{\dot{s}}{s} = \frac{d \ln s}{dt} = \xi \qquad \text{...(9.40}d\text{)}$$

Note that Eq. (9.40*d*), in fact, is redundant, because the other three equations form a closed set. Nonetheless, if we solve the equations of motion for s as well, we can use the following, $\overline{H}_{\text{Nose}}$, as a diagnostic tool, because this quantity has to be conserved during the simulation even though $\overline{H}_{\text{Nose}}$ is no longer a Hamiltonian:

$$\overline{H}_{\text{Nose}'} = \sum_{i=1}^{N} \frac{p_i^2}{2m_i} + U(q) + \frac{\xi 2Q}{2} + \frac{gs}{\beta} \qquad \text{...(9.41)}$$

where $g = 3N$ in this real-variable formulation.

Theoretically, static quantities are independent of the value chosen for the parameter Q. In practice, however, we observe that quantities are Q-dependent because of the finite number of MD simulation steps. Too high a value of Q results in slow energy flow between the system and reservoir, and in the limit $Q \to \infty$ we regain conventional MD. On the other hand, if Q is too low, the energy undergoes long-lived, weakly damped oscillations, resulting in poor equilibration. Nose suggested the choice of the adjustable parameter Q for the physical system. It may, however, be necessary to choose Q by trial and error in order to achieve satisfactory damping. Both Andersen and Nose-Hoover methods, indeed, have generated canonical distributions. However, sometimes the Nose-Hoover thermostat runs into ergodicity problems and the desired distribution cannot be achieved.

The Andersen method does not suffer from such problems, but its dynamics are less realistic than Nose-Hoover. To alleviate the ergodicity problems, Martyna et al. proposed a scheme in which the Nose-Hoover thermostat is coupled to another thermostat or, if necessary, to a whole chain of thermostates. The coupling ensures that the thermostats are allowed to fluctuate. In the original Nose-Hoover method, the thermostat variable does not fluctuate. This generalization of the original Nose-Hoover method is also shown to generate a canonical distribution, but this approach no longer faces the ergodicity problems. Details of the programming for this approach may be obtained from the book by Frenkel and Smit.

2. Constant-Pressure MD

In a normal molecular dynamics simulation with repeating boundary conditions (i.e., periodic boundary condition), the volume is held fixed, whereas at constant pressure the vol-ume of the system must fluctuate. In some simulation cases, such as simulations dealing with membranes, it is more advantageous to use'the constant-pressure MD than the regular MD. Various schemes for prescribing the pressure of a molecular dynamics simulation have also been proposed and applied. In all of these approaches it is inevitable that the system box must change its volume. To include the volume as a dynamic variable, the equations of motion are determined in the analysis of a system in which the positions and momenta of all particles are scaled by a factor proportional to the cube root of the volume of the system.

Andersen originally proposed a method for constant-pressure MD that involves coupling the system to an external variable, V, the volume of the simulation box. This coupling mimics the action of a piston on a real system. The piston has a "mass" M_v [which has units of (mass)(length)$^{-4}$]. From the Lagrangian for this extended system, the equations of motion for the particles and the volume of the cube are

$$\dot{r}_i = \frac{P_i}{m_i} + \frac{1}{3}\left(\frac{\dot{V}}{V}\right) r_i \quad \text{....(9.42a)}$$

$$\dot{p}_i = F_i - \frac{1}{3}\left(\frac{\dot{V}}{V}\right) p_i \quad \text{...(9.42b)}$$

$$\ddot{V} = \frac{1}{M_V}\left[P(t) - P_0\right] \quad \text{...(9.42c)}$$

where V is the volume, $P(t)$ is the instantaneous pressure, P_0 is the desired pressure, and r_i, p_i, m_i and Fi are the position, momentum, mass, and force, respectively, for each particle i. Andersen proved that the solution to these equations produces trajectories in the isobaric-isoenthalpic (*NPH*) ensemble where the particle number, pressure, and enthalpy of the system are constant. Here the choice of piston mass determines the decay time of the volume fluctuations.

It has been proven that equilibrium quantities are independent of M_v, but in practice M_v influences the dynamic properties in the simulations. Even though there is no precise formulation for choosing M_v, Andersen suggests that the piston mass may be chosen by trial and error to satisfy the length of time required for a sound wave to travel through the simulation cell.

An alternative procedure rescales the coordinates of each atom at each time step [24]. The atomic coordinate x and the characteristic distance for repeating boundary conditions, d, are rescaled to values μx and μd, respectively, where

$$\mu = \left[1 - \frac{\Delta t}{\tau_p}(P_0 - P)\right]^{1/3} \quad \text{...(9.43)}$$

Here, Δt is the size of the time step, τ_p is a characteristic relaxation time, and P_0 is the pressure of the external constant-pressure bath. The instantaneous pressure can be calculated as follows:

$$P = \frac{2}{3V}\left[E_k + \frac{1}{2}\sum_{i<j} r_{ij} \cdot F_{ij}\right] \qquad ...(9.44)$$

where V is the volume and E_k is the kinetic energy, r_{tj} is the vector from particle i to particle, j and F_{ij} is the force on particle j due to particle i. In simulations of water, values of $\tau_p = 0.01 - 0.1$ ps were found suitable. This method does not drastically alter the dynamic trajectories and is easy to program, but the appropriate ensemble has not been identified. Therefore, the meaning of fluctuations in any observed quantity cannot be determined. An algorithm for performing a constant-pressure molecular dynamics simulation that resolves some unphysical observations in the extended system (Andersen's) method and Berendsen's methods was developed by Feller et. al.

This approach replaces the deterministic equations of motion with the piston degree of freedom added to the Langevin equations of motion. This eliminates the unphysical fluctuation of the volume associated with thc piston mass. In addition, Klein and coworkers present an advanced constant-pressure method to overcome an unphysical dependence of the choice of lattice in generated trajectories. In the foregoing treatments of pressure feedback, the simulation volume retains its cubic form, so changes consist of uniform contractions and expansions. The method is readily extended to the case of a simulation region in which the lengths and directions of the edges are allowed to vary independently.

Parrinello and Rahman and Nose and Klein extended the Andersen method to the case of noncubic simulation cells and derived a new Lagrangian for the extended system. Though their equations of motion are different from Andersen's original equations, they produce an identical ensemble. This technique is particularly helpful in the study of crystal structures of proteins, because it allows for phase changes in the simulation, which may involve changes in the unit cell dimensions and angles. These constant-pressure MD methods can be combined with a suitable temperature control mechanism, as discussed in the previous section, to produce a more useful method to control both pressure and temperature simultaneously during the simulation. There are several approaches. The simplest approach is to use the scaling factors. The details of the algorithm are given by Berendsen et. al. Another approach is to define the appropriate extended Lagrangian for simultaneously coupling pressure and temperature to produce the isothermal-isobaric (*NPT*) ensemble.

Hoover presents a detailed description of the Nose-Hoover constant-temperature method with Andersen's constant-pressure method. Even though this extended system method is slightly more complicated to program, this is the best candidate for conducting *NPT* ensemble MD. Details on programming for this approach are available in the book by Rapaport. In addition, the new Langevin piston method for constant pressure can be easily extended to couple a Nose-Hoover thermostat to obtain a constant-pressure and constant-temperature method.

ADVANCED SIMULATION TECHNIQUES

Computer simulations have become a valuable tool for the theoretical investigation of biomolecular systems. Unfortunately, these simulations are often computationally demanding tasks owing to the large number of particles as well as the complex nature of their associated interactions. A longstanding problem, however, is that molecular dynamics is typically limited to a time scale of 10^{-6} s (1 μs) or less.

In an MD simulation, the most rapidly varying quantities, such as the bond lengths, limit the integration time step, while the more slowly varying molecular processes are of primary interest and determine the simulation length required. This would make the simulation of molecular substances very expensive. A variety of techniques have been introduced to increase the time step in molecular dynamics simulations in an attempt to surmount the strict time step limits in MD simulations so that long time scale simulations can be routinely undertaken. One such technique is to solve the equations of motion in the internal degree of freedom, so that bond stretching and angle bending can be treated as rigid. This technique is discussed in elsewhere of this book. Herein, a brief overview is presented of two approaches, constrained dynamics and multiple time step dynamics.

A. Constrained Dynamics

To avoid the situation in which high frequency motions, such as bond stretching and bond angle bending which limits the integration time step, it is customary to eliminate such degrees of freedom entirely by the simple expedient of replacing them with constraints. In general, the dynamics could satisfy many constraints simultaneously, *e.g.*, many bond lengths and bond angles. Assuming that a total of *n* distance constraints are imposed on a particular molecule, the constraint σ_k for a fixed distance d_{ij} between atom *i* and *j* can be expressed as

$$\sigma_k = r_{ij}^2 - d_{ij}^2 = 0, \; k = 1,..., n \qquad ...(9.45)$$

The equations of motion follow directly from the Lagrangian formulation containing all constraints. The result is

$$m_i \ddot{r}_i = F_i + G_i \qquad ...(9.46)$$

where F_i, is the usual force term, m_i the mass of the ith atom, and the additional term G_i expresses the effect of the constraints on atom *i*, G_i can be written

$$G_i = -\sum_{\alpha} \lambda_\alpha \frac{\partial \sigma_\alpha}{\partial r_i} \qquad ...(9.47)$$

Here α denotes the set of constraints that directly involve r_i and the $\{\lambda_\alpha\}$ are the Lagrange multipliers introduced into the problem. There are various ways to obtain the solutions to this problem. The most straightforward method is to solve the full problem by first computing the Lagrange multipliers from the time-differentiated constraint equations and then using the values obtained to solve the equations of motion. This method, however, is not computationally cheap because it requires a matrix inversion at every iteration.

In practice, therefore, the problem is solved by a simple iterative scheme to satisfy the constraints. This scheme is called SHAKE. Note that the computational advantage has to be balanced against the additional work required to solve the constraint equations. This approach allows a modest increase in speed by a factor of 2 or 3 if all bonds are constrained. Although constrained dynamics is usually discussed in the context of the geometrically constrained system described above, the same techniques can have many other applications. For instance, constant-pressure and constant-temperature dynamics can be imposed by using constraint methods. Car and Parrinello describe the use of the extended Lagrangian to maintain constraints in the context of their ab initio MD method.

B. Multiple Time Step Methods

According to the nature of the empirical potential energy function, described different motions can take place on different time scales, *e.g.*, bond stretching and bond angle bending vs. dihedral angle librations and non-bond interactions.

Multiple time step (MTS) methods [38-40,42] allow one to use different integration time steps in the same simulation so as to treat the time development of the slow and fast movements most effectively. Tuckerman et al. showed how to systematically derive time-reversible, area-preserving MD algorithms from the Liouville formulation of classical mechanics. Here, we briefly introduce the Liouville approach to the MTS method. The Liouville operator for a system of N degrees of freedom in Cartesian coordinates is defined as

$$iL = [..., H] = \sum_{i=1}^{N}\left[\dot{r}_i \frac{\partial}{\partial r_i} + F_i(r)\frac{\partial}{\partial p_i}\right] = \dot{r}\frac{\partial}{\partial r} + F(r)\frac{\partial}{\partial p} \qquad ...(9.48)$$

where [......] is the Poisson bracket, H is the Hamiltonian of the system, r_i and p_i, are the position and conjugate momentum at coordinate i, r_i is the time derivative of r_i, and F_i is the force acting on the ith degree of freedom. The state of the system, Γ, at time Δt is then given by

$$\Gamma\left[r(\Delta t), p(\Delta t)\right] = U(\Delta t)\cdot\Gamma\left[r(0), p(0)\right] \qquad ...(9.49)$$

where $U(t)$ is composed of a classical time evolution operator, $e^{iL\Delta t}$ and Γ could be any arbitrary function that depends on all coordinates and momenta of the system. We decompose the Liouville operator into two parts,

$$iL = iL_1 + iL_2 \qquad ...(9.50)$$

Unfortunately, we cannot replace $e^{iL\Delta t}$ by $e^{iL_1\Delta t}e^{iL_2\Delta t}$, because iL_1 and iL_2 are noncommutative operators. Applying Trotter's theorem [41], however, we can decompose the propagator, $U/(\Delta t)$:

$$U(\Delta t) = e^{iL\Delta t} = e^{iL_1\Delta t/2}e^{iL_2\Delta t}e^{iL_1\Delta t/2} \qquad ...(9.51)$$

The idea is now to replace the formal solution of the Liouville equation by the discretized version. The middle term of the propagator in Eq. (9.51) can be further decomposed by an additional Trotter factorization to obtain

$$e^{iL_2\Delta t} = (e^{iL_2\Delta\tau})^n + O(n\Delta\tau^3) \qquad ...(9.52)$$

where $\Delta t = n\ \Delta\tau$. Here the smaller time interval $\Delta\tau$ and the integer n determining the number of steps are chosen to guarantee stable dynamics for the system. Now Eq. (9.51) becomes

$$U(\Delta t) \approx e^{iL_1\Delta t/2}\left(e^{iL_2\Delta\tau}\right)^n e^{iL_1\Delta t/2} \qquad ...(9.53)$$

With the propagator written in this way, the equation of motion can be integrated by a multiple time step algorithm in Cartesian coordinates because Δt and $\Delta\tau$ are different integration time steps ($\Delta t > \Delta\tau$ when $n > 1$). As an example, the force terms are separated into two components

$$F(\mathbf{r}) = F_f(\mathbf{r}) + F_s(\mathbf{r}) \qquad ...(9.54)$$

where F_f associates with "stiff" degrees of freedom or fast-varying forces, such as forces from bond-stretching, angle-bending motions, and F_s is associated with the rest of the contributions (*i.e.,* slowly varying forces), such as forces from torsion motions and non-bond interaction. By introducing this decomposition into the Liouville operator we obtain

$$iL = \dot{r}\frac{\partial}{\partial r} + F_f(r)\frac{\partial}{\partial p} + F_s(r)\frac{\partial}{\partial p} \quad \text{...(9.55)}$$

In this separation, the two Liouville operators, iL_1 and iL_2 of Eq. (9.50) can now be defined:

$$iL_2 = \dot{r}\frac{\partial}{\partial r} + F_f(r)\frac{\partial}{\partial p};\ iL_1 = F_s(r)\frac{\partial}{\partial p} \quad \text{...(9.56)}$$

The propagator $U(\Delta t)$ defined in Eq. (9.53) can now be implemented algorithmically as follows:

1. Starting with the initial state $[r(0), p(0)]$, generate the motion by using the propagator $e^{iL_1 \Delta t/2}$.
2. Using the final state of step 1 as the initial state, generate the motion using the middle propagator $e^{iL_2 \Delta\tau}$. Repeat this step n times.
3. Finally, starting with the state generated in step 2 as the initial state, generate the motion using the propagator $e^{iL_1 \Delta t/2}$.

Corresponding implementations of the velocity Verlet operator can be easily derived for this Liouville propagator. It should also be realized that the decomposition of iL into a sum of iL_i and iL_2 is arbitrary. Other decompositions are possible and may lead to algorithms that are more convenient.

One example is that in a typical MD simulation, a large portion of the computer processing time is spent in examining the non-bond pair interactions. These non-bond forces, therefore, can be divided into fast and slow parts based on distance by using a continuous switching function. Applications of this MTS method to protein simulations have been shown to reduce the CPU time by a factor of 4-5 without altering dynamical properties. In addition, this MTS approach shows significantly better performance enhancement in systems where the separation of fast and slow motions is pronounced.

C. Other Approaches and Future Direction

There are other approaches to obtaining efficiency in MD simulations. Examples include eigenvector-based schemes implicit integration schemes, path optimization schemes and a transition state theory approach. Recently, a unique approach to overcome the time scale problems in MD was developed. Multibody order (N) dynamics [MBO(N)D] is based on aggregating atoms of a macromolecule into groupings of interacting flexible and rigid bodies. Body flexibility is modeled by a truncated set of body-based normal modes. This approach allows for the unimportant high frequency modes of vibration, such as bond and angle motions, to be eliminated, leaving only the important lower frequency motions. This results in the use of a larger integration time step size, substantially reducing the computational time required for a given dynamic simulation.

By coupling MBO(N)D with MTS described in the previous section, speed increases of up to 30-fold over conventional simulation methods have been realized in various MD simulations.

In addition to increasing computational efficiency, the approach also allows for a simplified analysis of dynamics simulations, as there are fewer degrees of freedom to consider. Additionally, continuous developments of computer architectures, such as the clock speed of CPU chips and massive parallel computers, also help to carry out simulations of large biomolecules that require enormous computing power. In recent years, distributed memory parallel computers have been offering cost-effective computational power to researchers. This approach shows a great advantage in the size of the system (it is possible to run a million atoms in the system), although the simulation length is not scaled as well as the size because of the nature of solving equations of motion sequentially in time. Finally, molecular modeling based on low resolution (coarse-grain) models has gained some attention in the field of biomolecular simulations.

This approach dramatically reduces the number of interaction sites by adapting a simple approach (*e.g.,* a single site per residue) or a multiple sites per residue approach (*e.g.,* one backbone and one side chain interaction site per residue). These coarse-grain potentials are described by two categories: those based on statistical contact information derived from high resolution protein structures, and those base on established molecular mechanics force fields. Coarse grain approaches are another way to gain a significant increase in speed and therefore begin to address large systems, such as protein-protein complexes, routinely. Despite recent developments in algorithms and computer hardware, to bridge the gap between the time and size scales accessible by computer simulations and those required by experimental observations we still need to develop noble approaches.

COMPUTATIONAL SEQUENCE AND ALIGNMENT

Proteomics is concerned with the analysis of the complete protein complements of genomes. Thus proteomics includes not only the identification and quantification of proteins, but also the determination of their localization, modifications, interactions, activities, and functions. This chapter focuses on protein sequences as the sources of biochemical information. Protein sequence databases are surveyed. Similarity search and sequence alignments using the Internet resources are described.

PROTEIN SEQUENCE: INFORMATION AND FEATURES

Proteome refers to protein complement expressed by a genome. Thus proteomics concerns with the analysis of complete complements of proteins. It is the study of proteins that are encoded by the genes of a cell or an organism. Such study includes determination of protein expression, identification and quantification of proteins as well as characterization of protein structures, functions and interactions. The functional classification of proteins in genomes (*i.e.,* proteomes) can be accessed from the Proteome Analysis Database. The unique characteristic of each protein is the distinctive sequence of amino acid residues in its polypeptide chain. The amino acid sequence is the link between the genetic message in DNA and the three-dimensional structure that performs a protein's biological function.

The sequence comparison among analogous proteins yields insights into the evolutionary relationships of proteins and protein function. The comparison of sequences between normal and mutant proteins provides invaluable information toward identification of critical residues involved in protein function as well as detection and treatment of diseases. The knowledge of amino acid sequences is essential to the protein's three-dimensional structure, mechanism of actions, and design.

The general strategy for determining the amino acid sequence of a protein involves several steps:

1. Separation and purification of individual polypeptide chain from a protein containing, subunits.
2. Cleavage of intramolecular disulfide linkage (*s*).
3. Determination of amino acid composition of purified polypeptide chain.
4. Analyses of N- and C-terminal residues.
5. Specific cleavage of the polypeptide chain reproducibly into manageable fragments and purification of resulting fragments.

6. Sequence analyses of the peptide fragments.
7. Repeat steps 5 and 6 using a different cleavage procedure to generate a different and overlapping set of peptide fragments and their sequence analyses.
8. Reconstruction of the overall amino acid sequence of the protein from the sequences in overlapping fragments.
9. Localization of the disulfide linkage(*s*).

Alternatively, protein sequence information can be derived from translating the nucleotide sequences of DNA that encode the proteins.

Protein Identity Based on Composition and Properties

A number of useful computational tools have been developed for predicting the identity of unknown proteins based on the physical and chemical properties of amino acids and vice versa. Many of these tools are available through the Expert Protein Analysis System (ExPASy) and other servers. Rather than using an amino acid sequence to search SWISS-PROT, AACompIdent of ExPASy Proteomic tools uses the amino acid composition of an unknown protein to identify known proteins of the same composition. The program requires the desired amino acid composition, the pI and molecular weight of the protein (if known), the appropriate taxonomic class, and any special keywords. The user must select from one of six amino acid constellations that influence how the analysis is performed.

For each sequence in the database, the algorithm computes a score based on the difference in compositions between the sequence and the query composition. The results, returned by e-mail, are organized as three ranked lists. Because the computed scores are a measure of difference, a score of zero implies that there is exact correspondence between the query composition and that sequence entry. AACompSim (ExPASy Proteomic tools) performs a similar type of analysis, but rather than using an experimentally derived amino acid composition as the basis of searches, the sequence of a Swiss-Prot protein is used instead. A theoretical pI and molecular weight are computed prior to computing the difference score using Compute pI/MW.

Physicochemical Properties Based on Sequence

Compute pI/MW (ExPASy Proteomic tools) is a tool that calculates the isoelectric point (pI) and molecular weight of an input sequence. Molecular weights are calculated by the addition of the average isotopic mass of each amino acid in the sequence plus that of one water molecule. The sequence can be furnished by the user in fasta format, or by a Swiss-Prot identifier. If a sequence is furnished, the tool automatically computes the pI and molecular weight for the entire length of the sequence. If a Swiss-Prot identifier is given, the user may specify a range of amino acids so that the computation is done on a fragment rather than on the entire protein.

The absorption coefficient and pI value are also calculated at aBi. If "courbe de titrage" is checked, the titration curve based on the query sequence is returned. PeptideMass (ExPASy Proteomic tools), which is designed for use in peptide mapping experiments, determines the cleavage products of a protein after exposure to a specific protease or chemical reagent. The enzymes and reagents available for cleavage via PeptideMass are trypsin, chymotrypsin, Lys C, cyanogen bromide, Arg C, Asp N, and Glu C. The AAindex database collects physicochemical properties of amino acids such as molecular weight, bulkiness, polarity, hydrophobicity, average area buried/ exposed, solvent accessibility, and secondary structure parameters (propensities for amino acid residues to form α-helix, β-strand, reverse turn and coil structures of proteins). The physicochemical properties can be plotted along the sequence using ProtScale at ExPASy. TGREASE calculates the hydrophobicity — that is, the propensity of the amino acid to bury itself in the core of the protein and away from surrounding water. This tendency, coupled with steric and other considerations, influence how a protein ultimately folds. As such, the program finds application in the determination of putative transmembrane sequences as well as in the prediction of buried regions of globular proteins. TGREASE can be downloaded and run on DOS-based computers.

The method relies on a hydropathy scale (Kyte and Doolittle, 1982). Amino acids with higher positive scores are more hydrophobic; those with more negative scores are more hydrophilic. A moving average, or hydropathic index, is then calculated across the protein. The window length is adjustable, with a span of 7 to 11 residues recommended to minimize noise and maximize information content. The results are then plotted as hydropathic index versus residue number. TMpred of EMB Net and TMAP predicts possible transmembrane regions and their topology from the amino acid sequence.

Sequence Comparison

When we compare the sequences of proteins, we analyze the similarities and differences at the level of individual amino acids with the aim of inferring structural, functional, and evolutionary relationships among the sequences under study. The most common comparative method is sequence alignment, which provides an explicit mapping between the residues of two or more sequences. *Sequence identity* refers to the occurrence of exactly the same nucleotide/amino acid in the same position in two sequences. *Sequence similarity* refers to the sequences aligned by applying possible substitutions scored according to the probability with which they occur. *Sequence homology* reflects the evolutionary relationship between sequences. Two sequences are said, to be homologous if they are derived from a common ancestral sequence. Thus, similarity is an observable quantity that refers to the presence of identical and similar sites in the two sequences while homology refers to a conclusion drawn from similarity data that two sequences share a common ancestor.

The changes that occur during divergence from the common ancestor can be characterized as substitutions, insertions, and deletions. Residues that have been aligned, but are not identical, would represent substitutions. Regions in which the residues of one sequence correspond to nothing in the other would be interpreted as either an insertion into one sequence or a deletion from the other (INDEL). *Analogy* refers to either protein structures that share similar folds but have no demonstrable sequence similarity or to proteins that share catalytic residues with almost exactly equivalent spatial geometry without sequence or structural similarity. It is useful to distinguish between homologous proteins as *orthologues* that perform the same function in different species, and as *paralogues* that perform different but related functions within one organism. Sequence comparison of orthologous proteins leads to the study of molecular systematics, and therefore the construction of phylogenetic trees has revealed relationships among different species. On the other hand, the study of paralogous proteins has provided insights into the underlying mechanisms of evolution because paralogous proteins arose from single genes via successive duplication events. In general, the three-dimensional structures are more likely to be conserved than are the corresponding amino acid sequences between distantly related proteins.

Apart from strict identity in pattern matching, one approach of tolerance in the sequence comparison can be introduced by considering amino acid residues as member of groups with shared biochemical properties as shown in Table 10.1.

The presence of the regular repeated structure regions in proteins permits the comparison of their structural characteristics that display common features. A search for structural similarity of proteins can be processed at various levels of complexity depending upon the field of application. Two approaches are considered in the similarity comparison of protein sequences :

1. Sequence comparison to detect common features. This may involve mapping profiles of conformational preferences, accessibility, hydrophilicity/hydrophobicity, and so on. The profiles are compared. Another useful topography is a dot-plot similarity matrix that compares sequences of two proteins according to the predefined criterion of similarity, and the similarities are identified as a succession of consecutive dots parallel to the diagonal.
2. Sequence alignment that defines the best one-to-one correspondence between amino acid sequences of two or more proteins. This scores correspondence *on the entire sequences with appropriate accommodations for insertions and deletions.*

Table 10.1. Classification of Amino Acids According to their Biochemical Properties

Property	*Residue*
Small	Ala, Gly
Hydroxyl/phenol	Ser, Thr, Tyr
Thiol/thioether	Cys, Met
Acidic/amide	Asp, Asn, Gln, Glu
Basic	Arg, His, Lys
Polar	Ala, Cys, Gy, Pro, Ser, Thr
Aromatic	Phe, Trp, Tyr
Alkyl hydrophobic	Ile, Leu, Met, Val

DATABASE SEARCH AND SEQUENCE ALIGNMENT

Primary Database

It is much easier and quicker to produce sequence information than to determine 3D structures of proteins in atomic detail. As a consequence, there is a protein sequence/structure deficit. In order to benefit from the wealth of sequence information, we must establish, maintain, and disseminate sequence databases; provide user-friendly software to access the information, and design analytical tools to visualize and interpret the structural/ functional clues associated with these data. There are different classes of protein sequence databases. Primary and secondary databases are used to address different aspects of sequence analysis.

Composite databases amalgamate a variety of different primary sources to facilitate sequence searching efficiently. The primary structure (amino acid sequence) of a protein is stored in primary databases as linear alphabets that represent the constituent residues. The secondary structure of a protein corresponding to region of local regularity (*e.g.*, α-helices, β-strands, and turns), which in sequence alignments are often apparent as conserved motifs, is stored in secondary databases as patterns. The tertiary structure of a protein derived from the packing of its secondary structural elements which may form folds and domains is stored in structure databases as sets of atomic coordinates.

Some of the most important protein sequence databases are PIR (Protein Information Resource), SWISS-PROT (at EBI and ExPASy), MIPS (Munich Information Center for Protein Sequences), JIPID (Japanese International Protein Sequence Database), and TrEMBL (at EBI). The PIR Protein Sequence Database developed at the National Biomedical Research Foundation (NBRF) has been maintained by PIR-International Protein Sequence Database (PSD), which is the largest publicly distributed and freely available protein sequence database.

The consortium includes PIR at the NBRF, MIPS, and JIPID. PIR-International provides online to numerous sequence and auxiliary databases. These include PSD (annotated and classified protein sequences), PATCHX (sequences not yet in PSD), ARCHIVE (sequences as originally reported in a publication), NRL_3D (sequences from 3D structure database PDB), PIR-ALN (sequence alignment of superfamilies, families, and homology domains), RESID (post-translational modifications with PSD feature information), iProClass (non-redundant sequences organized according to superfamilies and motifs), and ProtFam (sequence alignments of superfamilies). The PIR server offers three types of database searches:

1. Interactive text-based search that allows Boolean queries of text field.
2. Standard sequence similarity search including Peptide Match, Pattern Match, BLAST, FASTA, Pairwise Alignment, and Multiple Alignment.
3. Advanced search that combines sequence similarity and annotation searches or evaluate gene family relationships.

Each PIR entry consists of Entry (entry ID), Title, Alternate-names, Organism, Date, Accession (accession number), Reference, Function (description of protein function), Comment (*e.g.*, enzyme specificity and reaction, etc.), Classification (super-family), Keywords (*e.g.*, dimer, alcohol metabolism, metalloprotein, etc.), Feature (lists of sequence positions for disulfide bonds, active site and binding site amino acid residues, etc.), Summary (number of amino acids and the molecular weight), and Sequence (in PIR format). In addition, links to PDB, KEGG, BRENDA, WIT, alignments, and iProClass are provided. The Munich Information Center for Protein Sequences (MIPS) collects and processes sequence data for the PIR-International Protein Sequence Database project. Access to the database is provided through its Web server.

The implementation of PrIAn (Protein Input and Annotation) data input has greatly increased database entries of PIR-International. SWISS-PROT (Bairoch and Apweiler, 2000) is a protein sequence database that, from its inception in 1986, was produced collaboratively by the Department of Medical Biochemistry at the University of Geneva and the EMBL. The database is now maintained collaboratively by Swiss Institute of Bioinformatics (SIB) and EBI/ EMBL. SWISS-PROT provides high-level annotations, including descriptions of the function of the protein and of the structure of its domains, its post-translational modifications, its variants, and so on. The database can be accessed from http://expasy.hcuge.ch/sprot/sprot-top.html or numerous mirror sites. In 1966, Translated EMBL (TrEMBL) was created as a computer-annotated supplement to SWISS-PROT. Each SWISS-PROT entry consists of general information about the entry (*e.g.,* entry name and date, accession number), Name and origin of the protein (*e.g.,* protein name, EC number and biological origin).

References, Comments (*e.g.,* catalytic activity, cofactor, subuit structure, subcellular location and family class, etc.), Cross-reference (EMBL, PIR, PDB, Pfam, ProSite, ProDom, ProtoMap, etc.), Keywords, Features (*e.g.,* active site, binding site, modification, secondary structures, etc.), and Sequence information (amino acid sequence in Swiss-Prot format. To facilitate sequence efficiency, a variety of different primary sources are amalgamated to composite databases such as NRDB (PDB, Swiss-Prot, PIR, and GenPept), OWL (GenBank, PIR, Swiss-Prot, and NRL-3D), and MIPSX (PIR-MIPS, Swiss-Prot, NRL-3D, Kabat, and PseqIP). NRDB is built at the NCBI.

The database is comprehensive and up-to-date; however, it is not nonredundant but rather nonidentical. OWL (http://www.bioinf.man.ac.uk/dbbrowser/OWL/) is a non-redundant protein sequence database (Bleasby et al., 1994). The database is a composite of four major primary sources that are assigned a priority with regard to their level of annotations and sequence validation with SWISS-PROT being the highest priority. MIPSX is a merged database produced at MIPS.

Secondary Database

There are many secondary (or pattern) databases that contain the biologically significant information of sequence analyses from the primary sources. Because there are several different primary databases and a variety of ways to analyze protein sequences, the type of information stored in each of the secondary databases is different. However, the creation of most secondary databases is based on a common principle that multiple alignments may identify homologous sequences within which conserved regions or motifs may reflect some vital information crucial to the structure or function of these homologous proteins.

Motifs have been exploited in various ways to build diagnostic patterns for particular protein families. If the structure and function of the family are known, searches of pattern databases may offer a fast track to the inference of biological function of the unknown. Notwithstanding, the pattern databases should be used to augment primary database searches. The secondary database, PROSITE, which uses SWISS-PROT, is maintained collaboratively at the Swiss Institute of Bioinformatics (http://expasy.hcuge.ch/sprot/prosite.html) and Swiss

Institute for Experimental Cancer Research (http//www.isrec.isb-sib.ch/). It is rationalized that protein families could be simply and effectively characterized by the single most conserved motif observable in a multiple alignment of known homologous proteins. Such motifs usually associate with key biological functions such as enzyme active sites, ligand binding sites, structural signatures, and so on. Entries are deposited in PROSITE in two files: the data file and the documentation file.

In the data file, each entry contains an identifier (ID) and an accession number (AC). A title or description of the family is contained in the DE line, and the pattern itself resides on PA lines. The following NR lines provide technical details about the derivation and diagnostic performance of the pattern. Large numbers of false-positives and false-negatives indicate poorly performing patterns. The comment (CC) lines furnish information on the taxonomic range of the family, the maximum number of observed repeat, functional site annotations, and so on. The following DR lines list the accession number and SWISS-PROT identification codes of all the true matches to the pattern (denoted by T) and any possible matches (denoted by P). The last DO line points to the associated documentation file.

The documentation file begins with the accession number and cross-reference identifier of its data file. This is followed by a free format description of the family including details of the pattern, the biological role of selected motif(s) if known, and appropriate bibliographic references. Most protein families are characterized by several conserved motifs. The PRINTS fingerprint database was developed to use multiple conserved motifs to build diagnostic signatures of family membership. If a query sequence fails to match all the motifs in a given fingerprint, the pattern of matches formed by the remaining motifs allows the user to make a reasonable diagnosis.

The PRINTS can be accessed by keyword and sequence searches at http://www.bioinf.man.ac.uk/dbbrowser/PRINTS/. There are three sections in a PRINTS entry. In the first section, each fingerprint is given an identifying code, an accession number, number of motifs in the fingerprint, a list of cross-linked databases and bibliographic information. In the second section, the diagnostic performance is given in the summary and the table of results. The last section lists seed motifs and final motifs that result from the iterative database scanning. Each motif is identified by an ID code followed by the motif length.

The aligned motifs themselves are then provided, together with the corresponding source database ID code, the location in the parent sequence of each fragment (ST), and the interval between the fragment and its preceding neighbour (INT). In the BLOCKS database, the motifs or blocks are created by automatically detecting the most highly conserved regions of each protein family. Initially three conserved amino acids are identified. The resulting blocks are then calibrated against SWISS-PROT to obtain a measure of the likelihood of a chance match. The database is accessible by keyword and sequence searches via the BLOCKS Web server at the Fred Hitchinson Cancer Research Center.

The Block includes the SWISS-PROT IDs of the constituent sequences, the start position of the fragment, the sequence of the fragment, and a score or weight indicating a measure of the closeness of the relationship of that sequence to others in the block (100 being the most distant). Another resource derived from BLOCKS and PRINTS is IDENTIFY, which, instead of encoding the exact information at each position in an alignment, tolerates alternative residues according to a set of prescribed groupings of related biochemical properties. IDENTIFY and its search software, eMOTIF (Huang and Brutlag, 2001), are accessible for use from the protein function Web server of the Biochemistry Department at Stanford University, http://dna.Stanford.edu/identify/. PANAL (http://mgd.ahc.umn.edu/panal/rurL.panal.html) of Computational Biology Centers at the University of Minnesota is a combined resource for ProSite, BLOCKS, PRINTS, and Pfam.

Sequence Alignment and Similarity Search

Pairwise comparison of two sequences is a fundamental process in sequence analysis. It defines the concepts of sequence identity, similarity, and homology as applied to two proteins, DNA or RNA sequences. Database interrogation can take the form of text queries or sequence similarity searches. Typically, the user employs a query sequence to conduct sequence similarity search so that the relationships between the query sequence (probe) and another sequence (target) can be quantified and their similarity assessed. A sequence consists of letters selected from an alphabet. The complexity of the alphabet is defined as the number of different letters it contains.

The complexity is 4 for nucleic acids and 20 for proteins. Sometimes additional characters are used to indicate ambiguities in the identity of a particular base or residue—for example, B for aspartate or asparagine and Z for glutamate or glutamine (in nucleic acids: R for A or G, and Y for C, T, or U).

Similarity Scoring

A comprehensive alignment must account fully for the positions of all residues in the sequences under comparison. To achieve this, gaps may be inserted in a manner as to maximize the number of identical matches. However the result of such process is biologically meaningless. Thus scoring penalties are introduced to minimize the number of gaps that are opened and extension penalties are also incurred when a gap has to be extended. The total alignment score is a function of the identity between aligned residues and the incurred gap penalties. By doing this, only residue identities are considered, resulting in a unitary matrix that weights identical residue matches with a score of 1 and other elements being 0.

In order to improve diagnostic performance, the scoring potential of the biologically significant signals is enhanced so that they can contribute to the matching process without amplifying noise. A need to balance between high-scoring matches that have only mathematical significance and lower-scoring matches that are biologically meaningful is the essential requisite of sequence analysis. Scoring matrices that weight matches between nonidentical residues based on evolutionary substitution rates have been devised to address this need. Such tools may increase the sensitivity of the alignment, especially in a situation where sequence identity is low. One of the most popular scoring matrices is Dayhoff mutation data (MD) matrix (Dayhoff et al., 1978).

Table 10.2. PAM Scale for Amino Acid Residues

Observed % Difference	*Evolutionary Distance in PAM*
1	1
10	11
20	23
25	30
30	38
40	56
50	80
60	112
70	159
75	195
80	246
85	328

The MD score is based on the concept of the *point accepted mutation* (PAM). Thus, one PAM is a unit of evolutionary divergence in which 1 % of the amino acids have been changed. If these changes were purely random, the frequencies of each possible substitution would be determined simply by the overall frequencies of the different amino acid (background frequencies). However, in related proteins, the observed substitution frequencies (target frequencies) are biased toward those that do not seriously disrupt the protein's function. In other words, these are mutations that have been accepted during evolution. Thus substitution scores in the mutation probability matrix are proportional to the natural log of the ratio of target frequencies to background frequencies.

Mutation probability matrices corresponding to longer evolutionary distance can be derived by multiplication of this matrix of probability values by itself the appropriate number of times (*n* times for a distance of *n* PAM). Table 1.2 shows the values connecting the overall proportion of observed mismatches in amino acid residues with the corresponding evolutionary distance in PAM (PAM scale). The 250 PAM (250 accepted point mutations per 100 residues) matrix gives similarity scores equivalent to 80% difference between two protein sequences. Because the sequence analysis is aimed at identifying relationships in the twilight zone (20% identity), the MD for 250 PAM has become the default matrix in many analysis packages.

The appropriate similarity score for a particular pairwise com-parison at any given evolutionary distance is the logarithm of the odds that a particular pair of amino acid residues has arisen by mutation. These odds can be derived via dividing the values in the mutation probability matrix by the overall frequencies of the amino acid residues. In the matrix, residues likely undergoing mutations, neutral (random mutations), and unlikely undergoing mutations have values of greater than 0, equal to 0, and less than 0, respectively. There are two approaches of sequence alignments: A global alignment compares similarity across the full stretch of sequences, while a local alignment searches for regions of similarity in parts of the sequences.

Global Alignment

In this approach (Needleman and Wunsch, 1970), a 2D matrix is constructed by comparing two sequences that are placed along the x-and the y-axes, respectively (Table 10.3). Cells representing identities are scored 1, and those with mismatches are scored 0 to populate the 2D array with 0's and 1's. An operation of successive summation of cells begins at the last cell (S, T) in the matrix progressing to the next row or column by adding the maximum number from two constituent subpaths to the cell.

For the cell (R, R) in the example, the original score is 1. The maximum value of the two subpaths is 5, thus the new value for the cell (R, R) becomes 5 + 1 = 6. This process is repeated for every cell in the matrix. An alignment is generated by working through the completed matrix, starting at the highest-scoring element at the N-terminal and following the path of high scores through to the C-terminal. For example:

Local Alignment

Two sequences that are distantly related to each other may exhibit small regions of local similarity, though no satisfactory overall alignment can be found. The Smith and Waterman algorithm is designed to find these common regions of similarity and has been used as the basis for many subsequent algorithms. The FASTA and BLAST programs are local similarity search methods that concentrate on finding short identical matches, which may contribute to a total match. FASTA uses *ktups* (1 or 2 amino acids for proteins and 4 or 6 nucleotides for nucleic acids) while BLAST uses words (3 amino acids for proteins and 11 nucleotides for nucleic acids) of short sequences in the database searching. In a typical output from a FASTA

Table 10.3. Unitary Scoring Matrix

	C	*H*	*E*	*M*	*I*	*C*	*A*	*L*	*R*	*E*	*A*	*G*	*E*	*N*	*T*
C	1	0	0	0	0	1	0	0	0	0	0	0	0	0	0
H	0	1	0	0	0	0	0	0	0	0	0	0	0	0	0
E	0	0	1	0	0	0	0	0	0	1	0	0	1	0	0
M	0	0	0	1	0	0	0	0	0	0	0	0	0	0	0
I	0	0	0	0	1	0	0	0	0	0	0	0	0	0	0
S	0	0	0	0	0	0	0	0	0	0	0	0	0	0	0
T	0	0	0	0	0	0	0	0	0	0	0	0	0	0	1
R	0	0	0	0	0	0	0	0	1	0	0	0	0	0	0
Y	0	0	0	0	0	0	0	0	0	0	0	0	0	0	0
A	0	0	0	0	0	0	1	0	0	0	1	0	0	0	0
G	0	0	0	0	0	0	0	0	0	0	0	1	0	0	0
E	0	0	1	0	0	0	0	0	0	1	0	0	1	0	0
N	0	0	0	0	0	0	0	0	0	0	0	0	0	1	0
T	0	0	0	0	0	0	0	0	0	0	0	0	0	0	1
S	0	0	0	0	0	0	0	0	0	0	0	0	0	0	0

Table 10.4. Completed Maximum-Match Matrix

	C	*H*	*E*	*M*	*I*	*C*	*A*	*L*	*R*	*E*	*A*	*G*	*E*	*N*	*T*
C	11	9	8	7	6	7	6	6	6	5	4	3	2	1	0
H	9	10	8	7	6	6	6	6	6	5	4	3	2	1	0
E	8	8	9	7	6	6	6	6	5	6	4	3	2	1	0
M	7	7	7	8	6	6	6	6	5	5	4	3	2	1	0
I	6	6	6	6	7	6	6	6	5	5	4	3	2	1	0
S	6	6	6	6	6	6	6	6	5	5	4	3	2	1	0
T	6	6	6	6	6	5	6	6	5	5	4	3	2	1	1
R	5	5	5	5	5	5	5	5	6	5	4	3	2	1	0
Y	5	5	5	5	5	5	5	5	5	5	4	3	2	1	0
A	4	4	4	4	4	4	5	4	4	4	5	3	2	1	0
G	3	3	3	3	3	3	3	3	3	3	3	4	2	1	0
E	2		3	2	2	2	2	2	2	3	2	2	3	1	0
N	1	1	1	1	1	1	1	1	1	1	1	1	1	2	0
T	0	0	0	9	0	0	0	0	0	0	0	0	0	0	1
S	0	0	0	0	0	0	0	0	0	0	0	0	0	0	0

```
CHEMICALREAGEMNT–
|||||   | |||||
CHEMIST-RYAGENTS
```

search, the name and version of the program are given at the top of the file with the appropriate citation. The query sequence together with the name and version of the search database are indicated. A range of parameters used by the algorithm and the run-time of the program are printed. Following these statistics come the results of the database search with a list of a user-defined number of matches with the query sequence and information for the source database (identifier, ID code, accession number, and title of the matched sequences). The length in amino acid residues of each of the retrieved matches is indicated in brackets.

Various initial and optimized score and *E* (expected) value are given. After the search summary, the output presents the complete pairwise alignments of a user-defined number of hits with the query sequence. Within the alignments, identities are indicated by the character while similarity is indicated by the character. The BLAST (Basic Local Alignment Search Tool) searehes two sequences for subsequences of the same length that form an ungapped alignment. The algorithm searches and calculates all segment pairs between the query and the database sequences above a scoring threshold. The resulting high-scoring pairs (HSPs) form the basis of the ungapped alignments that characterize BLAST output. In the gapped BLAST, the algorithm seeks only one ungapped alignment that makes up a significant match.

Dynamic programming is used to extend a central pair of aligned residues in both directions to yield the final gapped alignment to speed up the initial database search. In the typical BLAST output, the name and version of the program together with the appropriate citation are given at the top of the file. The query sequence and the name of the search database are then indicated. Next come the results of the database search with a list of a user-defined number of matches with the query sequence and information for the source database (identifier, ID code, accession number, and title of the matched sequences).

The highest score of the set of'matching segment pairs for the given sequence with the number of HSPs denoted by the parameter N is given. This is followed by a p (probability) value. After the search summary, the output presents the pairwise alignments of the HSPs for each of the user-defined number of hits with the query sequence. For each aligned HSP, the beginning and end locations within the sequence are marked and identities between them are indicated by the corresponding amino acid symbols.

Multiple Sequence Alignment. The goal of multiple sequence alignment is to generate a concise summary of sequence data in order to make an informed decision on the relatedness of sequences to a gene family. Alignments should be regarded as models that can be used to test hypothesis. There are two main approaches on the construction of alignments. The first method is guided by the comparison of similar strings of amino acid residues, taking into account their physicochemical properties, mutability data, and so on. The second method results from comparison at the level of secondary or tertiary structure where alignment positions are determined solely on the basis of structural equivalence.

Alignments generated from these rather different approaches often show significant disparities. Each is a model that reflects a different biochemical view of sequence data. The multiple sequence analysis draws together related sequences from various species and expresses the degree of similarity in a relatively concise format. The ClustalW algorithm (Higgins et al., 1996) is one of the most widely used multiple sequence alignment program (where W stands for weighting). This program takes an input set of sequences (the simplest format is fasta format) and calculates a series of pairwise alignments, comparing each sequence to every other sequence, one at a time. Based on these comparisons, a distance matrix is calculated, reflecting the relatedness of each pair of sequences. The distance matrix, in turn, serves in the calculation of a phylogenetic tree that can be weighted to favor closely related alignment of the most closely related sequences, then realigning with the addition of the next sequence, and so on. The Clustal X is the portable windows interface Clustal program for multiple sequence alignment.

PROTEOMIC ANALYSIS USING INTERNET RESOURCES: SEQUENCE AND ALIGNMENT

Sequence Databases

The Protein Information Resources (PIR) of NBRF in collaboration with MIPS and JIPID produces the annotated protein sequence database in the PIR-MIPS International Protein Sequence Database (PSD). The PSD is a comprehensive annotated and nonredundant protein sequence database. Its annotation includes concurrent cross-references to other sequence, structure, genomic and citation databases, as well as functional descriptions and structural features. The PIR-International database is accessible at the PIR site and the MIPS site. To open the search protein sequence database page at PIR, select Text/Entry Search under Search Databases.

Enter the search keyword string by using Boolean expression or by using one of the five options of textboxes marked Search string and then click the Submit button. After an initial return of hits, you may modify the number of entries on the list, and view of the output. Pick the desired entry and the annotated sequence to receive the return in the default PIR format. An output in the fasta format can be requested. Links to the structure (PDB), alignment, and protein classification (at PIR and MIPS) are provided on the output page.

The PIR site also provides facilities for sequence similarity search (BLAST or FASTA), alignment (ClustalW), and analysis (ProClass, ProtFam, RESID, SSEARCH) of proteins. To perform the similarity search, select Blast or Fasta to open the search form.

Paste the query sequence and click the Submit button. The search returns a list of hits (PIR Search results with option (*s*) for modifying the entries (modify query by choosing refine, add to or remove from the list). The Modify option(*s*) enable the user to design the list of desired entries.

DB	ID	Title	Score	c-value	#aa	%idn	Ov.lap
PIR	A42343	alcohol dehydrogenasc – Baltic cod	703	0.0	375	100	375
PIR	JC4967	alcohol dehydrogenasc class III - -gil...	450	c-126	376	62	375
PIR	S68061	alcohol dehydrogenasc class III – Ind...	447	c-125	373	64	370
PIR	A33419	alchol dehydrogenasc class III – horsc	440	c-123	373	62	371
PIR	DEHUC2	alcohol dehydrogenasc 5 [validatcd]	438	c-122	374	63	372
PIR	DERTA	alcohol dehydrogenasc – rat	437	c-122	373	63	372
PIR	A56643	alcohol dehydrogenasc 2 – mouse	431	c-120	374	62	372
PIR	S51187	alcohol dehydrogenasc class III – All...	430	c-120	376	61	373
PIR	A49662	alchol dehydrogenasc class III – com...	412	c-114	378	58	373
PIR	T04164	formaldehyde dehydrogenasc (glutathione)	401	c-111	381	56	378
PIR	S51357	alcohol dehydrogenasc Fdh – fruit fly...	400	c-111	379	58	376
PIR	S71244	alcohol dehydrogenasc class III – Ara...	394	c-109	379	56	375
PIR	DEMSAA	alcohol dehydrogenasc A – mouse	391	c-108	375	57	375
PIR	A35837	alcohol dehydrogenase – Japanesc – Japanese quail	390	c-108	375	57	376
PIR	DECHAL	alcohol dehydrogenasc – chicken	387	c-107	375	56	376
PIR	A26468	alcohol dehydrogenasc I – rat	386	c-107	376	56	376
PIR	A49107	alcohol dehydrogenase I – deer mouse	385	c-5106	375	55	381
PIR	DEHOAL	alcohol dehydrogenasc E – horse	385	c-106	375	55	375
PIR	S62640	alcohol dehydrogenasc – Indian cobra	385	c-106	375	56	374
PIR	T03289	formaldehyde dehydrogenasc (glutathione)	384	c-106	381	55	377

DEHOAL alcohol dehydrogenase (EC 1.1.1.1) E – horse

Length = 375

Score = 385 bits (1036), Expect = e-106

Identities = 196/374 (52%), Positives = 260/374 (69%), Gaps = 2/374 (0%)

```
Query 1  ATVGKVIKCKAAVAWENKPLVIEEIEVDVPHANEIRIKIATGVHTDLYHILFEGKHKD 60
          +T GKVIKVKCAV WE KP IEE+EV P A+E+RIK++ATG+C +D H+ G
```

```
Sbjct 2     STAGKVIKCKAAVLWEEKKPFSIEEVEVAPPKAHEVRIKMVATGICRSD-DHVVSGTLVT 60
Query 61     GFPVVLGHEGAGIVESVGPGVTEFQPGEKVIPLFISQCGECRFCQSPKTNQCVKGWANES 120
            PV+ GHE AGIVES+G GVT +PG+KVIPLF QCG+CR C+ P+ N C+K +
Sbjct 61    PLPVIAGHEAAGIVESIGEGVTTVRPGDKVIPLFTPQCGKCRVCKHPEGNFCLKNDLSMP 120
Query 121   PDVMSPKETRFTCKGRKVLQFLGTSTFSQYTVVNQTAVAKIDPSAPLDTVCLLGCGVSTG 180
            M +RFTC+G+ FLGTSTFSQYTVV++I+VAKID ++PL+ VCL+GCG STG
Sbjct 121   RGTMQDGTSRFTCRGKPIHHFLGTSTFSQYTVVDEISVAKIDAASPLEKVCLIGCGFSTG 180
Query181    FGAAVNTAKVEPGSTCxxxxxxxxxxxCHSAGAKRIIAVDLNPDKFEKAKVFGATD 240
            +G+AV AKV GSTC     C +AGA RII VD+N DKF KAK GAT+
Sbjct 181   YGSAVKVAKVTQGSTCAVFGLGGVGLSVIMGCKAAGAARIIGVDINKDKFAKAKEVGATE 240
Query 241   FVNPNDHSEPISQVLSKMTNGGVDFSLECVGNVGVMRNALESCLKGWGVSVLVGW – TDLH
            299 VNP D+ +PI IVL++M+NGGVDFS E +G + M AL C + + GVSV + VG D
Sbjct 241   CVNPQDYKKPIQEVLTEMSNGGVDFSFEVIGRLDTMVTALSCCQEAYGVSVIVGVPPDSQ 300
Query 300   DVATRPIQLIAGRTWKGSMFGGFKGKDGVPKMVKAYLDKKVKLDEFITHRMPLESVNDAI 359
            +++ P+ L++GRTWKG++FGGFK KD VPK+V ++ KK LK ITH +P E + N+
Sbjct 301   NLSMNPMLLLSGRTWKGAIFGGFKSKDSVPKLVADFMAKKFALDPLITHVLPFEKINEGF 360
Query 360   DLMKHGKCIRTVLS 373
            DL++ G+ IRT+L+
Sbjct 361   DLLRSGESIRTILT 374
```

Fig. 10.1. PIR BLAST Search results for Baltic cod alcohol dehydrogenase. A list of hits (only top 20 hits are shown; the EC numbers are deleted from the entries) is followed by pairwise alignments of similar sequences (only the sequence alignment between the query protein and horse liver E isozyme is shown).

SWISS-PROT is a curated protein sequence database maintained by the Swiss Institute of Bioinfornmatics and is a collaborative partner of EMBL. The database consists of SWISS-PROT and TrEMBL, which consists of entries in SWISS-PROT-like format derived from the translation of all CDS in the EMBL Nucleotide Sequence Database. SWISS-PROT consists of core sequence data with minimal redundancy, citation and extensive annotations including protein function, post-translational modifications, domain sites, protein structural information, diseases associated with protein deficiencies and variants. SWISS-PROT and TrEMBL are available at EBI site, http://www.ebi.ac.uk/swissprot/, and ExPASy site, http://www.expasy.ch/sprot/. From the SWISS-PROT and TrEMBL page of ExPASy site, click "Full text search" (under Access to SWISS-PROT and TrEMBL) to open the search page (Fig. 10.3). Enter the keyword string (use Boolean expression if required), check SWISS-PROT box, and click the Submit button.

Select the desired entry from the returned list to view the annotated sequence data in Swiss-Prot format. An output in the fasta format can be requested. Links to BLAST, feature table, some ExPASy proteomic tools (*e.g.,* Compute pI/Mw, ProtParam, ProfileScan, ProtScale, PeptideMass, ScanProsite), and structure (SWISS-MODEL) are provided on the page. The amino acid sequences can be searched and retrieved from the integrated retrieval sites such as Entrez (Schuler et al., 1996), SRS of EBI (http://srs.ebi.ac.uk/),and DDBJ (http://srs.ddbj.nig.ac.jp/index-e.html). From the Entrez home page (http://www.ncbi.nlm.nih.gov/Entrez), select Protein to open the protein search page. Follow the same procedure described for the Nucleotide sequence to retrieve amino acid sequences of proteins in two formats; GenPept and fasta.

The GenPept format is similar to the GenBank format with annotated information, reference(s), and features. The amino acid sequences of the EBI are derived from the SWISS-PROT database. The retrieval system of the DDBJ consists of PIR, SWISS-PROT, and DAD, which returns sequences in the GenPept format.

Sequence Information Tools

The comprehensive molecular biology server, ExPASy (Expert Protein Analysis System), provides Proteomic tools (http://www.expasy.ch/tools) that can be accessed directly or from the ExPASy home page by selecting Identification and Characterization under Tools and Software Packages. The Protein identification and characterization tools of the Proteomics tools (Fig. 10.4) provide facilities to:

- Identify a protein by its amino acid composition (AACompIdent)
- Compare the amino acid composition of an entry with the database (AACompSim)
- Predict potential post-translational modifications (FindMod)
- Calculate the mass of peptides for peptide mapping (PeptideMass)
- Evaluate physicochemical parameters from a query sequence (ProtParam)
- Scan for PROSITE (ScanProsite).
- Calculate/plot amino acid scales (ProtScale)

The identification of a protein from its amino acid composition and physicochemical properties (*e.g.,* pI and molecular weight) is a valuable step prior to initiating sequence determination. On the AACompIdent page, select a Constellation to be used (e.g., Constellation 0 for all amino acids) to open the request form. Enter pI value, molecular weight, biological source of the protein (OS or OC), the ExPASy recognizable keyword (*e.g.,* dehydrogenase, kinase), your e-mail address, and amino acid composition (in molar %). Click the Run AACompIdent button. PROPSEARCH (http://www.embl-heidelberg.de/prs.html) is designed to find the putative protein family if alignment methods fail.

The program uses the amino acid composition of a protein taken from the query sequence to discern members of the same protein family by weighing 144 different physical properties including molecular weight, the content of bulky residues, average hydrophobicity, average charge, and so on, in performing the analysis. This collection of physical properties is called the query vector, and it is compared against the same type of vector precomputed for every sequence in the target databases (SWISS-PROT and PIR). The search results are returned by e-mail.

Table 10.5. Search Result of AACompldent for Lipoamide Dehydrogenase from Schizosaccharomyces pombe

The closest SWISS-PROT entries (in terms of AA composition) for the keyword FLAVOPROTEIN and the species SCHIZOSACCHAROMYCES POMBE:

Rank	Score	Protein	(pI	Mw)	Description
1	2	DLDH_SCHPO	8.29	50793	DIHYDROLIPOAMIDE DEHYDROGENASE.
2	16	GSHR_SCHPO	6.90	49999	GLUTATHIONE REDUCTASE (EC 1.6.4.2) (GR)
3	25	YDGE_SCHPO	6.20	62103	PUTATIVE FLAVOPROTEIN C26F1.14C.
4	31	DCP1_SCHPO	6.11	63294	PUTATIVE PYRUVATE DECARBOXYLASE C13A11
5	34	ODPA_SCHPO	6.72	41421	PYRUVATE DEHYDROGENASE E1 COMPONENT
6	40	ILVB_SCHOP	7.85	57643	ACETOLACTATE SYNTHASE.
7	43	PYRD_SCHPO	9.44	44526	DIHYDROOROTATE DEHYDROGENASE.
8	46	DCP2_SCHPO	5.71	64787	PROBABLE PYRUVATE DECARBOXYLASE C1F8..07C
9	47	ETFD_SCHPO	8.78	65780	PROBABLE ELECTRON TRANSFER FLAVOPROTEIN
10	53	ETFA_SCHPO	5.48	32424	PROBABLE ELECTRON TRANSFER FLAVOPROTEIN
11	54	TRXB_SCHPO	5.19	34618	THIOREDOXIN REDUCTASE (EC 1.6.4.5).
12	56	MTHS_SCHPO	5.91	42140	METHYLENETETRAHYDROFOLATE REDUCTASE 2
13	63	ODPB_SCHPO	5.29	35852	PYRUVATE DEHYDROGENASE E1 COMPONENT BETA
14	67	OYEA_SCHPO	8.77	43813	PUTATIVE NADPH DEHYDROGENASE C5H10.04
15	67	GPDM_SCHPO	6.24	68229	GLYCEROL-3-PHOSPHATE DEHYDROGENASE
16	77	OYEB_SCHPO	5.83	44983	PUTATIVE NADPH DEHYDROGENASE C5H10.10
17	79	MTHR_SCHPO	5.17	69012	PROBABLE METHYLENETETRAHYDROFOLATE
18	80	NCPR_SCHPO	5.41	76775	NADPH-CYTOCHROME P450 REDUCTASE (EC 1.6
19	91	MT10_SCHPO	5.69	111353	PROBABLE SULFITE REDUCTASE [NADPH]
20	92	COQ6_SCHPO	9.29	52160	PUTATIVE UBIQUINONE BIOSYNTHESIS

An amino acid scale refers to a numerical value assigned to each amino acid such as polarity, hydrophobicity, accessible/buried area of the residue, and propensity to form secondary structures (e.g., amino acid parameter of AAIndex). For example, selecting (clicking) ProtScale from the list of ExPASy Proteomics tools opens the query page with a list of predefined amino acid scales. Paste the query sequence into the sequence box, select the desired scale and window size, set your choice (yes/no) for normalizing the scale from 0 to 1, and click the Submit button.

The query returns with a list of scales used for amino acids and the profile plot (score versus residue position as shown in Fig. 10.5), which can be saved as image file (protscal.gif or protscal.ps). In addition, numerous links are made to various servers for similarity searches, pattern and profile searches, post-translational modification predictions, structural analyses, and sequence alignments.

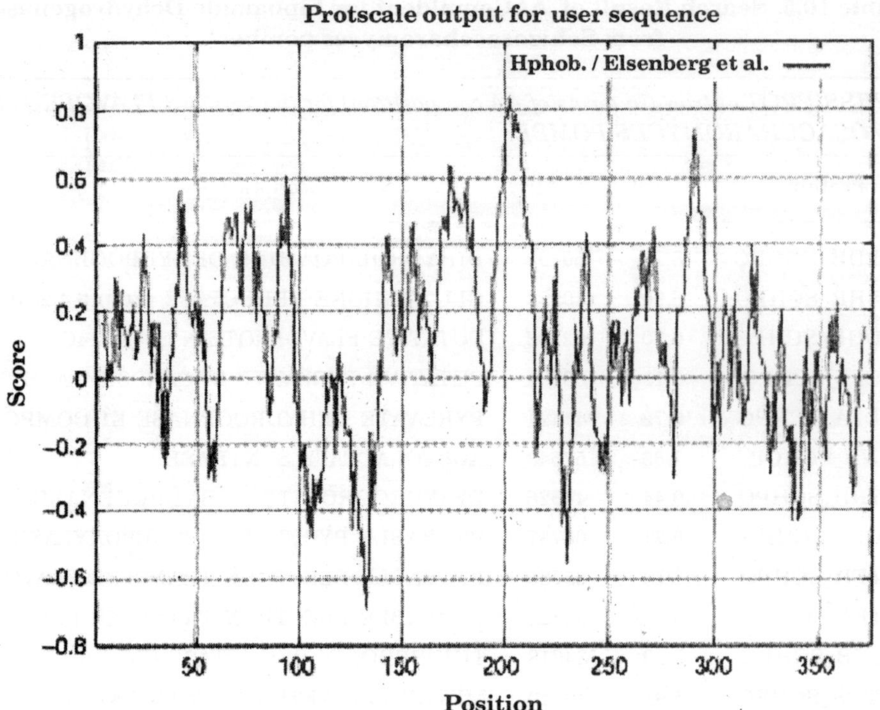

Fig. 10.2. ProtScale (ExPASy Proteomic Tools) output of hydrophobicity parameter. The hydrophobicity profile of cod alcohol dehydrogenase is generated with ProtScale using Eisenberg hydrophobic scale. The individual values for the 20 amino acids used in the profile are also listed in the report.

Sequence Alignment

One of the most common analyses of protein sequences is the similarity/homology search. It allows a mapping of information from known sequences to novel ones, especially the functional sites of the homologous proteins. Dynamic programming that is a general. optimization technique has been applied to sequence alignment and related computational problems. A dynamic programming algorithm finds the best solution by breaking the original problem into smaller sequentially dependent subproblems that are solved first. Each intermediate solution is stored in a table along with a score and the sequence of solutions that yields the highest score is chosen. The search for similarity/homology is well supported by the Internet accessible tools.

A query sequence may be entered to conduct homology_search using the BLAST server at ExPASy or NCBI. Launch Basic BLAST at NCBI (http:// www.ncbi.nlm.hih.gov/BLAST/) as previously shown in Fig. 10.1. Paste or upload the query sequence (fasta format is preceded by >seqname). Choose the program (blastn for nucleic acids and blastp for proteins) and database (nr for non-redun-dant). Set to either Perform ungapped alignment or Perform CDD Search (Protein only). You may choose the formatting options now or later. Enter the address to receive reply by e-mail.

After clicking the Search button, you receive a confirmation message, "Your request has been successfully submitted and put into the Blast Queue. Query = seqname" with the request ID which is used to retrieve the search results. The output includes alignment scores plot, list of sequences producing significant alignments (accession ID, name, score, and *E* value), and pairwise alignments (if this Alignment view is chosen). Similarity search can be performed with the FASTA program, though BLAST is easier to use. Both FASTA and BLAST homology searches are also available at EBI, PIR and DDBJ. The small *P* values indicate homologous sequences. The multiple sequence alignment of homologous proteins can be accomplished with ClustalW, which is one of the most commonly used programs.

A number of ClustalW servers are available. These include EBI of EMBL (http://www2.ebi.ac.uk/ clustalw/), PIR (http://pir.georgetown.edu/), IBC of Washington University in St. Louis (http://www.ibc.wustl.edu/msa/clustal.html), and PredictProtein of Columbia University (http://cubic.bioc.columbia.edu/predictprotein/). BCM server (http://dot.imgen.bcm.tme.edu.9331/multi-align/multi-align.html) and DDBJ (http://www. ddbj.nig.ac.jp/). The returned alignment via e-mail can be saved as alignfile.ps (*e.g.*, Fig. 10.6 for lysozyme C from EBI). To perform multiple sequence alignment online (at DDBJ), select (click) Search and Analysis to access DDBJ Homology Search System (http://spiral.genes.nig.ac.jp/homology/tope.html). Select ClustalW to open the query form (Fig. 10.7). Select the desired options, upload the file containing multiple sequences, pick WWW, and click Send. The returned output includes pairwise alignment score, multiple sequence alignment and tree file.

WORKSHOPS

1. A purified mammalian protein with a molecular weight of 40,000 + 1000 and a pI of 8.0 (±0.5 is composed (an amino acid composition in mol %) of Ala, 7.5; Arg, 3.2; Asn, 2.1; Asp, 4.5; Cys, 3.7; Gin, 2.1; Glu, 5.6; Gly, 10.2; His, 1.9; Il, 6.4; Leu, 6.7; Lys, 8.0; Met, 2.4; Phe, 4.8; Pro, 5.3; Ser, 7:0; Thr, 6.4; Trp, 0.5; Tyr, 1.1; and Val, 10.4. Submit these data to the Internet server that searches its database to identify the protein.
2. Analyze peptide fragments (PeptideMass of ExPASy) produced by treating a protein with the following amino acid sequence with chymotrypsin, proteinase K, and trypsin,. Deduce the specificities of these enzymes.
   >gi | 35497 | emb | CAA78820.1 | protein kinase C gamma [Homo sapiens]
   QLEIRAPTADEIHVTVGEARNLIPMDPNGLSDPYVKLKLIPDPRNLTKQKTRTVKATLNPVWNETFVFNL
   KPGDVERRLSVEVWDWDRTSRNDFMGAMSFGVSELLKAPVDGWYKLLNQEEGEYYNVPVADADNCSLLQK
   FEACNYPLELYERVRMGPSSSPIPSPSPSPTDPKRCFFGASPGRLHISDFSFLMVLGKGSFGKVMLAERR
   GSDELYAIKILKKDVIVQDDDVDCTLVEKRVLALGGRGPGGRPHFLTQLHSTFQTPDRLYEVMEYVTGGD
   LMYHIQQLGKFKEPHAAFYAAEIAIGLFFLHNQGIIYRDLKLDNVMLDAEGHIKITDFGMCKENVFPGTT
   TRTFCGTPDYIAPEIIAYQPYGKSVDWWSFVLL YE M LAGQPPFDGEDEELFQAIMEQTVTYPKSLSRE
   AVAICKGFLTKHPGKRLGSGPDGEPTIRAHGFFRWIDWERLERLEIPPPFRPRPPCGRSGENFDKFFTRAA
   PALTPPDRLVLASIIDQADFQGFTYVNPDFVHPDARSPTSPVPVPVM
3. Given the following amino acid sequence, estimate (ProtParam of ExPASy) its amino acid composition, numbers of charged residues, extinction coefficient, estimated half-life, and instability index of the protein. Elaborate briefly how extinction coefficient, half-life and instability index are estimated.
   >gi | 68532 | pir | | SYBYDC aspartate—tRNA ligase (EC 6.1.1.12), cytosolic – yeast
   MSQDENIVKAVEESAEPAQVILGEDGKPL'SKKALKKLQKEQEKQRKKEERALQLEAEREAREKKAAAEDT
   AKDNYGKLPLIQSRDSDRTGQKRVKFVDLDEAKDSDKEVLFRARVHNTRQQGATLAFLTLRQQASLIQGL

VKANKEGTISKNMVKWAGSLNLESIVLVRGIVKKVDEPIKSATVQNLEIHITKIYTISETPEALPILLED
ASRSEAEAEAAGLPWNLDTRLDYRVIDLRTVTNQAIFRIQAGVCELFREY.LATKKFTEVHTPKLLGAPS
EGGSSVFEVTYFKGKAYLAQSPQFNKQQLIVADFERVYEIGPVFRAENSNTHRHMTEFTGLDMEMAFEEH
YHEVLDTLSELFVFIFSELPKRFAHEIELVRKQYPVEEFKLPKDGKMVRLTYKEGIEMLRAAGKEIGDFE
DLSTENEKFLGKLVRDKYDTDFYILDKFPLEIRPFYTMPDPANPKYSNSYDFFMRGEEILSGAQRIHDHA
LLQERMKAHGLSPEDPGLKDYCDGFSYGCPPHAGGGIGLERVVMFYLDLKNIRRASLFPRDPKRLRP

4. ProtScale tool of ExPASy computes amino acid scale (physicochemical properties/ parameters) and presents the result in a profile plot. Perform ProtScale computations to compare the hydrophobicity/polarity profiles with %buried residues/%accessible residues profiles for human serine protease with the following amino acid sequence.

>gi|2318115|gb|AAB66483.1| serine protease [Homo sapiens]

MKKLMVVLSLIAAAWAEEQNKLVHGGPCDKTSHPYQAALYTSGHLLCGGVLIHPLWVLTAAHCKKPNLQV
FLGKHNLRQRESSQEQSSVVRAVIHPDYDAASHDQDIMLLRLARPAKLSELIQPLPLERDCSANTTSCHI
LGWGKTADGDFPDTIQCAYIHLVSREECEHAYPGQITQNMLCAGDEKYGKDSCQGDSGGPLVCGDHLRGL
VSWGNIPCGSKEKPGVYTNVCRYTNWIQKTIQAK

5. Scan the relative mutability of the protein with the following amino acid sequence:

>gi|67414|pir||LZPY lysozyme (EC3.2.1.17) c - pigeon

KDIPRCELVKILRRHGFEGFVGKTVANWVCLVKHESGYRTTAFNNNGPNSRDYGIFQINSKYWCNDGKTR
GSKNACNINCSKLRDDNIADDIQCAKKIAREARGLTPWVAWKKYCQGKDLSSYVRGC

Is there any correlation between the relative mutability and polarity, average flexibility and/or average buried area of the amino acid residues?

6. Search the Web site to predict transmembrane topology of rhodopsin with the following sequence and compare the membrane-spanning regions with the hydrophobicity profiles of ProScale

>gi|10720173|sp|Q9YH05OPSD_DIPAN RHODOPSIN

MNGTEGPFFYVPMVNTTGIVRSPYEYPQYYLVNPAAYAALGAYMFLLILVGFPINFLTLYVTIEHKKLRT
PLNYILLNLAVADLFMVLGGFTTTMYTSMHGYFVLGRLGCNIEGFFATLGGEIALWSLVVLAIERWVVVC
KPISNFRFGENHAIMGLAFTWTMAMACAAPPLVGWSRYIPEGMQCSCGIDYYTRAEGFNNESFVIYMFIC
HFTIPLTVVFFCYGRLLCAVKEAAAAQQESETTQRAEKEVTRMVIMMVIAFLVCWLPYASVAWYIFTHQG
SEFGPVFMTIPAFFAKSSSIYNPMIYICLNKQFRHCMITTLCCGKNPFEEEEGASTASKTEASSVSSSSVSPA

7. Perform a similarity search of PIR database for a protein with the following amino acid sequence:

>gi|67428|pir||LZGSG lysozyme (EC 3.2.1.17) g - goose

RTDCYGNVNRIDTTGASCKTAKPEGLSYCGVSASKKIAERDLQAMDRYKTIIKKVGEKLCVEPAVIAGII
SRESHAGKVLKNGWGDRGNGFGLMQVDKRSHKPQGTWNGEVHITQGTTILINFIKTIQKKFPSWTKDQQL
KGGISAYNAGAGNVRSYARMDIGTTHDDYANDWARAQYYKQHGY

8. Perform a pairwise sequence alignment for the following analogous proteins:

>gi|539057|pir||A61024 alcohol dehydrogenase (EC 1.1.1.1) -wheat

MATAGKVIECKAAVAWEAGKPLSIEEVEVAPPHAMEVRVKILYTALCHTDVYFWEAKGQTPVFPRILGHE
AGGIVESVGEGVTELVPGDHVLPVFTGECKDCAHCKSEESNLCDLLRINVDRGVMIGDGQSRFTINGKPI
FHFVGTSTFSEYTVIHVGCLAKINPEAPLDKVCVLSCGISTGLGATLNVAKPKKGSTVAIFGLGAVGLAA

MEGARMAGASRIIGVDLNPAKYEQAKKFGCTDFVNPKDHTKPVQEVLVEMTNGGVDRAVECTGHIDAMIA AFECVHDGWGVAVLVGVPHKEAVFKTYPMNFLNERTLKGTFFGNYKPRTDLPEVVEMYMRKELELEKFIT HSVPFSQINTAFDLMLKGEGLRCIMRMDQ

>gi | 82347 | pir | | S01893 alcohol dehydrogenase (EC 1.1.1.1) 1 - barley

MATAGKVIKCKAAVAWEAGKPLTMEEVEVAPPQAMEVRVKILFTSLCHTDVYFWEAKGQIPMFPRIFGHE AGGIVESVGEGVTDVAPGDHVLPVFTGECKECPHCKSAESNMCDLLRINTDRGVMIGDGKSRFSIGGKPI YHFVGTSTFSEYTVMHVGCVAKINPEAPLDKVCVLSCGISTGLGASINVAKPPKGSTVAIFGLGAVGLAA AEGARIAGASRIIGVDLNAVRFEEARKFGCTEFVNPKDHTKPVQQVLADMTNGGVDRSVECTGNVNAMIQ AFECVHDGWGVAVLVGVPHKDAEFKTHPMNFLNERTLKGTFFGNFKPRTDLPNVVEMYMKKELEVEKFIT HSVPFSEINTAFDLMAKGEGIRCIIRMDN

9. Retrieve amino acid sequences (in fasta format) of human alcohol dehydrogenase isozymes and perform multiple alignment with ClustalW to evaluate their homology. Identify the amino acid substitutions among the seven isozymes.
10. Perform a BLAST similarity search for a human enzyme with the following amino acid sequence:

    MALEKSLVRLLLLVLILLVLGWVQPSLGKESRAKKFQRQHMDSDSSPSSSSTYCNQMMRRRNMTQGRCKP VNTFVHEPLVDVQNV.CFQEKVTCKNGQGNCYKSNSSMHITDCRLTNGSRYPNCAYRTSPKERHIIVACEG SPYVPVHFDASVEDST

Identify its isozymes and subject these isozymes to multiple alignment with ClustalW to evaluate their homology.

PROTEOMICS: PREDICTION OF PROTEIN STRUCTURES

Computation proteome annotation is the process of proteome database mining, which includes structure/fold prediction and functionality assignment. Methodologies of secondary structure prediction and problems of protein folding are discussed. Approaches to identify functional sites are presented. Protein structure databases are surveyed. Secondary structure predictions and pattern/fold recognition of proteins using the Internet resources are described.

PREDICTION OF PROTEIN SECONDARY STRUCTURES FROM SEQUENCES

The enormous structural diversity of proteins begins with different amino acid sequences (primary structure) of polypeptide chains that fold into complex 3D structures. The final folded arrangement of the polypeptide chain is referred to as its conformation (secondary and tertiary structures). It appears that the information for folding to the native conformation is present in the amino acid sequences; however, a special class of proteins known as *chaperons* is required to facilitate *in vivo* folding of a protein to form its native conformation. Large-scale sequencing projects produce data of genes; hence they produce protein sequences at an amazing pace.

Although experimental determination of protein three-dimensional structure has become more efficient, the gap between the number of known sequences and the number of known structures is rapidly increasing. Protein structure prediction aims at reducing this sequence-structure gap. Methods for abstracting protein structures from their sequences can be aimed at secondary structures or 3D structures (folding problems).

There are four approaches to secondary structure prediction.

1. Empirical statistical methods that use parameters derived from known 3D structures.
2. Methods based on physicochemical properties of amino acid residues (Lim, 1974) such as volume, exposure, hydrophobicity/hydrophilicity, charge, hydrogen bonding potential, and so on.
3. Methods based on prediction algorithms that use known structures of homologous proteins to assign secondary structures.
4. Molecular mechanical methods that use force field parameters to model and assign secondary structures.

The requirements for hydrogen bond preservation in the folded structure result in the cooperative formation of regular hydrogen bonded secondary structure regions in proteins. The secondary structure specifies regular polypeptide chain-folding patterns of helices, sheets, coils, and turns which are combined/folded into tertiary structure. The main chain of an α helix forms the backbone with the side chains of the amino acids projecting outward form the helix.

The backbone is stabilized by the formation of hydrogen bonds between the CO group of each amino acid and the NH group of the residue four positions forward $(n + 4)$, creating a tight, rodlike structure. Some residues form a helices better than others: Alainine, glutamine, leucine, and methionine are commonly found in a helices, whereas proline, glycine, tyrosine, and serine are not. Proline is commonly thought of as α helix breaker because its pyrrole ring structure disrupts the formation of hydrogen bonds, α Helices vary considerably in length (4-40 residues) in proteins with the average length of. about 10 residues corresponding to 3 turns.

The most common location for an a helix in a protein is along the outside of the protein with one side of the helix facing the solvent and the other side toward the hydrophobic interior of the protein. The β-sheet structure is built up from a combination of several regions of polypeptide chain usually 5 to 10 residues long in an almost fully extended conformation and stabilized by hydrogen bonding between adjacent strands. The overall structure formed through the interaction of these individual β strands is known as a β-pleated sheet, which is generally right-twisted. The regular secondary structures, α helices and β sheets, are connected by coil or loop regions of various lengths and irregular shape.

A combination of secondary structure elements forms the stable hydrophobic core of the protein molecule. The loop regions are at the surface of the molecule and frequently participate in forming binding/catalytic sites. Loop regions joining two adjacent secondary structures that change directions are known as turns or reverse turns (or β bends or hairpins for connecting two adjacent antiparallel β strands). Most reverse turns involve four successive amino acid residues (often with Gly or Pro at the second or third position). Almost all proteins of larger than 60 amino acid residues contain one or more loops of 6 to 16 residues referred to as Ω loops (necked-in shape), which may contain reverse turns located on the surface of protein molecules. The propensities of the 20 amino acid residues over three structural states (α, *β*, turn) predict segments of a secondary structure when a cluster of consecutive preferences along the sequence show a mean preference greater than the threshold value in the empirical statistical method of Chou and Fasman.

An example of the physicochemical approach is the use of polar and nonpolar properties of residue along the sequence. In an a helix for example, successive side chains occur at rotation of nearly 100° to form helical wheel. Polar residues are usually on one side of the wheel to face the external aqueous environment, and hydrophobic residues are usually on the other side to pack against the protein's internal core. This pattern of hydrophilic/ hydrophobic residues can be searched for to predict a helix. The information approach uses known structures and collects information on the significant pairwise dependence of an amino acid in a given position with number of residues defined by the window (*e.g.,* 7-12 residues) on either side of it in a particular secondary structural setting.

The method of Gamier, Osguthorpe, and Robson (GOR) considers the effect that residues of a given position within the region eight residues N-terminal to eight residues *C*-terminal have on the structure of that position. Thus a profile that quantifies the contribution of the residue type toward the probability of one of four states — α helix (H), β sheet (E), turn (T), and coil (C)—exits for each residue type. Four probabilities are calculated for each residue in the sequence by summing information from the 17 local residues, $i \pm 8$. The statistical information derived from proteins of known structure is stored in four (17 × 20) matrices for H, E, T, and C. For any residue, the predicted state is the one with the largest probability value with reference to the known structures. The refinement of this approach includes the information from the alignment of homologous sequences. An important approach that has been applied frequently in structure predictions is the *neural network.*

Basically a neural network, a realm of artificial intelligence (Al), gives computational processes the ability to "learn" in an attempt to approximate human learning, whereas most computer programs execute their instructions blindly in a sequential manner. The neural networks have been applied extensively to solve problems that require analysis of patterns and trends, such as secondary structure prediction of proteins. Every neural network has an *input layer* and an *output layer.* In the case of secondary structure prediction, the input layer would be information from the sequence itself, and the output layer would be the probabilities of whether a particular residue could form a particular structure. Between the input and output layers would be one or more hidden layers in which the actual *learning* takes place. This is accomplished by providing a training data set for the network. Here, an appropriate training set would be all sequences for which three-dimensional structures have been deduced.

The network can process this information to look for what are possibly relationships between an amino acid sequence and the structures they can form in a particular context. Kneller et al. (1990) discussed an application of neural networks to protein secondary structure prediction. Rost and Sander (1993) combined neural networks with multiple sequence alignments to improve the success rate of the prediction in the prediction program (PHD at PredictProtein server). Molecular mechanical method for predicting protein structures is the topic. A number of servers offer various methods to predict secondary structures of proteins.

Secondary structure prediction of ExPASy Proteomic tools (http://www.expasy.ch/tools) lists various prediction methods and pointers to their servers. Some of prediction servers and their URL are listed in Table 10b.1. Results are returned either online (Jpred, BCM, nnPredict, and NPS@) or by e-mail (BMERC, Predator, PredictProtein, and PSIpred).

Table 10.6. Web Servers for Protein Secondary Structure Prediction

Server	*URL*
BCM	http://www.hgsc.bcm.tmc.edu/search launcher/
MBERC	http://bmerc-www.bu.edu/psa/index.html
Jpred	http://jura.ebi.ac.uk:8888/index.html
nnPredict	http://www.cmpharm.ucsf.edu/ ~ nomi/nnpredict.html
NPS@	http://npsa-pbil.ibcp.fr
Predator	http://www.embl-heidelberg.de/cgi/predator serv.pl
PredictProtein	http://cubic.bioc.columbia.edu/predictprotein/
PSIpred	http://insulin.brunel.ac.uk/psipred/
SSThread	http://www.ddbj.nig.ac.jp/E-mail/ssthread/www service.html

PROTEIN FOLDING PROBLEMS AND FUNCTIONAL SITES

Fold Recognition and Structure Alignment

It is generally accepted that there are a finite number of different protein folds, and it is very likely for a newly determined structure to adopt a familiar fold rather than a novel one. This limitation to protein folds presumably derives from the relatively few secondary structure elements in a given domain and the fact that the arrange-ment of these elements is greatly - constrained by folding environments, protein functions and probably evolution. Because proteins sharing at least 30% of their amino acid sequence will adopt the same fold and will often exhibit similar functions, protein homology has been an important quality in the identification of related proteins with a common fold during database searches. Thus a sequence can be compared with the known 3D structure of homologous protein using one of two basic strategies.

In the first approach, one-dimensional sequences are compared with one-dimensional templates or profiles derived from known 3D structures. The template provides a set of scores for matching a position within the structure with those in the sequence. Thus threading methods fit a probe sequence onto a backbone of a known structure and evaluate the compatibility by using pseudo-energy potential and residue environmental information. The second strategy involves an evaluation of the total energy of a sequence of the molecule which is folded similarly to the known 3D structure. Each amino acid has distinct attributes such as size, hydrophobicity/hydrophilicity, hydrogen bonding capacity, and conformational preferences that allow it to contribute to a protein fold.

Attempts are being made to interpret the protein fold in terms of amino acid descriptors. Structural alignments are more accurate than sequence alignments, and the local physicoehemical environment of every residue within each 3D structure is directly obtainable (*e.g.*, AAindex at http://www.genome.ad.jp/aaindex/ or ProtScale of ExPASy Proteomic tools at http://www.-expasy.ch/tools/). In sequence alignments, a gap penalty is typically assigned as either a constant value or one that consists of two components: a constant component and a length-dependent component. However, gaps do not occur with equal frequency everywhere within the protein fold. They are rarely found within elements of regular secondary structure, but rather most often within loop regions that are exposed to solvent at the surface of proteins.

Likewise, a hydrogen-bonded polar amino acid buried in the protein core is highly conserved, whereas these same residues at the surface of a protein that is exposed to solvent are not. When representative structures for most of the common folds become available, the structure alignment will certainly be the most powerful approach to protein fold recognition. Thus, computer graphics with its interactive capabilities emerged as an attractive tool for displaying molecular shapes or superimposing active molecules to detect their common substructures. For two related structures, the main question is finding the best geometric transformation in order to superimpose their frameworks together as closely as possible and evaluate their degree of similarity. If a protein with a known sequence but unknown structure can be matched to one of the common folds, it is possible to model the 3D structure of the protein and learn something about its function.

Protein Folding Classes

A study of the amino acid sequences in homologous proteins proves very informative not only in establishing the evolutionary relationships of different species but also in deciphering the complicated question of how amino acid sequence is translated into a specific 3D structure. Programs to predict secondary structures can now distinguish helix, sheet, and turn residues with over 60% accuracy. Thus, a hierarchical classification of proteins (Richardson, 1981) can be employed to assess the quality of our ability to predict protein structures from amino acid sequences.

Most protein classifications refer to structure domains. The domain is a region of protein that has relatively little interaction with the rest of the protein, and therefore its structure is essentially independent. Typically, domains are collinear in sequence, but occasionally one domain will have another inserted into it, or two homologous domains will intertwine by swapping some topologically equivalent parts of their chains. A classification system, such as SCOP, categorizes structure domains based on secondary structural elements within a protein into α structure (made up primarily from α helices), β structure (made up primarily from β strands), α/β structure (comprised of primarily β strands alternating with α helices), and α + β structure (comprised of a mixture of isolated α helices and β strands).

In this classification, only the core of the domain is considered. Therefore, it is possible for an all-α structure to have very small amount of β strand outside the α-helical core. Similarly, an all-β protein may have a small presence of α or 3_{10} helix. The SCOP (http://scop.mrc-lmb.cam.ac.uk/scop/) can be summarized as follows:

[1] *α-Helical Proteins* : Most contain some helix. In a significant number of proteins, the secondary structure is limited to α helices. For example:

- There is a class of small proteins (*e.g.*, haemerythrin, Fig. 10b.1a) that have the form of a four-helix bundle.
- Globins (myoglobin, hemoglobin) are rich in helices and all but one (3_{10}) are α helices.
- The cytochrome Cs are another family of haem proteins with exclusively helical structures.

- Very large proteins (*e.g.*, citrate synthase) can also have a purely helical secondary structure.
- Membrane attached receptor proteins contains seven transmembrane helices (*e.g.*, bacteriorhodopsin, Fig. 10.1).

For individual pairs of interacting helices, certain regularities in the patterns of tertiary-structural interactions between pairs of helices have been observed.

[2] β-*Sheet Proteins.* Domains in which the secondary structure is almost exclusively β-sheet tend to contain two sheets packed face to face. There are two major classes: those in which the strands are almost parallel and those in which the strands are almost perpendicular. The strands of sheet can vary in direction (parallel and antiparallel) and in connectivity, giving great topological variety.

(2.1) Parallel β-Sheet Proteins. The most common arrangement in this class contains four strands for each sheet to form the natural twist such as prealbumin. However, the directions of the strands are not all equivalent and the connectivities of the strands are different, such as transthyretin (Fig. 10b.1c).

(2.2) Orthogonal β-Sheet Proteins. An alternative way of packing two β-sheets together is with the strands in the two sheets almost perpendicular. Each domain of the serine proteases (*e.g.*, trypsin, Fig. 10b.1d) shows this arrangement.

(2.3) Other β-Sheet Proteins. Influenza neuraminidase contains a very unusual β-sheet propellor-like structure. Ascorbate oxidase (Fig. 10b.1e) is a large β-sheet protein that contains parallel β-sheet domains, and interleukin-1β may be described as containing both a barrel and a propeller.

[3]α + β *Proteins.* Many proteins such as lysozyme (Fig. 10b.1f) contain both α helices and β sheets, but do not have the special structures created by alternating *(β-α-β* patterns. In the sulfhydryl proteases, such as papain and actinidin (Fig. 10b.1g), the strands of sheets and the helices tend to be segregated in different regions of space.

[4] α/β *Proteins.* The supersecondary structure consisting of a β-α-β unit with the hydrogen-bonded parallel β strands forms the basis of many enzymes, especially those that bind nucleotides (*e.g.,* adenylate kinase, Fig. 10b.1h) or related molecules. The strands form a parallel β sheet. In some cases, there is a linear β-α-β-α-β-...arrangement, but in other cases the β sheet closes on itself with the last strand hydrogen-bonded to the first.

(4.1) Linear or Open β-α-β Proteins. Many proteins that bind nucleotides contain a domain made up of six β-α units with a special topology. The long loop between β_C and β_D tends to create a natural pocket for the nucleotide ligand. The N—H groups in the last turn of the α_A helix are well-positioned to form hydrogen bonds to phosphate oxygen atoms of the ligand. The NAD-binding domains of horse liver alcohol dehydrogenase and lactate dehydrogenase (Fig. 10b.1i) are typical. Other dehydrogenases have very similar domains. Flavodoxin and adenylate kinase contains a variation on the theme. They have five strands instead of six. Dihydrofolate reductase has eight strands.

(4.2) Closed β-α-β Barrel Structures. Chicken triose phosphate isomerase (Fig. 10b.1J) is typical of a large number of structures that contain eight β-α units in which the strands form a sheet wrapped around into a closed structure, cylindrical in topology. The helices are on the outside of the sheet.

[5] *Irregular Structures.* There are structures that contain very few of their residues in helices and sheets. These tend to be stabilized by disulfide bridges or by metal ligands. For instance: (a) In the case of wheat germ agglutinin, there are numerous disulfide bridges, (b) In the case of ferridoxin, there are ion-sulfur clusters, (c) The kringle structure occurring in many proteins contains disulfidc bridges as well as several short stretches of two-stranded β sheet.

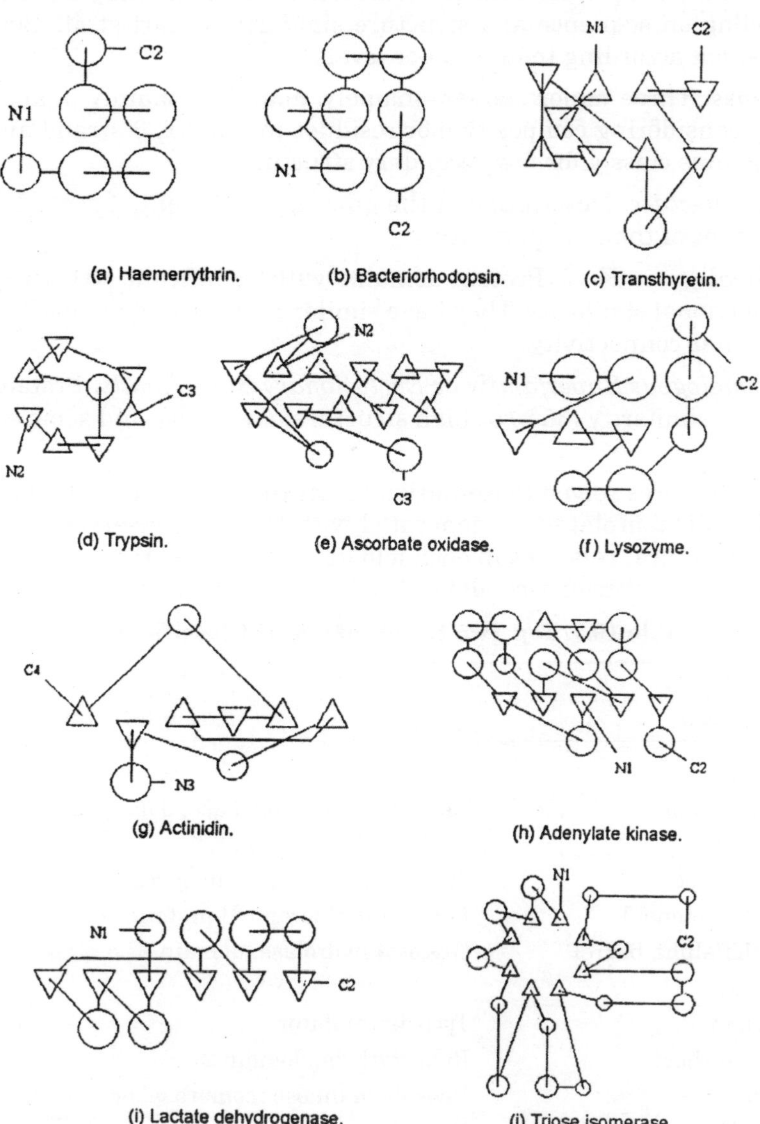

Fig. 10.3. Protein topology cartoons of representative protein domains. The 2D schematic drawings of some representative domains are illustarated in TOPS cartoons (originally retrieved from TOPS server) in which the domain start at N_i and ends at C_{i+1}. α-(and 3_{10}) Helices and β strands are represented by circles and triangles, respectively, according to the symbolisms described.

[6] ***Multidomain Structures.*** Proteins may have multiple domains, but the different domains of these proteins have never been seen independently of each other, therefore accurate determination of their boundaries is not possible and perhaps not meaningful.

The CATH protein domain database (http://www.biochem.ucl.ac.uk/bsm/cath) is a hierarchical classification of protein domain structures into evolutionary families and structural groupings depending on sequence and structure similarity (Pearl et al., 2000). The protein domains are classified according to four major levels.

Level 1: *Class.* Three major classes—namely, mainly α, mainly β, and both α and β – are recognized by considering composition or residues in α helix, β-strand and the secondary structure packing, plus class 4 for few secondary structures.

Level 2: *Architecture.* Description of the gross arrangement of secondary structures in 3D space independent of their connectivity.

Level 3: *Topology* or *Fold.* Protein domains with significant structural similarity but no sequence or functional similarity. They have similar number and arrangement of secondary structures and similar connectivity.

Level 4: *Homologous Superfamily* or *Evolutionary Superfamily.* Proteins either having significant sequence similarity (35%) or high structural similarity and some sequence identity (20%).

Only protein structures solved to resolution better than 3.0 Å from the Protein Data Bank are considered. The 3D templates are generated with CORA (Conserved Residue Attributes) for recognition of structural relatives in each fold group. The CATH Architectural descriptions that denote the arrangements of secondary structures are given in Table 10b.2.

Table 10.7. Top Two Levels of CATH Classification

I.D. No.	*Architecture Description*	*Example*
Class 1, Mainly Alpha:		
10	Non-bundle	Aldehyde dehydrogenase, domain 2
20	Bundle	ATP Synthase, domain 2
25	Horseshoe	70-kD soluble lytic transglycosylase, domain 1
40	Alpha solenoid	Peridinine chlorophyll protein, chain M
50	Alpha/alpha barrel	Glycosyl hydrolase, domain 2
Class 2, Mainly Beta:		
10	Ribbon	Trypsin inhibitor
20	Single sheet	Rubrerythrin, domain 2
30	Roll	Phosphomannose isomerase, domain 1
40	Barrel	Endoglucanase V
50	Clam	Bacteriochlorophyll-a protein
60	Sandwich	Galactose oxidase, domain 1
70	Distroted sandwich	Topoisomerase I, domain 3
80	Trefoil	Acidic fibroblase growth factor, subunit A
90	Othogonal prism	Agglutinin, subunit A

	100	Aligned prism	Vitelline membrane outer layer protein I, subunit A
	102	3 Layer sandwich	Rieske iron-sulfur protein
	110	4 Propeller	Hemopexim
	120	4 Propeller	Neuraminidase
	130	7 Propeller	Methylamine dehydrogenase, chain H
	140	8 Propeller	Methanol dehydrogenase
	150	2 Solenoid	Alkaline protease, subunit P, domain 1
	160	3 Solenoid	Alkaline protease, subunit P, domain 1
	170	Complex	Phosphoenol pyruvate carboxykinase, domain 1
Class 3, Alpha and beta:			
	10	Roll	Elastase, domain 1
	15	Superroll	Bactericidal permeability incr. Protein, domain 1
	20	Barrel	Aconitase, domain 4
	30	2 Layer sandwich	α-Amylase, domain 2
	40	3 Layer (aba) sandwich	Lysozyme
	50	3 Layer (bba) sandwich	Restriction endonuclease, domain 2
	60	4 Layer sandwich	Deoxyribonuclease I, subunit A
	65	Alpha-beta prism	UDP-NAG-Cbov, transferase, chain A, domain 1
	70	Box	Proliferating cell nuclear antigen
	75	5 Stranded propeller	$_{L}$-Arg/Gly aminotransferase, chain A
	80	Horseshoe	Ribonuclease, inhibitor
	85		Sulfite reductase hemoprotein, domain 2
	90	Complex	Glutamine synthetase, domain 1
Class 4, Few secondary structures:			
	10	Irregular	Glucose oxidase, domain 2

Table 10.8. Web Servers for Protein Structure Analysis

Web Site	*Service*	*URL*
3D_PSSM	Protein folding recognition	http://www.bmm.icnet.uk/ ~ 3dpssm/
ASTRAL	Protein structure analysis	http://astral.stanford.edu/
CATH	Domain classes, 3D fold templates	http://www.biochem.ucl.ac.uk/bsm/cath/
Doe-Mbi	Fold recognition	http://fold.doe-mbi.ucla.edu/Login/
Enzyme DB	Enzyme structures, active sites	http://www.biochem.ucl.ac.uk/bsm/enzyme/index.html
LIBRA I	Folding, accessibility	http://www.ddbj.nig.ac.jp/E-mail/libra/LIBRA_I.html
ModBase	Comparative protein structure model	http://guitar.rockefeller.edu/modbase/
PDBSum	Summary and analysis of PDB	http://www.biochem.ucl.ac.uk/bsm/pdbsum/
Pfam	Protein domain families	http://www.sanger.ac.uk/Software/Pfam/
PredictProtein	PHDs and globularity	http://cubic.bioc.columbia.edu/predictprotein/
ProDom	Protein domain analysis	http://www.toulouse.inra.fr/prodom.html
Reliabase	Receptor ligand interaction	http://relibase.ebi.ac.uk/
SCOP	Protein structure classification	http://scop.mrc-lmb.cam.ac.uk/scop/

The Protein Data Bank (PDB) is the primary structure database serving as the international repository for the processing and distribution of 3D structures of biomacromolecules. The database is operated by the Research Collaboratory for Structural Bioinformatics (RCSB) and is accessible from the primary RCSB site at http://www.rcsb.org/pdb/. Most of the structure fold/motif/domain databases and analysis servers utilize 3D-structure information from PDB and sequence information from primary sequence databases. Some of these databases/analysis servers and their URL are listed in Table 10b.3.

Functional Sites

An important aspect in the sequence analysis of proteins is the detection of functional sites such as the active site, ligand binding site, signal sequence/cleavage site, and post-translational modification sites. Generally, a prior knowledge concerning the chemical nature of these sites is available. For example, the active triad (Ser-His-Asp/Glu) of serine proteases, the Arg residue of $NADP^+$ binding site, and the Asn/Gln residue of glycosylation site are known, and detection of these functional sites can be accomplished by the alignment of a query sequence against sequences with known candidate sites.

When studying the specificity of molecular functional sites, it has been common practice to create consensus sequences from alignments and then to choose the most common amino acid residue (s) representative at the given position(*s*). The known functional sites are described in the FEATURE lines of the annotated sequence files such as PIR, GenPept, and Swiss-Prot. The PROSITE database (http://www.expasy.ch/sprot/prosite.html) consists of biologically significant patterns/profiles/signatures that may aid in identifying the query sequence to the known family of proteins (Hofmann et al., 2000).

The database that is composed of two ascii (text) files, PROSITE.dat and PROSITE.doc, can be downloaded using FTP from ExPASy (ftp://www.expasy.ch/databases/prosite/) or Swiss Institute for Experimental Cancer Research (ISREC) at ftp://ftp.isrec.isbsib.ch/sib-isrec/profiles or EBI (ftp://ftp.ebi.ac.uk/pub/databases/profiles/). The residue types that are identified as ProSites are generally enzyme active sites, prosthetic group attachment sites, ligand/metal binding sites, post-translational modification sites, and protein family signature residues.

The similarity search for the highly conserved sequences/regions of each protein family is available from the BLOCKS database, which serves to identify the functional/signature regions of the query sequence. The database is accessible at the BLOCKS Web server (http://www.blocks.fhcrc.org/) of the Fred Hitchinson Cancer Research Center. Pfam is a protein families database containing functional annotation. The database is available at http://www.sanger.ac.uk/Pfam. The active site topology of enzymes can be displayed/saved by accessing the Enzyme Structure Database (http://www.biochem.ucl.ac.uk/bsm/enzymes/index.html), and the binding site of receptors can be analyzed at Relibase (http://relibase.ebi.ac.uk/reli-cgi/rll?/reli-cgi/query/form_home.pl) as discussed earlier.

The center for Biological Sequence Analysis (CBS) at http://www.cbs.dtu.dk/ offers methods for predicting post-translational modifications including signal peptide cleavage, *O*-glycosylation, and phosphorylation sites. Prediction of mitochondrial targeting sequences is available at http://www.mips.biochem.mg.de/cgi-bin/proj/medgen/mitofilter. TargetP (http://www.cbs.dtu.dk/services/TargetP/) assigns the subcellular location of proteins based on the predictions of the N-terminal chloroplast transit peptide, mitochondrial targeting peptide, or secretory signal peptide.

Proteins with intracellular half-life ($t_{1/2} = 0.693/k$, where k is the first-order rate constant for the proteolytic degradation of a cellular protein) of less than two hours are found to contain region rich in proline, glutamic acid, serine, and threoinine, termed PEST. The PEST region is generally flanked by clusters of positively charged amino acids. The PEST sequence search is available at http://www.icnet.uk/LRITu/projects/pest/.

PROTEOMIC ANALYSIS USING INTERNET RESOURCES: STRUCTURE AND FUNCTION

Structure Database

Protein Data Bank (PDB) at http://www.rcsb.org/pdb/) is the worldwide archive of structural data of biomacromolecules. PDB was established at Brookhaven National Laboratories (BNL) in 1971. In October 1998, the management of the PDB became the responsibility of the Research Collaboratory for Structural Bioinformatics (RCSB). Two methods, SearchLite (simple keyword search) and SearchFields (advanced search), are available to search and retrieve atomic coordinates (pdb files). If you know the PDB identification code (4-character identifier of the form [0-9][A-Z or 0-9] [A-Z or 0-9] [A-Z or 0-9], *e.g.*, 1LYZ for hen's egg-white lysozyme and 153L for goose lysozyme), enter the identification code into the PDB ID box and hit the Explore button.

The Structure Explore page for this entry is returned. If the PDB ID is not known, clicking SearchLite opens the query page. Enter the keyword (name of ligand/biomacromolecule or author) and click the Search button. In the advanced search, clicking SearchField opens the query form (Fig. 10.2). Construct combination of your query options including PDB ID, citation author, chain type (for protein, enzyme, carbohydrate,.DNA or RNA), compound information, PDB header, and experimental technique used. Clicking the Search button returns you to the search page with a list of hits from which you select the desired entry to access the Summary information of the selected molecule. From the Summary information, the user can choose one of many options including View Structure, Download/Display File, Structural Neighbors (links to CATH, CE, PSSP, SCOP and VAST), Geometry, or Sequence Details. Select Download/Display File, then choose PDB text and PDB noncompression format to retrieve the pdb file in text format.

To view the structure online, select View Structure followed by choosing one of the 3D display options. The display can be saved as image.jpg/image.gif. The Structure Neighbors provide links to CATH (fold classification by domain), CE (representative structure comparison, structure alignments, structure superposition tool), FSSP (fold tree, domain dictionary, sequence neighbors, structure superposition), SCOP (Class, fold, superfamily, and family classification), and VAST (representative structure comparison, structure alignments, structure superposition tool). The Geometry option provides tables of dihedral angles, bond angles, and bond lengths. Molecular modeling database (MMDB) of Entrez is a subset of 3D structures from PDB recorded in asn.l format.

The database that provides the link between 3D structures and sequences can be accessed at http:// www.ncbi.nlm.nih.gov/Entrez/structure.html Enter author's last name or text words/ keywords (*e.g.*, zinc finger, DNA protein complex, or topoisomerase) or PDB ID (if it is known) and click the Search or Go button. A list of hits is returned. Select the desired entries and then click the PDB ID to receive the MMDB Structure summary. Choose options (radios) to view or save the structure file as structure.cgi (Cn3D), structure.kin (KineMage), or structure.pdb

(RasMol). Sequence similarities or structure similarities can be viewed/saved by clicking the respective chain designation of Sequence neighbors or Structure neighbors. This returns the sequence display summary or VAST structure neighbors, respectively. The structural information about nucleic acids can be obtained from PDB or Nucleic Acid Database (NDB) at http://ndbserver.rutgers.edu/NDB/ndb.html On the NDB home page, choose Search→Nucleic acid database search→Quick search to open the search form for the database (Fig. 10c.4). Activate Classification list box (listing of the database by classes: DNA, DNA/RNA, Peptide nucleic acid, Peptide nucleic acid/DNA, Protein/DNA, Protein/RNA, Ribosome, Ribozyme, RNA, and tRNA), highlight the desired class, and click the Execute Selection button to open a list of database for the class. Highlight an entry, and click the Display Selection button.

An information page providing NDB ID, Compound name, Sequence, Citation, Crystal information, Coordinates, and Views is returned. Click the link, coordinates for the asymmetric unit to open the coordinate file and save it as nuclacid.pdb. WPDB, which can be downloaded from http:// www.sdsc.edu/pb/wpdb, is a protein structural data resource. It is a Microsoft windows-based, data management and query system that can be used locally to interrogate the 3D structures of biomacromolecules found in PDB.

The program provides a query ability to find structures and performs sequence alignment, property profile analyses, secondary structure assignment, 3D rendering, geometric calculation, and structure superposition (based on PDB atomic coordinates without computation). The accompanying WPDB loader (WPDBL) permits the user to build a subset of database from selected pdb files for use as tutorials on protein structures. The summaries and analyses of PDB structures can be accessed from PDBsum at http://www.biochem.ucl.ac.uk/pdbsutn. Search the structure by entering the four-character PDB code (*e.g.*, llyz, this returns the summary page) or a string of the protein name (*e.g.*, lysozyme, this returns a list of hits; you have to select the desired entry to open the summary page).

The summary page contains brief descriptions of the structure, hyperlinks to PDB, GRASS (Virtual graphic server at Colombia University), MMDB, CATH, SCOP, PROCHECK (Ramachandran plot statistics), PROMOTIF (summary of protein secondary struc-tures at the site), and clickable presentations of CATH classification, secondary structures, PROMITIF, TOP, SAS (annotated FASTA alignment of related sequences in the PDB), PROSITE, and LIGAND (ligand molecules and ligand binding residues).

In addition to the graphical representation of secondary structures along the sequence, the structural information can be viewed by selecting PROMO-TIF → sheet/strands/helices/beta turns/gamma turn/bulge/hairpins/disulfide bridges as exemplified in Table 10c.4 for helical structure of human lysozyme. Click on Helical wheels and nets to view diagrams of helical wheels. The residues are color coded for hydrophobic (purple), polar (blue) and charged (red) amino acid types.

Secondary Structure Predictions

A number of Web sites offer services in the secondary structure predictions of proteins using different approaches with varied accuracy. The Secondary structure prediction of ExPASy Proteomic tools (http://www.expasy.ch/tools/) provides pointers to different Web servers for predicting secondary structures of proteins. The ProtScale of ExPASy Proteomic tools produces conformatiorial profiles by plotting statistical scales of various parameters (e.g., Chou and

Fasman's conformational propensities, Levitt's conformational parameters) against residue positions. Numerous prediction methods including SOPM, SOPAMA, HNN, MLRC, DPM, DSC, GOR I, GOR III, GOR IV, PHD, PREDATOR, and SIMPA96 are available at Network Protein Sequence Analysis (NPS@). The server can be accessed via http://npsa-pbil.ibcp.fr. For example, the prediction method of Gamier, Osguthorpe, and Robson (GOR I,

Table 10.9. Secondary (Helical) Structure Descriptions from PROMOTIF Individual Helices

Helix Number	*Start*	*End*	*Type*	*Number of Residues*	*Unit Residues*					
					Length	*Rise*	*per Turn*	*Pitch*	*Deviation*	*Sequence*
1	5	14	H	10	15.52	1.51	3.62	5.47	3.5	RCELAAAMKR
2	25	36	H	12	17.82	1.49	3.74	5.56	9.6	LGNWVCAAKFES
3	80	84	G	5	9.27	1.80	3.18	5.71	17.0	CSALL
4	89	99	H	11	17.05	1.51	3.65	5.51	9.1	TASVNCAKKIV
5	104	107	G	4	7.11	1.95	3.06	5.95	41.7	GMNA
6	109	114	H	6	8.66	1.43	3.57	5.09	8.9	VAWRNR
7	120	124	G	5	9.19	1.84	3.13	5.77	15.5	VQAWI

Helix Interactions Involving This Chain

Helix	*Numbers*	*Helix*	*Types*	*Distance*	*Angle*	*Interaction*	*Type*	*No. of Residues* total	*Helix 1*	*Helix 2*	
1	2	H	H	9.6	123.5	I	I	9	5	4	Intrachain
1	7	H	G	7.7	−109.8	N	C	4	2	3	Intrachain
2	4	H	H	13.4	110.0	I	I	2	1	2	Intrachain
2	5	H	G	11.8	− 4.7	C	C	3	3	1	Intrachain
2	6	H	H	8.2	118.7	I	I	8	5	4	Intrachain
2	7	H	G	6.9	77.2	I	I	11	7	3	Intrachain
3	4	G	H	6.1	−109.5	c	n	3	2	2	Intrachain
4	5	H	G	7.2	114.7	C	N	4	2	3	Intrachain
5	6	G	H	5.0	−120.3	c	n	5	3	3	Intrachain

Note: The helix types are H for α helix and G for 3_{10} helix. The helical geometry is followed by a measure of the deviation from an ideal helix (this value should be 0 for a perfect helix). For the interacting pairs, the interaction type describes where in each of the helices the distance of closest approach occurs (C, beyond the C terminus of the helix; N, beyond the N terminus of the helix; I, internal to the helix).

IIII, or IV) can be initiated by selecting GOR IV under Secondary structure prediction tool to open the query form. Paste the query sequence into the sequence box and click the Submit button. The result is returned showing the query sequence with the corresponding predicted conformations (c, h, e, and t for coil, helix, extended strand, and turn, respectively), summarized content (%) of conformations, and conformational profiles (Fig. 10c.6).

Click Secondary Structure Consensus Prediction enabling the simultaneous execution of a number of selected prediction methods. The predicted secondary structures including the consensus secondary structure is returned (Fig. 10c.7).

The BCM (Baylor College of Medicine) at http://www.hgsc.bcm.tmc.edu/Search/ Launcher/ offers segment-oriented prediction (PSSP/SSP) in which the most prob-able secondary structure segments (a for α helix, b for β strand, and c for the remainder) are assigned based on the probability for a, b, or c (0 to 9). The PredSS assignments of a and b are made. The nnPredict (http://www.cmpharm.ucsf.edu/~ nomi/nnpredict.html) uses the neural network in the structure prediction and reports both the predicted secondary structures (H = helix, E = strand, and – = no prediction) and tertiary structure class (all-α, all-β, and α / β). The Protein Sequence

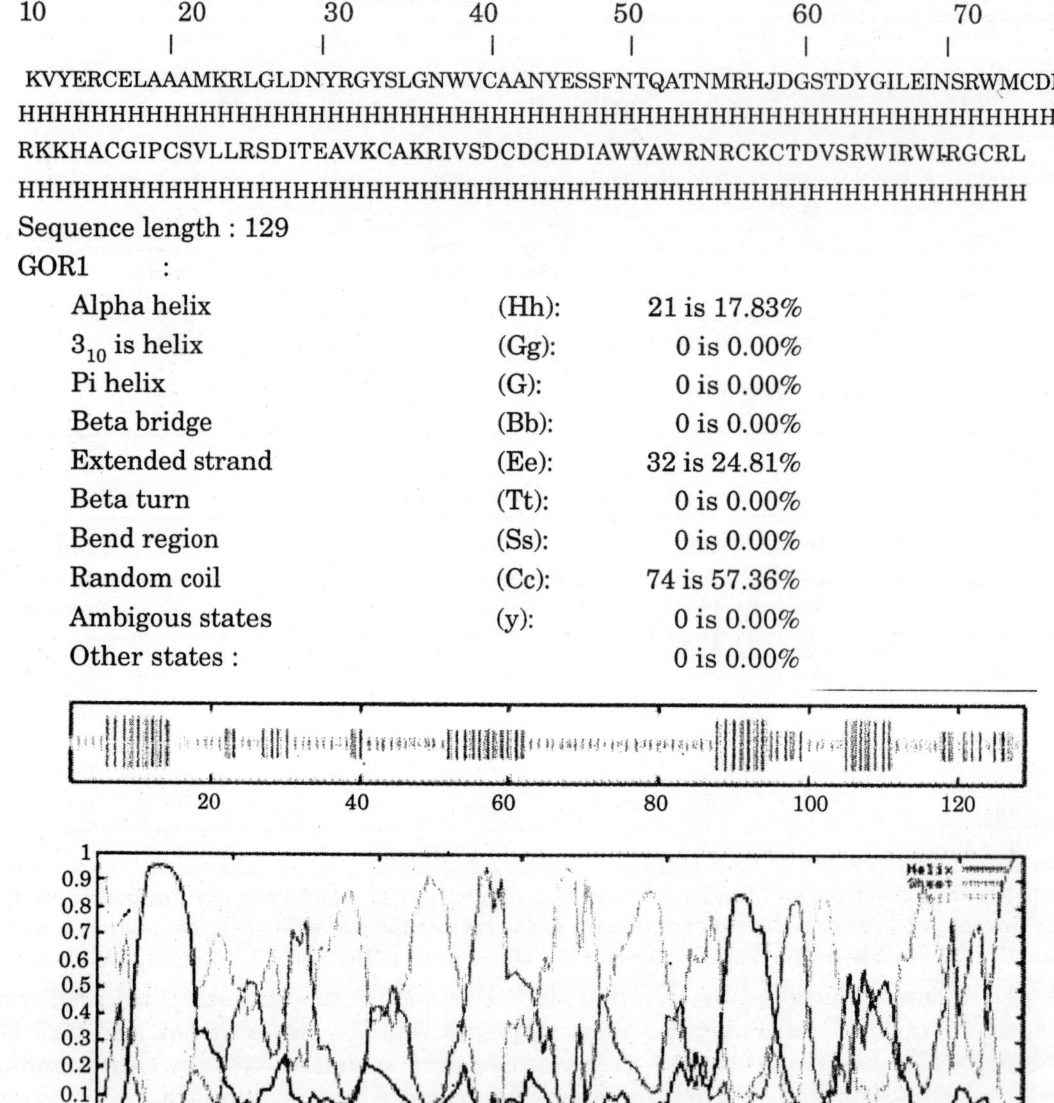

```
10         20         30         40         50         60         70
           |          |          |          |          |          |          |
KVYERCELAAAMKRLGLDNYRGYSLGNWVCAANYESSFNTQATNMRHJDGSTDYGILEINSRWMCDNGKTP
HHHHHHHHHHHHHHHHHHHHHHHHHHHHHHHHHHHHHHHHHHHHHHHHHHHHHHHHHHHHHHHHHHHHHHHH
RKKHACGIPCSVLLRSDITEAVKCAKRIVSDCDCHDIAWVAWRNRCKCTDVSRWIRWIRGCRL
HHHHHHHHHHHHHHHHHHHHHHHHHHHHHHHHHHHHHHHHHHHHHHHHHHHHHHHHHHHHHHHHH
```

Sequence length : 129

GOR1 :

Alpha helix	(Hh):	21 is 17.83%
3_{10} is helix	(Gg):	0 is 0.00%
Pi helix	(G):	0 is 0.00%
Beta bridge	(Bb):	0 is 0.00%
Extended strand	(Ee):	32 is 24.81%
Beta turn	(Tt):	0 is 0.00%
Bend region	(Ss):	0 is 0.00%
Random coil	(Cc):	74 is 57.36%
Ambigous states	(y):	0 is 0.00%
Other states :		0 is 0.00%

Fig. 10.4. Secondary structure prediction of duck lysozyme at NPS@. The predicted secondary structures of duck lysozyme at Network Protein Sequence Analysis (NPS@) with GOR IV method are depicted in different representations.

Analysis (PSA) server of BMERC predicts secondary structures and folding classes from a query sequence. On the PSA home page at http://bmerc-www.bu.edu/psa/ index.html, select Submit a sequence analysis request to submit the query sequence and your e-mail address. The returned results include (a) probability distribution plots (conventional X/Y and contour plots) for strand, turn, and helix and (b) a list of structure probabilities for loop, helix, turn, and strand for every amino acid residues. The method of Rost and Sander, which combines neural networks with multiple sequence alignments known as PHD, is available from the PredictProtein server of Columbia University (http://cubic.bioc. columbia.edu/predictprotein/). This Web site offers the comprehensive protein sequence analysis and structure prediction.

For the secondary structure prediction, choose Submit a protein sequence for prediction to open the submission form. Enter e-mail address, paste the sequence, choose options, and then

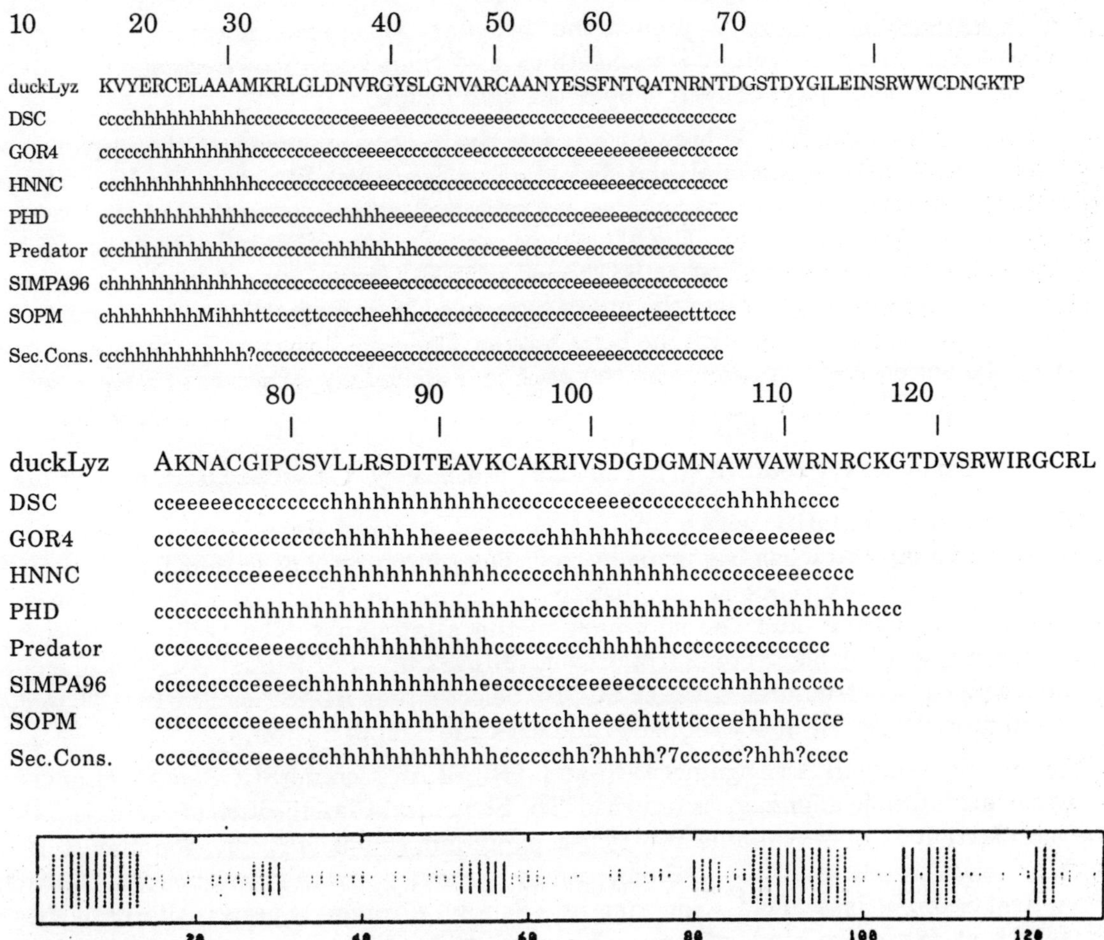

Fig. 10.5. Secondary structure consensus prediction at NPS@. Network Protein Sequence Analysis (NPS@) offers numerous methods for secondary structure prediction of proteins. The secondary structure consensus prediction of duck lysozyme is derived from simultaneous execution of predictions with more than one methods.

click the Submit/Run Prediction button. The predicted results for the secondary structure (PHDsec), solvent accessibility (PHDacc), and helical transmembrane regions (PHDhtm) are returned by e-mail. The prediction reports the multiple alignment (with statistical summary) of homologous proteins, and it reports the query sequence (AA) with

- PHD_sec: predicted H for helix and E for strand
- Rel_sec: reliability index for PHDsec prediction with 0 = low and 9 = high
- P_3_acc: predicted relative solvent accessibility in *b* = 0-9%, i = 9-36%, and *e* = 36-100%
- Rel_acc: reliability index for PHDacc prediction with 0 = low and 9 = high

The prediction of the secondary structures can be made by the structure similarity search of PDB collection at the site. Several servers provide such prediction method. The Jpred, which aligns the query sequence against PDB library, can be accessed at http://jura.ebi.ac.uk:8888/index.html To predict the secondary structures, however, check Bypass the current Brookhaven Protein Database box and then click Run Secondary Structure Prediction on the home page of Jped to open the query page.

Upload the sequence file via browser or paste the query sequence into the sequence box. Enter your e-mail address (optional) and click the Run Secondary Structure Prediction button. The results with the consensus structures are returned either online (linked file) or via e-mail (if e-mail address is entered). The 3D-ID compatibility algorithm is applied to predict the secondary structures by threading at SSThread of DDBJ (http://www.ddbj.nig.ac.jp/E-mail/ssthread/www_service.html). Paste the query sequence (fasta format) into the sequence box, enter your e-mail address, and click the Send button. The e-mail returns the threading result reporting the amino acid sequence with the predicted secondary structures (H for α helix, E for β strand, and C for coil or other).

Toward 3D Structure Prediction: Structural Similarity, Classification, and Folds

The structural similarity search can be conducted at Jpred if the Bypass on the current Brookhaven Protein Database box is unchecked. The search returns output for the pairwise alignments of the query sequence against the sequences of PDB files with the consensus sequence between them and the summary of the alignments. The G To P site (http://spock.genes.nig.ac.jp/~genome/gtop.htrm)provides BLAST and PSI-BLAST searches of a query sequence against 3D structures (PDB or SCOP). Select either BLAST or PSI-BLAST to open the search form. Paste the query sequence and click the Search button.

The search result (BLAST against PDB and. PSI_BLAST againt SCOP) with summary of the search and multiple alignment is returned. The Structural Classification of Proteins (SCOP) database (Section 12.2) is available at http://scop.mrc-lmb.cam.ac.uk/scop/. You may start the search by selecting (clicking) Top of the hierarchy to move down the classification hierarchy, that is, Root (scop)→Class→Fold→Superfamily→Family→Protein→Species. Alternatively, you may search at any stage of classification by going to http://scop.mrc-lamb.cam.ac.uk/scop/search.cgi and enter the PDB ID or the keyword at the Root (Fig. 10c.11). Click the Search button to view the list of related entries (matching the keyword). Choose the desired entry to display the lineage and PDB entry domains with pointers.

The SCOP is augmented by the ASTRAL compendium at http://astral.stanford.edu/ that provides sequence databases corresponding to the domain structures. The CATH database (http://www.biochem.ucl.ac.uk/bsm/cath new/index.thml) is another hierarchical protein structure classification that classified protein domains into four Classes. Each class is subclassified into several structural Architectures (Section 12.2), which is further classified into fold family (Topology) and then Homologous/evolutionary superfamily. The representative domain for each architecture description can be displayed by selecting Navigation from Top of the hierarchy. A search for the domain structure can be conducted by specifying PDB code or general text. To search the database, go to DHS (Dictionary of Homologous Superfamilies) at http://www.biochem.ucl.ac.uk/bsm/dhs/. You may click one of the four classes to initiate a search that returns the CATH table (classification) of the class.

Alternatively, you may search category by entering PDB code (PDB ID) or CATH number (c.aa.tt.ss, where c stands for class, a for architecture, t for topology, and s for superfamily, *e.g.*, 3.40.80.10 for lysozyme) or Enzyme classification (EC number) and click the Search button. To request 3D structure analysis by multiservers via META-PP submission at the PredictProtein. site, click Submit Requests to META-PP to open the submission form. Enter your e-mail address, paste the query sequence, and pick (check boxes) of the desired 3D analyses (homology modeling and threading) then click the Submit/Run prediction button.

The analytical results are returned via e-mail. As mentioned above, the PredictProtein server also carries out the prediction of relative solvent accessibility (PHDacc), which is reported in three states (b, i, and e). The threading prediction of the query sequence against the structural library chosen from PDB at LIBRA I (http://www.ddbj.nig.ac.jp/E-mail/libra/LIBRA I.html) reports a list of compatible structures and 3D-ID alignment results recording the secondary structures and accessibility. The 3D-PSSM (Kelley et al., 2000) server at http://www.bmm.icnet.uk/~3dpssm/ offers online protein fold recognition. On the submission form, enter your e-mail address and a one-line description of the query protein, then paste the query sequence into the sequence box and click the Submit button.

The query sequence is used to search the Fold library for homologues. You will be informed of the URL where the result is located for 4 days. The output includes a summary table (hits with statistics; models that can be viewed with RasMol; classifications; and links) and fold recognition by 3D-PSSM with a printout as exemplified in Fig. 10c.13. The alignment displays consensus sequence, secondary structures (C for coil, E for extended, and H for helix), and core score (0 for exterior to 9 for interior core). The FUGUE site (Shi et. al., 2001) at http://www-cryst.bioc.cam.ac.uk/~fugue/prfsearch.html is a protein fold recognition server. On the request form, enter your e-mail address and the name of the query sequence (optional), upload or paste the query sequence, and select "search options" then click the Search button.

The query sequence is used to search structurally homologous proteins in the Fold library for hits (representatives). You will be informed with a short summary of the result and the URL where the detailed result can be examined for 5 days. The fold recognition results (Fig. 10c.14) are summarized in a view ranking listed according to Z-score with the following recommended cutoffs.

```
    Recommended cutoff :      ZSCORE >  = 6.0 (CERTAIN 99% CONFIDENCE)
    Other cutoff :            ZSCORE >  = 5.0 (LIKELY 95% confidence)
    Other cutoff:             ZSCORE > =4.7 (MARGINAL 90% confidence)
    Other cutoff:             ZSCORE >  = 3.5 (GUESS 50% confidence)
    Other cutoff:             ZSCORE<      3.5 (UNCERTAIN)
Rk StrC Protein                                   Lsr    Lal    Rsc      SD      Rs/N    ID%
1 2ohxA ALCOHOL DEHYDROGENASE (HOLO FO             374    375   –211.8   –7.34   –0.565   53.3
2 1kevA NADP–DEPENDENT ALCOHOL DEHYDRO            351    372   –148.0   –4.54   –0.398   21.8
3 1bouB 4.5–DIOXYGENASE ALPHA CHAIN;              298    291   –09.5    –2.86   –0.366    9.4
4 2cnd–NADH–DEPENDENT NITRATE REDUCTA             260    396   –107.8   –2.78   –0.364   10.8
5 2bbkH METHYLAMINE DEHYDROGENASE (MAD            355    357   -106.6   -2.72   -0.298    8.7
 ....    ....    ....    ....    ....    .....    ....  ....  ..      ..      ..        ..
    3D-1D alignments are sequences (only 1 2ohxA structure vs codADH is shown):
    2631138164      9768413133      95335293852     6413697949      6-69998889
    egeleeBBBB      BBBegeelee      eBBBBBBee       eee1BBBBBB      B-BBBeeAAA
    STAGKVIKCK      AAVLWEEKKP      FSIEEVEVAP      PKAHEVRIKM      V-ATGICRSD
    :  :::::::      ::: ::   ::     :::     ::      :  :  :  :::
    ATVGKVIKCK      AAVWENKP        LVIEEIEVDV      PHANEIRIKI      IATGVHTDL
    . .. .  .  .  .  .  .  .  .  .  .  .  .  .  .  .  .
    9899986998      8966719249      898562383       1892499754      7724244123
    geAAAAAAAA      AAAAglegBB      BBBegegggA      AAAAAAlegB      BBegggegee
    LGGVGLSVIM      GCKAAGAARI      IGVDINKDKF      AKAKEVGATE      CVNPQDYKKP
    ::  :::  :   ::  :::  ::  : :: ::  ::           ::: :::  :      :::    :    :
    LGAVGLAAVM      GCHSAGAKRI      IAVDLNPDKF      EKADVFGATD      FVNPNDHSEP

    8425942892      4698999896      5251884789      6996586868      9997683424
    AAAAAAAAgl      leegBBBBgg      leAAAAAAAA      AgeegggeBB      BBgeeeeele
    IQEVLTEMSN      GGVDFSFEVI      GRLDTMVTAL      SCCQEAYGVS      VIVGVPPDSQ
    :  ::  : :  : : : : : :         :    :  : :        :    :::     : : :
    ISQVLSKMTN      GGVDFSLECV      GNVGVMRNAL      ESCLKGWGVS      VLVGWTDLHD

    2718294775      7777995999      8695954892      399344213       941896564
    eeeeegAAAA      gleBBBBeel      lleeAAAAAA      AAAAAAAglg      eeggggBBB
    NLSMNPMLLL      SGRTWKGAIF      GGFKSKDSVP      KLVADFMAKK      FALDPLITHV
           :        ::::::    :     : : : : :: ::   : :       ::        :::.::
    VATRPIQLI-AGRTWKGSMF            GGFKGKDGVP      KMVKAYLDKK      VKLDEFTHR

    5171494486      2674243679      78838-
    BBgggAAAAA      AAAAAlgegB      BBBBe-
    LPFEKINEGF      DLLRSGESIR      TILTF-
    :  :  :         ::     :  ::   : :
    MPLESVNDAI      DLMKHGKCIR      TVLSLE
```

Fig. 10.6. Protein structure prediction by LIBRA I server. The amino acid sequence of Baltic cod alcohol dehydrogenase is used for structure predictions against the structure library by LIBRA I. The results consist of a list (a partial list is shown) of compatible structures and aligned sequences with predicted structural features (only one set is shown). The abbreviations used for the list are Rk, rank position; StrC, structural code; Lsr, length of the structural template; Lal, length of the aligned region; Rsc, raw score of the structural template; SD, standardized score; Rs/N, raw score (Rsc) normalized by the alignment length (Lal); and ID%, sequence identity. The structural features in the aligned sequences are as follows. First line: accessibility of the aligned site with scores from 1 (exposed to solvent) to 9 (buried, in protein); second line: local conformation of the aligned site with A(alpha), B(beta), gel(coils classified by the dihedral angles); third line: sequence of the template structure, fourth line: match site of the alignment and fifth line: query sequence.

Sawted

	Domain	E-value	Score	Class		Fold	Superfamily	Family
	=====		=======		=====	===== ====	===========	======
:	clteha_	1.07e-05	0.00	no	SCOP	3D-PSSM pseudeo-family	founder id childa_	-
:	clhlda_	3.73e-02	0.11	no	SCOP	3D-PSSM pseudo-family	founder id clhlda_	-
:	clcdoa_	5.42e-D2	0.07	no	SCOP	3D-PSSM pseudo-family	founder id clhlda_	-
:								
:	dlteha2	6.04e+00	0.00	α and β		NAD (P)-bd Rossmann-fold	NAD(P) bd-Rosmann-fold	Alc/glc DH
:	dlcdoa2	7.85e+00	0.07	α and β		NAD(P)-bd Rossmann-fold	NAD(P) bd-Rossmann-fold	Alc/glc DH
:	d3dbtoa2	7.85e+00	0.11	α and β		NAD(P) Rosmann-fold	NAD(P) bd-Rosmann-fold	Alc/glc DH
:								

```
              150                                                                       200
fish_al_pass  EEEEEEEEEE    ECCCCCCCCE    EEEECCHHHC    CCEEEECCCC    CCCCEEEEEC
fish_al_Seq   TVVNQIAVAK    IDPSAPLDTV    CLLGCGVSTG    FGAAVNTAKV    EPGSTCAVFG
————————                                      G+STG    +GAAVNTAK+    EPGS-CAVFG
dlteha2_Seq   ............  ...........   .......GISTG  YGAAVNTAKL    EPGSVCAVFG
dlteha2_SS    ............  ...........   ......HHHHH   HHHHHHCCCC    CCCCEEEEEC
CORE          ............  ...........   ........00013 0098001305    0001257932

              200                                                                       250
fish_al_PSS   CHHHHHHHHH    HHHHHCCCEE    EEEECCHHHH    HHHHHCCCCC    CCCHHHCCCC
fish_al_Seq   LGAVGLAAVM    GCHSAGAKRI    IAVDLNPDKF    EKAKVFGATD    FVNPNDHSEP
————————      LG-VGLA—M     GC++AGA+RI    I+VD+N+DKF    ++AK+FGAT+    ++NP+D+S+P
dlteha2_Seq   LGGVGLAVIM    GCKVAGSRI     IGVDINKDKF    ARAKEFGATE    CINPQDFSKP
dlteha2_SS    CCHHHHHHHH    HHHHHCCCEE    EEECCCHHHH    HHHHHHCCCE    EECHHHCCCC
CORE          3003211960    2700104006    5560000000    0050012600    2501000000

              250                                                                       300
fish_al_PSS   HHHHHHCCCC    CCCCEEEEEE    CCHHHHHHHH    HHHHCCCCEE    EEEEEC.CHH
fish_al_Seq   ISQVLSKMTN    GGVDFSLECV    GNVGVMRNAL    ESCLKGWGVS    VLVGWT.DLH
————————      I++VL++MT+    GGVD+S+EC+    GNV+VMR+AL    E+C+KGWGVS    V+VG++ +++
dlteha2_Seq   IQEVLIEMTD    GGVDYSFECI    GNVKVMRAAL    EACHKGWGVS    VVVGVAASGE
dlteha2_ss    HHHHHHHHHC    CCCCEEEECC    CCHHHHHHHH    HCCCCCCCEE    EECCCCCCCC
CORE          3001700110    0050038051    0000100020    0410000001    5080000000

              300                                                                       350
fish_al_PSS   HCCCCCEEEE    CCCEEEEEEE    ECCCCHHHHH    HHHHHHHCCC    CCCCCCEEEC
fish_al_Seq   DVATRPIQLI    AGRTWKGSMF    GGFKGKDGVP    KMVKAYLDKK    VKLDEFITHR
————————      ++ATRP+QL+    +GRTWKG++F    GG+K+
dlteha2_Seq   EIATRPFQLV    TGRTWKGTAF    GGWKS.....    ..........    ..........
dlteha2_SS    CCCCCHHHHH    HCCEEEECCH    HHCCH.....    ..........    ..........
CORE          0000000000    0000001300    20000.....    ..........    ..........
```

Fig. 10.7. Protein fold recognition by 3D_PSSB. The summary fold recognition results of cod alcohol dehydrogenase analyzed by 3D_PSSB (secondary structure prediction by PsiPred is also included) are sorted by E value, most confident assignment first. E values below 0.05 are highly confident, and E values up to 1.0 are worthy of attention. Following the list (shown partial, abbreviations used are no SCOP, not in SCOP v1.50; NAD(P)-bd, NAD(P)-binding; and Alc/glc DH, Alcohol/glucose dehydrogenase) is the pairwise top 20 alignments (only partial of one example is shown).

The keys linking to the files for alignments with structures/fold assignments are available from the View Alignments table. The query sequence is aligned against the sequence(s)/ structure(s) of either all representatives or a selection of the group according to the following alignment options (keys):

View Ranking (Click on a profile hit will bring you to the corresponding HOMSTRAD family)

Profile Hit	PLEN	RAWS	RVN	ZSCORE	ZORI	AL	
adh	375	1713	2043	76.58	77.55	00	CERTAIN
hsdlkeval	187	-2	210	11.11	12.08	02	CERTAIN
hsdlqoral	196	-35	185	9.11	10.08	02	CERTAIN
LRR	485	-382	26	3.58	4.54	00	GUESS
hsldova	346	-182	76	3.40	4.37	00	UNCERTAIN
ATP-synt	535	-503	82	3.33	4.30	00	UNCERTAIN
hsdlayl	532	-485	63	2.50	3.47	00	UNCERTAIN
hsdlordal	107	4	17	2.33	3.30	22	UNCERATAIN
hslfbna	225	-147	180	2.31	3.28	00	UNCERTAIN
kex	498	-3-537	108	2.31	3.28	00	UNCERTAIN

View Alignments (aa option is shown)

```
                      10          20          30          40          50
3huda     ( 1 )   stagkvikckAAVLwevkkpFsiedVeVappkayEVRIkMvAVGICrtDD
1cdoa     ( 1 )   atvgkvIkckAAVAweankpLvieeIeVdvPhaneIRIkIiATGVChtDL
1teha     ( 3 )     anevikckAAVAweagkpLsieeIeVapPkahEVRIkIiATAVChtDa
2ohxa     ( 1 )   stagkvikckAAVLweekkpFsieeVeVapPkahEVRIkMvATGICrSDd
1dlta     ( 1 )   dtagkvikckAAVLweqkqpFsieeIeVapPktkEVRIkIlATGICrtDD
gi|482344|        ATVGKVIKCKAAVAWEANKPLVIEEIEVDVPHANEIRIKIIATGVCHTDL
                        bbbbbbb bbbbbbb        bbbbbbbbbb     aaaa
                         60          70          80        90    100
3huda     ( 51 )  hVvsgn-lvTplpVILGHEAAGiVesvgegVttVkpgdkVIPLFTPqCgk
1cdoa     ( 51 )  YHLfegkhkdGFpVVLGHEGAGiVesvGpqVteFqpgekVIPLfisqcge
1teha     ( 51 )  yTlsqadpegcfpVILGHEGAGiVesvGegVtkLkagdTVIPLyiPgCge
2ohxa     ( 51 )  hVvsgt-lvTplpVIAGHEAAGiVesigegVttVrpgdkVIPLFTPqCgk
1dlta     ( 51 )  hVikgt-mvSkfpVIVGHEATGiVesigegVttVkpqDkVIPLfLPQcre
gi|482344|        YHLFEGKHKDGFPVVLGHEGAGIVESVGPGVTEFQPGEKVIPLFISQCGE
                        aaa           bbbbbbb                bbbb
                         110         120         130        140     150
3huda     ( 100 ) crvCknpesnyClkndlgnprGtLqdgtrrFtCrgkpIhHFLgTSTFSqy
1cdoa     ( 101 ) crfCqspktnqCvkgwanespdvmSpkeTrFtCkgrkVlOFLgTSTFSqy
1teha     ( 100 ) ckfClnpktNlCgkirvtggkglMpdgtSRFtCkgktIlHmgTSTFSey
2ohxa     ( 100 ) crvCkhpeqnfClkNdlsmprGtMqdgtsRFtCrgkpIhHFLgTSTFSqy
1dlta     ( 100 ) cnaCrnpdgnlCirsDit-grgvLadyttRFtCkgkpVHHFlnTSTFTey
gi|482344|        CRFCQSPKTNQCVKGWANESPDVMSPKETRFTCKGRKVLQFLGTSTFSQY
                                                  bb          bb           b
                              160         170         180         190   200
3huda     ( 150 ) TVVdenaVAkIdaaSpLekVCLIGCGFSTGyGSAvnvAkVtpgstCAVFG
1cdoa     ( 151 ) TVVnqiaVAkIdpsApldTVCLLGCGVSTGfGAAvntAkVepgstCAVFG
1teha     ( 150 ) TVVadiSVAkIdplApldkVCLLGCGTSTGYGAAvntAkLepgsvCAVFG
```

```
 2ohxa     ( 150 )  TVVdeiSVAkIdaaSpLekVCLIGCGFSTGyGSAvkvAkVtqgstCAVFG
 1dlta     ( 150 )  TVVdesSVAkIddaAppekVCLIGCGFStGyGAAvktGkVkpgstCVVFg
Gi I 482344 I        TVVNQIAVAKIDPSAPLDTVCLLGCGVSTGFGAAVNTAKVEPGSTCAVFG
                           bbbb333bbb       bbbb3333          aaaaaaaaa       bbbbb
                              210       220    230       240    250
3huda      ( 200 )  LGGVGLSAVmGCkaagAarIIAVDinkdkfakAkelgAteCinpqdykkp
1coda      ( 201 )  LGAVGLAAVMGChsagAkrIIAVDlnpdkfekAkvFgAtdfvnpndhsep
1teha      (200 )   LGGVGLAVIMGCkvagAsrIIGVDinkdkfarAkeFgAteCinpqdfskp
2ohxa      ( 200 )  LGGVGLSVImGCkaagAarIIGVDinkdkfakAkevGAteCvnpqdykkp
1dlta      ( 200 )  LgGvGLSVImGCksagAsrIIGIdlnkdkfekAmaVgAteCiSpkdstkp
gi I 482344 I       LGAVGLAAVMGCHSAGAKRIIAVDLNPDKFEKAKVFGATDFVNPNDHSEP
                          aaaaaaaaaaaa       bbbbb    333aaaaaaa       bbb      333
                              260   270          280         290   300
3huda      ( 250 )  iqevLkemtdgGVdfSFEViGrldtMmaSLlccHEACgTsvIvgVPPASQ
lcdoa      ( 251 )  isqvLskmtngGVdfSLECvGnvgvMrnALeSClkgwGvSVlVGwT-dlh
lteha      ( 250 )  iqevLicmtdgGVdySFECiGnvkvMrAALeAChkgwGvSVvVGvaasge
2ohxa      ( 250 )  iqevLtemSngGVdfSFEVIGrldtMvtALsCCqeayGvSViVGvppdsq
ldlta      ( 250 )  isevLsemTgnnVgyTFEViGhletMidALaSChmnyGtSVvVgvppsak
gi I 482344 I       ISQVLSKMTNGGVDFSLECVGNVGVMRNALESCLKGWGVSVLVGW-TDLH
                         aaaaaaa bbbb       aaaaaaa        bbbb
                         310           320          330          340          350
3huda      ( 300 )  nlsinpmillltgRtwkgavYGgfkSkegIpklVadfmakkfsLdaLIthv
lcdoa      ( 300 )  dvatrpiqlragrtwkgSmFGgfkGkdgVpkmVkayldkkvkLdeFIthr
lteha      ( 300 )  eiaTrpfqlvtgrtwkgtaFGgwkSvesVpkLVseymskkIkVdeFVthn
2ohxa      ( 300 )  nlsmnpmlllsgrtwkgaiFGgfkSkdsVpklVadfmakkfaLdpLithv
ldlta      ( 300 )  mltydpmllftgrtwkgCvFGgLkSrddVpklVteflakkfdLdqLIthv
gi I 482344 I       DVATRPIQLTAGRTWKGSMFGGFKGKDGVPKMVKAYLDKKVKLDEFITHR
                         aaaa        bbbb        333         aaaaaaaaaaaaa                    333bbbb
                    360        370
3huda      ( 350 )  lpFekIneGFdlLhsgkSiRTVLtf
lcdoa      ( 350 )  mpLesVndAIdlMkhgkCIRTVLsl
lteha      ( 350 )  LsFdeInkAfelmhsgksIRTVVki
2ohxa      ( 350 )  lpFekIneGFdlLrsgeSIRTILtf
ldlta      ( 350 )  lpfkkIseGfellnsgqSIRTVLtf
gi I 482344 I       MPLESVNDAIDLMKHGKCIRTVLSLE
                    bb333aaaaaaaaabbbbb
```

Fig. 10.8. Protein fold recognition by FUGUE. The recognition results for cod alcohol dehydrogenase performed by FUGUE are summarized in view ranking and view alignment tables that provide links to pertaining structural information. The view ranking lists information on: PLEN, profile length; RAWS, raw alignment score; RVN, (Raw score)-(Raw score for NULL model); ZSCORE, Z-score normalized by sequence divergence; ZORI, original Z-score (before normalization); and AL, alignment algorithm used for Z-score/alignment calculation and comments on the confidence of the analysis. The view alignments provide links to sequence/structure information of the compatible (hit) structures with keys (aa, ma, mh, and hh) for alignment options of which the aa structure alignment is presented.

aa: query sequehces (including PSI-BLAST homologues) aligned against all the representative structures from a HOMSTRAD family

ma: master sequence aligned against all the representative structures from a HOMSTRAD family

mh: master sequence aligned against a single structure of highest sequence identity from a HOMSTRAD family

hh: single sequence/structure pair with highest sequence identity in 'aa'

Sequences of the representative proteins are displayed in JOY protein sequence/ structure representation. The representa-tions are uppercase for solvent inaccessible, lowercase for solvent accessible, red for α helix, blue for *β* strand, maroon for 3_{10} helix, **bold** for hydrogen bond to main chain amide, underline for hydrogen bond to main chain carbonyl, cedilla for disulfide bond, and *italic* for positive ϕ angle. The query sequence is displayed in all capital letters. The consensus secondary structure (a for α helix, b for β strand, and 3 for 3_{10} helix) as defined, if greater, than 70% of the residues in a given position in that particular conformation, is given underneath.

Post-translational Modifications and Functional Sites

The annotated amino acid sequence files normally provide information concerning the sites and types for post-translational modifications and functions. These descriptions can be found from Features line(s) in all of the three sequence formats, namely, PIR, GenPept, and Swiss-Prot. The ExPASy Proteomics tools provide numerous links for such services.

The prediction of signal peptide cleavage site is performed at CBS prediction server, SignalP (http://www.cbs.dtu.dk/services/SignalP/). On the submission form, paste the query sequence, enter the sequence name, select one of the organism groups, and request Postscript graphics then click the Submit Sequence button. The SignalP returns three scores (C for raw cleavage site, S for signal peptide, and Y for combined cleavage site) between 0 to 1 for each residue position in a table and a profile graph (Fig. 10c.15)

The CBS prediction server also provides services for predicting *O*-glycosylation sites (NetGly) in mammalian proteins and phosphorylation sites (NetPhos) in eukaryotic proteins. Paste the query sequence and click the Submit Sequence button to receive the predicted results. NetOGly returns tables of potential *versus* threshold assignments for threonine and serine residues as well as a plot of *O*-glycosylation potential versus sequence position. NetPhos returns tables of context (nanopeptides, [S,T,Y] ±4 residues) and scores for serine/threonine/ tyrosine predictions.

The ProSite (http://www.expasy.ch/sprot/prosite.html) is the collection of sequence information for biologically significant patterns, motifs, signatures, and fingerprints, each of

KESPAKKEQR CHMDPDESES NESMYCMENH SRRHIQORC KPVNIFVHES LADVQAVCSQ INVNCKNGOT NCYOENSTNH TEDCROTGSE KYPMCAYKAS QEQTHIIVAC EGNPPVVVHF DASV

[1] PDOC00001 FEUD001 ASN_GLYCOSYLATION
N–glycocylation site
Number of matches : 3
1 21–24 NSEN
2 34–37 EMTQ
3 76–79 NSTM
[2] PDOC00005 PS00005 PKC_PHOSPHO_SITE
Protein kinase C phosphorylation site
Number of matches : 2
1 31–33 SRR
2 89–91 SSK
[3] PDOC00006 PSO0006 CK2_PHOEPH0_SITE
Case in kinase II phosphorylation site
50–53 SLAX
[4] PDOCDO008 PS00008 NYRISTYL
N-myristoylation site G
68–73 OTNCY
[5] PDCOC00118 PE00127 PNASE_PANCREATIC
Pancreatic ribonuclease family signature
40–46 PVNIF
PPSearch output :
Sequence/net/nfs0/voll/production/w/nobody/tmp/132934.ppsearches (124 residues) :
Matching pattern PS00001 ASN_GLYCOSYLATION :
21 : NSSN
34 : NMTQ
76 : NSTM
Total matches : 3
Matching pattern PS00005 PKC_PHOSPHO_SITE :
31 : SRR
89 : SSK
Total matches : 2
Matching pattern PS00006, CK2_PHOSPHO_SITE :
50 : SLAD
Total matches : 1
Matching pattern PS00008 MYRISTYL :
68 : GOTNCY
Total matches : 1
Matching pattern PS00127 PNASE_PANCREATIC :
40 : CKPVNTF
Total matches : 1
Total no of hits in this sequences : 8

Fig. 10.9. ProSite search by different servers. The ProSite searches for Pig ribonuclease are performed by ScanProsite of ExPASy, PPSearch of EBI, and ProfileScan. Identical results with slightly different outputs are obtained.

Query=Unknown (cod alcohol dehydrogenase)
Size=375 Amino Acids
Blocks Searched=11117
Alignments Done= 4440928
Cutoff combined expected value for hits = 5
Cutoff block expected value for repeats/other = 5

Family		Strand	Blocks	Combined E-value
IPB002328	Zinc-containing alcohol dehydrogena	1	5 of 5	1.2e-92
IPB002364	Quinone oxidoreductase/zeta-crystal	1	2 of 4	4.9e-12
IPB002162	D-isomer specific 2-hydroxyacid deh	1	1 of 4	0.3
IPB000464	Fumarate reductase/succinate dehy	1	1 of 10	0.033
IPB001625	Glutamate/leucine/phenylalanine/val	1	1 of 4	0.42
PR01211	Synuclein signature	1	1 of 5	0.45
BP00240	COLLAGEN GLYCOPROTEIN PRECUR	1	1 of 2	0.89
IPB001557	L-lactate dehydrogenase	1	1 of 6	1.2
PR00302	Lupus La protein signature	1	1 of 5	1.7
PF001873	Domain found in IF/IF5	1	1 of 4	2.1
DM01023	2 GLYCOSYL HYDROLASES FAMILY	1	1 1 of 6	2.3
PF00718	Polyomavirus coat protein	1	1 of 8	3.1
IPB000171	Bacterial-type phytoene dehydrogena	1	1 of 5	3.2
IPB001636	SAICAR synthetase	1	1 of 7	3.9
PR01259	Na+/Ca^{2+} exchanger signature	1	1 of 4	4.9

```
>IPB002364        2/4      blocks Combined E-value= 4.9e-12 : Quinone oxidoreductase/
zeta-crystallin
Block          Frame   Location (aa)   Block E-value
IPB002364B     0       63-88           2.6e-11
IPB002364BB    0       260-271         53
Up to 1 repeats expected :
Other reported alignments :
|—2364 amino acids—|
   IPB002364 AB:.......C....................................D
         Unknown :B: :D
IPB002364B          <->b    (52, 4627) : 62
QOR_PSEAE|P43903        58          PSGLGSEGAGEVEAVGSEVTRFKVGD
                                    | ||||| || ||   |||| | |
Unknown                 63          PVVLGEGAGIVESVGPGVTeFqPGE
IPB002364D              B<->D       (110, 5614) : 171
QOR_LEIAM|P42865        212         PKGVDVVYECVG
                                      |||   ||||
Unknown                 260         nGGVDfsLECVG
```

Fig. 10.10. Search a sequence against Blocks database. The search of cod alcohol dehydrogenase sequence against Blocks database returns a list of hits and pairwise alignments of blocks (only one set is shown).

which associated with an accession number (PDOCxxxxx). Some of the descriptors are: Post-translational modification sites, Domain profiles/ sequences, DNA or RNA associated protein profiles/signatures, Enzyme active sites/signatures (for oxidoreductases, transferases, hydrolases, lyases, isomerases, and ligases), Electron transport protein signatures, Structural protein signatures, Cytokine and growth factor signatures, Hormone and active peptide signatures, Toxin signatures, Inhibitor signatures, and Secretive protein and chaperone signatures.

The ProtSite search can be conducted at ScanProsite tool (selecting Scan a sequence for the occurrence of PROSITE patterns) of ExPASy Proteomic tools (http://www.expasy:ch/tools/scnpsite.html), or PPSearch of EBI (http://www2.ebi.ac. uk/ppsearch/) or ProfileScan server (http://www.isrec.isb-sib.ch/software/PFSCAN_form.html). On the search form, paste the query sequence, select options (e.g., display in PPSearch and databases in ProfileScan) and click the Run/Start the Scan button. The search results are returned (after clicking SEView applet button in ProfileScan) with different outputs as shown in Fig. 10c.16.

To search amino acid sequence against the BLOCKS database at http://www.blocks.fhcrc.org/, select Block Searcher to open the search form. Choose the options, paste the query sequence, and click the Perform Search button. The search returns lists of matched blocks (I.D. descriptions, number of blocks, and combined *E* values), locations (amino acid positions of individual blocks and its E value), and pairwise alignment of the blocks with the matched query sequences (Fig. 10c.17).

The protein domain families database, Pfam, contains curated multiple sequence alignments for each family with linked functional annotation and profile hidden Markow models (HMMs) for finding the domains in the query sequences (Bateman et. al. 2000). The database can be accessed interactively at http://www.sanger.ac.uk/Pfam/ by *Keyword search* (enter keywords, *e.g.,* enzyme name), Protein search (paste or upload query sequence) or *DNA search* (paste query sequence). Select *Protein search* to paste or upload the query sequence and click the Search Pfam button. Pfam returns tabulated search results and alignments of the matched domains (Fig. 10c.18). Select the hyperlinked match (domain) to view the functional description, linked database, and references for the domain. Each family provides two multiple alignments that can be viewed/saved by clicking the option button. The seed alignment that contains a relatively small number of representative family members is used to build profile HMM. The family alignment contains all family members in the database. Select *Browse Pfam* on the home page to search/view the functional description (linked databases, references, alignments, and organization) of Pfam domains.

The PANAL/MetaFam server (http://mgd.ahc.umn.edu/panal/run_panal.html) analyzes protein sequence for Prosite patterns, Prosite profiles, BLOCKS, PRINTS, and Pfam. Check the options, paste the query sequence, and click the Submit button. The analytical results are returned with a graphical sketch of the predicted features, and a summary list is shown in Table 10c.5 The detailed listings of the results are also available by clicking *Full Report.*

The PEST search can be conducted at http://www.incnet.uk/ORITu/projects/pest/runpest.html. Paste the query sequence, enter options, and click Submit button. Candidate PEST sequences with hydrophobicity index and PEST score are returned.

Table 10.10. Summary List of PANAL for Cod Alcohol Dehydrogenase

Program	*Domain*	*Start*	*End*	*Motifs*	*E Value*	*Description*
PROSITE	PS00013	201	211			Membrane lipprot.lipid attachment site
PROSITE	PS00059	66	80			Copper fist DNA binding domain profile
BLOCKS	IPB002328	20	243	5 of 5	4.592e-91	Zinc-containing alcohol dehydrogena
BLOCKS	IPB002364	63	271	2 of 4	1.87e-10	Quinone oxidoreductase/zeta-crystal
PFAM	adhzinc	20	374		2.3e-134	Zinc-binding dehydrogenases

WORKSHOPS

1. Conduct the structural similarity search for a protein with the following sequence:
MSQDENIVKAVEESAEPAQVILGEDGKPLSKKALKKLQKEQEKQRKKEERALQLEAEREAREKKAAAEDT
AKDNYGKLPLIQSRDSDRTGQKRVKFVDLDEAKDSDKEVLFRARVHNTRQQGATLAFLTLRQQASLIQGL
VKANKEGTISKNMVKWAGSLNLESIVLVRGIVKKVDEPIKSATVQNLEIHITKIYTISETPEALPILLED
ASRSEAEAEAAGLPVVNLDTRLDYRVIDLRTVTNQAIFRIQAGVCELFREYLATKKFTEVHTPKLLGAPS
EGGSSVFEVTYFKGKAYLAQSPQFNKQQLIVADFERVYEIGPVERAENSNTHRHMTEFTGLDMEMAFEEH
YHEVLDTLSELFVFIFSELPKRFAHEIELVRKQYPVEEFKLPKDGKMVRLTYKEGIEMLRAAGKEIGDFE
DLSTENEKFLGKLVRDKYDTDFYILDKEPLEIRPFYTMPDPANPKYSNSYDFFMRGEEILSGAQRIHDA
LLQERMKAHGLSPEDPGLKDYCDGFSYGCPPHAGGGIGLERVVMFYLDLKNIRRASLFPRDPKRLRPCGV
PSFPPNLSARVVGGEDARPHSWPWQISLQYLKDDTWRHTCGGTLIASNFVLTAAHCISNTWTYRVAVGKN
NLEVEDEEGSLFVGVDTIHVHKRWNALLLRNDIALIKLAEHVELSDTIQVACLPEKDSLLPKDYPCYVTG
WGRLWTNGPIADKLQQGLQPVVDHATCSRIDWWGFRVKKTMVCAGGDGVISACNGDSGGPLNCQLENGSW
EVFGIVSFGSRRGCNTRKKPVVYTRVSAYIDWINEKMQL

2. Predict the secondary structure of the following protein by GOR IV method.

CGVPSFPPNLSARVVGGEDARPHSWPWQISLQYLKDDTWRHTCGGTLIASNFVLTAAHCISNTWTYRVAV
GKNNLEVEDEEGSLFVGVDTIHVHKRWNALLLRNDIALIKLAEHVELSDTIQVACLPEKDSLLPKDYPCY
VTGWGRLWTNGPIADKLQQGLQPVVDHATCSRIDWWGFRVKKTMVCAGGDGVISACNGDSGGPLNCQLEN
GSWEVFGIVSFGSRRGCNTRKKPVVYTRVSAYIDWINEKMQL

3. Perform the consensus secondary structure prediction of pea alcohol dehydrogenase at NSP@ by selecting six different methods.

>gi|81891|pir||S0012 alcohol dehydrogenase (EC 1.1.1.1) 1-graden pea

MSNTVGQIIKCRAAVAWEAGKPLVIEVEVAPPQAGEVRLKILFTSLCHTDVYFWEAKGQTPLFPRIFGH
EAGGIVESVGEGVTHLKPGDHALPVFTGECGECPHCKSEESNMCDLLRINTDRGVMLNDNKSRFSIKGQP
VHHFVGTSTFSEYTVVHAGCVAKINPDAPLDKVCILSCGICTGLGATINVAKPKPGSSVAIFGLGAVGLA
AAEGARISGASRIIGVDLVSSRFELAKKFGVNEFVNPKEHDKPVQQVIAEMTNGGVDRAVECTGSIQAMI
SAFECVHDGWGVAVLVGVPSKDDAFKTHPMNFLNERTLKGTFYGNYKPRTDLPNVVEKYMKGELELEKFI
THTVPFSEINKAFDYMLKGESIRCIIKMEE

4. List SCOP lineages for the following proteins:
   a. Pig D-amino acid oxidase, N-terminal domain
   b. Human carbonic anhydrase
   c. Human cyclin A
   d. Nuclear estrogen receptor, DNA-binding domain
   e. Goose lysozyme, active site domain
   f. HIV-1 reverse transcriptase, palm domain

5. Furnish CATH architecture descriptions for the following proteins:
   a. Tyrosyl-tRNA synthetase
   b. Serine/threonine protein phosphatase
   c. ATP Synthase, domain 1
   d. Glysosyl transferase
   e. Phenylalanine-tRNA synthetase
   f. Phenylalanine-tRNA synthetase
   g. Carbonic anhydrase II
   h. Dihydrolipoamide transferase
   i. Inorganic pyrophosphatase
   j. Dihydrofolate reductase, subunit A
6. Submit the following sequence to a structure prediction by displaying the secondary structure and core index in the multiple alignment of the query sequence, and with the representative sequences of proteins whose 3D structures are known.

KVYERCELAAAMKRLGLDNYSRGYLGNWVCAANYESSFNTQATNRNTDGSTDYGILEINSRWWCDNGKTP
RAKNACGIPCSVLLRSDITEAVKCAKRIVSDGDGMNAWVAWRNRCKGTDVSRWIRGCRL

Retrieve the PDB file of the protein which is structurally the most homologous to the query protein.

7. Submit the following sequence to structure prediction by the multiple alignment of the query sequence with the reference sequences of proteins whose 3D structures are known in Joy sequence/structure representation.

RTDCYGNVNRIDTTGASCKTAKPEGLSYCGVPASKTIAERDLKAMDRYKTIIKKVGEKLCVEPAVIAGII
SRESHAGKVLKNGWGDRGNGFGLMQVDKRSHKPQGTWNGEVHITQGTTILTDFIKRIQKKFPSWTKDQQL
KGGISAYNAGAGNVRSYARMDIGTTHDDYANDVVARAQYYKQHGY

Retrieve the PDB file of the protein which is structurally most homologous to the query protein.

8. What is the most likely structural domain/motif of a protein with the following sequence?.

MIKLAKFPRDFVWGTATSSYQIEGAVNEDGRTPSIWDTFSKTEGKTYKGHTGDVACDHYHRYKEDVEILK
EIGVKAYRFSIAWPRIFPEEGKYNPKGMDFYKKLIDELQKRDIVPAATIYHWDLPQWAYDKGGGWLNRES
IKWYVEYATKLFEELGDAIPLWITHNEPWCSSILSYGIGEHAPGHKNYREALIAAHHILLSHGEAVKAFR
EMNIKGSKIGITLNLTPAYPASEKEEDKLAAQYADGFANRWFLDPIFKGNYPEDMMELYSKIIGEFDFIK
EGDLETISVPIDFLGVNYYTRSIVKYDEDSMLKAENVPGPGKRTEMGWEISPESLYDLLKRLDREYTKLP
MYITENGAAFKDEVTEDGRVHDDERIEYIKEHLKAAAKFIGEGGNLKGYFVWSLMDNFEWAHGYSKRFGI
VYVDYTTQKRILKDSALWYKEVILDDGIED

9. Identify the signal peptide and family signature sequences of big pancreatic ribonuclease with the given amino acid sequence:

>gi | 67344 | pir | NRPG pancreatic ribonuclease (EC 3.1.27.5) – pig

KESPAKKFQRQHMDPDSSSSNSSNYCNLMMSRRNMTQGRCKPVNTFVHESLADVQAVCSQINVNCKNGQT
NCYQSNSTMHITDCRQTGSSKYPNCAYKASQEQKHIIVACEGNPPVPVHFDASV

10. Identify the ProSite residues and protein domain families for the following proteins of given amino acid sequences:

>gi|11373714|pir||CCZP cytochrome c – Schizosaccharomyces pombe

MPYAPGDEKKGASLFKTRCAQCHTVEKGGANKVGPNLHGVFGRKTGQAEGFSYTEANRDKGITWDEETLF
AYLENPKKYIPGTKMAFAGFKKPADRNNVITYLKKATSE

>gi|7430909|pir||552316 flavodoxin-Escherichia coli

MNMGLFYGSSTCYTEMAAEKIRDIIGPELVTLHNLKDDSPKLMEQYDVLILGIPTWDFGEIQEDWEAVWD
QLDDLNLEGKIVALYGLGDQLGYGEWFLDALGMLHDKLSTKGVKCVGYWPTEGYEFTSPKPVIADGQLFV
GLALDETNQYDLSDERIQSWCEQILNEMAEHYA

>gi|2318115|gb|AAB66483.1| serine protease [Homo sapiens]

MKKLMVVLSLIAAAWAEEQNKLVHGGPCDKTSHPYQAALYTSGHLLCGGVLIHPLWVLTAAHCKKPNLQV
FLGKHNLRQRESSQEQSSVVRAVIHPDYDAASHDQDIMLRLARPAKLSELIQPLPLERDCSANTTSCHI
LGWGKTADGDFPDTIQCAYIHLVSREECEHAYPGQITQNMLCAGDEKYGKDSCQGDSGGPLVCGDHLRGL
VSWGNIPCGSKEKPGVYTNVCRYTNWIQKTIQAK

>gi|7489456|pir||T03296 beta-glucosidase (EC 3.2.1.21), rice chloroplast

MADLFTEYADFCFKTFGNVKHWFTFNEPRIVALLGYDQGTNPPKRCTCKCAAGGNSATEPYIVAHNFLLS
HAAAVARYRTKYQAAQQGKVGIVLDFNWYEALSNSTEDQAAQRARDFHIGWYLDPLINGHYSQIMQDLV
KDRLPKFTPEQARLVKGSADYIGINQYTASYMKGQQLMQQTPTSYSADWQVTYVAKNGKPIGPQANSNW
LYIVPWGMYGCVNYIKQKYGNPTVVITENGMDQPANLSRDQYLRDTTRVHFYRSYLTQLKKAIDEGANVA
GYFAWSLLDNFEWLSGYTSKFGIVYVDFNTLERHPKASAYWFRDMLKH

>gi|6319698|ref|NP_009780.1|beta subunit of pyruvate dehydrogenase – Saccharomyces cerevisiae

MFSRLPTSLARNVARRAPTSFVRPSAAAAALRFSSTKTMTVREALNSAMAEELDRDDDVFLIGEEVAQYN

GAYKVSKGLLDRFGERRVVDTPITEYGFTGLAVGAALKGLKPIVEFMSFNFSMQAIDHVVNSAAKTHYMS
GGTQKCQMVFRGPNGAAVGVGAQHSQDFSPQYGSIPGLKVLVPYSAEDARGLLKAAIRDPNPVVFLENEL
LYGESFEISEEALSPEFTLPYKAKIEREGTDISIVTYTRNVQFSLEAAEILQKKYGVSAEVINLRSIRPL
DTEAIIKTVKKTNHLITVESTFPSFGVAEIVAQVMESEAFDYLDAPIQRVTGADVPTPPYAKELEDFAFP
DTPTIVKAVKEVLSIE

TRANSMISSION OF SIGNALS

Orchestrating the activity of cell populations for physiological functioning of the brain, organs, and musculature depends on transmission of signals, learning and memory devices, and feedback control systems. By what biophysical mechanisms do cells communicate in order to coordinate their activity as local ensembles, as multimodal circuits, and across system levels? Here we only scratch the surface of this fascinating topic. We will focus on electrically active cells; for this, you can have in mind, for example, cardiac cells, many types of secretry cells, and neurons. We know that ionic currents underlie cellular electrical activity. Hence, one way that cells can interact is by directly passing ionic current between each other.

Perhaps the simplest mechanism for such communication is the analogue of resistive coupling between units, *i.e.,* with the intercellular current being proportional to the voltage difference between cells. Heart cells, and many other types of cells, communicate in this manner, with the ions flowing directly between two coupled cells. In this case, referred to as electrical coupling (and sometimes, in the neural context, electrotonic coupling), the current flows through channel proteins that span the plasma membranes of both cells.

The clusters of such channel proteins that are found at cell-to-cell contacts are called gap junctions. While gap junctions are occasionally found in neural circuits, they seem to be more common during development or they may not constitute an exclusive means of interaction. Neurons have a rich repertoire of other ways for exciting or inhibiting other target neurons, via the indirect means of chemical synaptic transmission depicted in Fig. 11.5. At a terminal of a neuron's axon the neurotransmitter of one type or another is bundled in vesicles that are released with increased probability when the terminal is active (depolarized). The transmitter, released into the extracellular space in the terminal region, diffuses and binds to receptors on the postsynaptic cell's nearby membrane. These receptors may be part of a receptor-channel

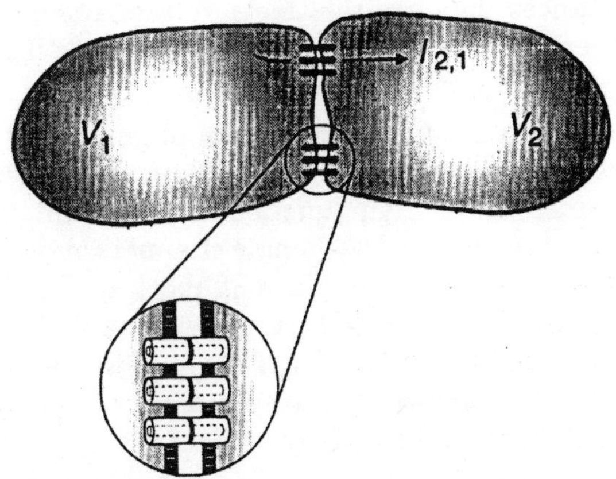

Fig. 11.1. Schematic of gap junctions between two cells. Coupling current from cell #1 to #2 is $I_{2,1} = g_c \langle V_1 - V_2 \rangle$, the exact negative of $I_{1,2}$.

complex or be linked indirectly through second messengers to a nearby channel. The activated channels then lead to the postsynaptic current and action. The variety of transmitters and receptors enable many different time scales and "sign" of the input that is being delivered to the target cell. We find synapses that are excitatory or inhibitory; they can be fast or slow; they can be shunting; their synaptic parameters can change with usage, either, depressing or facilitating; they can be voltage gated or not on the postsynaptic side. Gap-junctional coupling is typically localized, certainly for cells without spatially extended processes, to nearby neighbors.

Prime examples include the heart and islets of Langerhans in the pancreas. In contrast, neurons can interact across distances that are many times greater than a cell body diameter, by means of their potentially far-reaching axonal and dendritic arbors. The synaptic interactions enable them to participate in local calculations and with distant assemblies, on selective time scales, fast or slow and to various degrees.

For long-distance communication between cells propagated action potentials typically mediate signal transfer to the synapses via axons. Given the variety of coupling mechanisms and connectivity patterns there are many possible behavioral modes, spontaneous and/or stimulus-driven, that such circuits may exhibit. Obviously we cannot elaborate on many of these behaviors; we will restrict attention to a few examples and focus on pairwise interactions. Also we choose not to present here some other mechanisms for intercellular interaction such as coupling by cell-generated electrical fields, diffusion and exchange of ions (that effect Nernst equilibrium potentials) and second messengers, mechanical and hormonal effects.

Electrical Coupling and Gap Junctions

The proteins that constitute the channels at gap junctions are of the connexin family. Connexin molecules in the plasma membrane of one cell link up those in an adjacent cell to form the channels. These channels pass most ions as well as various molecules up to molecular weights of tens of kilodaltons, including those involved in second messenger systems such as IP_3. The single channel conductance can be in the range of 75-150 pS, and in most known instances they are relatively voltage-independent. The variety of connexins provides a rich repertoire for diverse modulatory possibilities by changes in intracellular pH, second messengers, neurotransmitters, or voltage.

For example, the reduction of gap-junctional coupling due to increasing levels of Ca^{2+} is described as a protection mechanism. For cell-level modeling we typically represent the net conductance of a gap junction as a constant, the product of the mean number of open channels at the junction and the single channel conductance. Simultaneous pairwise recordings, at least for geometrically simple neighboring cells, can be used to estimate this conductance. For spatially extended cells, like neurons with branching dendrites, such measurements would be confounded especially if the gap junctions are at electrically remote sites. Gap junctions may also be detected by dye coupling, although this method is problematic without good controls to test how readily the dye passes.

Synchronization of Two Oscillators

Consider the case of two cells idealized as isopotential compartments as shown in Fig. 11.1. The gap-junctional current $I_{2,1}$ that flows from cell 1 into cell 2 is written as

$g_c(V_1 - V_2)$, where g_c is the net coupling conductance of all the gap junctions formed between the two cells. This current appears as a source term in the current balance equation for cell 2. Defined as such, with the current as leaving from cell 1, it appears as a sink term for cell 1 (or equivalently, $g_c(V_2 - V_1)$ is a source term into cell 1). Thus we have

$$C_m \frac{dV_1}{dt} = -I_{ion,1} + I_{app,1} + g_c(V_2 - V_1), \quad ...(11.1)$$

$$C_m \frac{dV_2}{dt} = -I_{ion,2} + I_{app,2} + g_c(V_1 - V_2), \quad ...(11.2)$$

where $I_{ion,1}$ is a function of V_1 and the set of gating variables w_1 in cell 1, and $I_{ion,2}$ is similarly dependent on V_2 and w_2. Let us first predict the behavior, say in the case where the gap junction conductance is large. This means that the cells are very tightly electrically coupled. We would then expect them to have approximately the same voltage. Indeed if the cells are identical, then $V_1 = V_2$ is always a solution: The coupling current would be zero in this case. Of course, a perturbation (*e.g.*, brief current pulse) to just one cell would make their *V*'s differ, transiently. By subtracting the two equations and then dividing by g_c we see, after dropping the term of order $1/g_c$, that

$$\tau_c \frac{d(V_1 - V_2)}{dt} \approx -(V_1 - V_2), \quad ...(11.3)$$

where $\tau_c = C_m/g_c$. This shows that the cells will re-establish uniformity with an effective time constant of τ_c, that is very short when g_c is large.

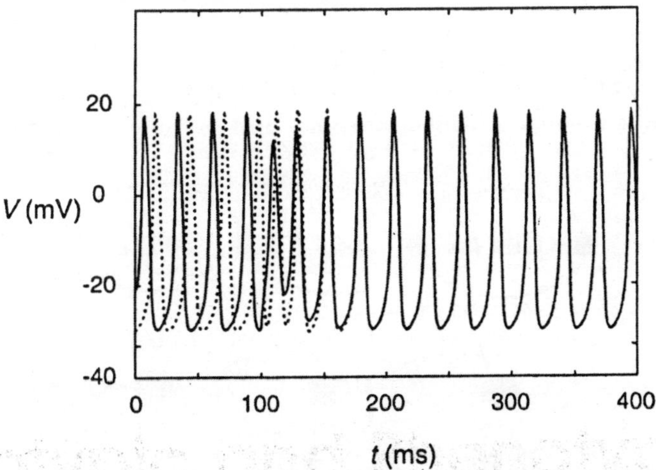

Fig. 11.2. Two identical Morris-Lecar neuron models, auto-rhythmic, synchronize with electrical coupling (turned on at t = 100 ms). Parameter values for the model are those of with the following exceptions: $v3 = -5$ mV, $v4 = 10$ mV, $\phi = 0.5$, $g_{Ca} = 8$ mS/cm^2, $I_{app} = -10\mu A/cm^2$. Net gap junction coupling conductance is $g_C = 2$ mS/cm^2. Initial conditions: $V_1 = -20$ mV, $w_1 = 0.2$, $V_2 = -30$ mV, $w_2 = 0$.

This is illustrated in Fig. 11.2 for two Morris-Lecar cells that are tuned into an oscillatory regime. In this example the cells are initially out of phase and uncoupled. Synchronization occurs promptly after the coupling is introduced at $t = 100$ ms.

Asynchrony Between Oscillators

While electrical coupling is typically considered as a mechanism for uniformizing cells, one should be aware that the outcome depends considerably on coupling strength and dynamic features of the individual cells. For example, the same cells used above when coupled weakly with gap junctions do not synchronize. Instead, they establish a rhythm with the two cells in a stable antiphase-locked pattern: just the opposite of togetherness (Fig. 11.3). Also, for this situation the period of the network oscillation depends on the value of g_c. This is not true for the in-phase pattern. When the coupling strength is large enough, the period (for a pair of identical cells) is that of the isolated cell and is therefore independent of g_c.

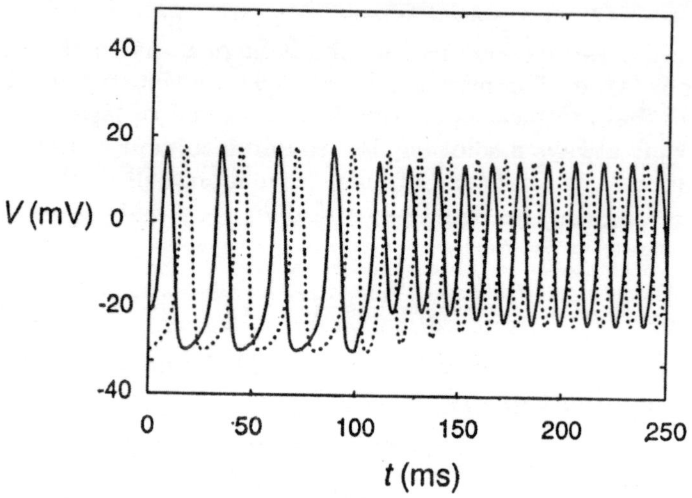

Fig. 11.3 Time courses of V_1, V_2 showing antiphase behavior when weak gap-junctional coupling is turned on at t = 100 ms. Same initial conditions and parameters as in Fig. 11.2, except that here g_c = 1 mS/cm^2.

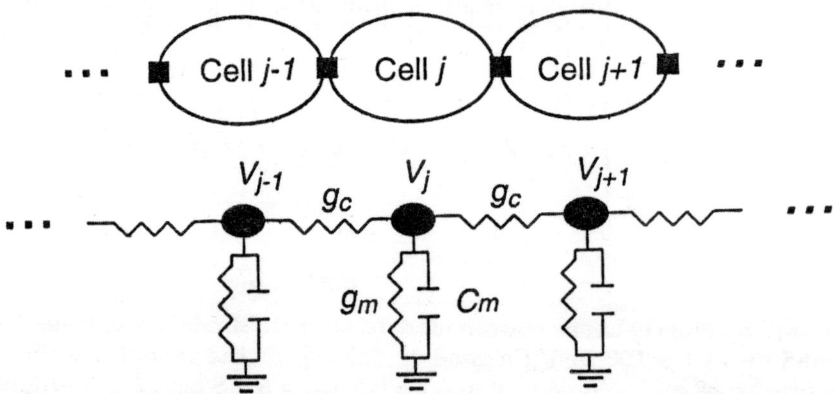

Fig. 11.4. Schematic of linear array of passive cells coupled (nearest neighbor) by gap junctions.

Cell Ensembles, Electrical Coupling Length Scale

Such synchronization also occurs for spatially distributed multicell systems; even though electrical coupling is localized it can, when large enough, synchronize a population that might be spread over a sizable spatial region. In order to formalize this, one can introduce the concept of an electrical length scale that involves coupling conductance as well as leakage conductance. This emphasizes the relativeness of intercellular current to the current flowing across a cell's plasma membrane. By using this notion we can refer to the "electrical size" of a cellular array; it could be quite compact even though the spatial extent might not be.

In order to formalize this notion we consider the simple case of Fig. 11.4, where we have a chain of gap-junction-coupled cells that have passive membrane properties, *i.e.*, $I_{\text{ion}} = g_m V$. Thus we have the equation for a typical cell, cell j:

$$\frac{dV_j}{dt} = \frac{g_c\left(V_{j-1} - V_j\right) - g_m V_j + g_c\left(V_{j+1} - V_j\right)}{C_m}. \quad \text{...(11.4)}$$

Suppose that we have a very long chain of cells (and for now, ignore end effects) and suppose we voltage clamp the cell in the middle for which $j = 0$ ($j > 0$ refers to cells to the right and $j < 0$ is for cells to the left). In this passive system the voltage distribution will go to a steady state after some transient, and the voltage will decrease from V_0 as $|j|$ increases. This attenuation occurs as current spreads from one cell to the next away from $j = 0$ and as current is lost through the "leaky" membrane of each cell.

The steady voltage decays in an exponential fashion with distance. The decay "rate" y-gan can be found by considering the steady-state case of (11.4) (*i.e.*, setting $dV_j/dt = 0$) and seeking a solution of the form $V_j = V_0 \gamma^j$. This leads to a quadratic equation for γ:

$$\gamma^2 - \left(2 + g_m / g_c\right)\gamma + 1 = 0. \quad \text{...(11.5)}$$

Hence, there are two values of γ :

$$\gamma\pm = 1 + g_m/(2g_c) \pm \sqrt{\left[1 + g_m/(2g_c)\right]^2 - 1}. \quad \text{...(11.6)}$$

Note that both roots are positive with $\gamma_+ > 1$ and $\gamma_- < 1$.

For a voltage profile that decays with distance to the right of $j = 0$ we must disallow γ_+ (such a term would grow with j). Thus, if we ignore end effects due to termination of the chain, the steady state profile for $j \geq 0$ has the form

$$V_j = V_0 \gamma_- \quad \text{...(11.7)}$$

The solution for; $j \leq 0$ is $V_j = V_0 \gamma_j = V_0 \gamma^{-1}{}_-$, since $\gamma_+ \gamma_-^{-1} = 1$.

We can test our intuition by considering some limiting cases. When g_m/g_c is large, the leakage conductance is dominant and the voltage should attenuate rapidly with distance from $j = 0$. Indeed, $\gamma_- \approx g_c / g_m$, so that V_j decays steeply. On the other hand, when g_m/g_c is small, then $\gamma_- \approx 1 - \sqrt{g_m / g_c}$ and the spatial decay is gradual.

Pursuing this latter case a bit more, if cell-to-cell attenuation is small and the spatial profile changes smoothly, we might treat the one-dimensional cellular array approximately as a continuum with position $x \approx j\Delta x$, where Δx is a cell-size parameter, considered small. The leakage conductance g_m is proportional to a cell's surface area, which is the perimeter of a cross section times Δx. We will take some care with the representation of g_c, Let's think about the reciprocal, $r_c = 1/g_c$, the resistance to longitudinal current flow between cells, say from one cell center to the next. Then, r_c involves resistance due to the gap junctions as well as cytoplasmic resistance along the cell's axial direction. If there are many gap junctions between adjacent cells, then the major contribution to r_c will be the cytoplasmic term, which is proportional to the area of a cross section times Δx. Consequently, the ratio g_m/g_c $(= g_m r_c)$ is proportional to Δx^2 and may be written as $(\Delta x/\lambda)^2$. Finally, combining this with our expression for γ_- (from (11.6) for small g_m/g_c) and noting that $\log(\gamma_-) \approx -\sqrt{g_m / g_c}$, we see that the solution form for the continuum approximation is

$$V(x) = V_j = V_0 \exp(j \log \gamma-) \approx V_0 \exp(-j\Delta x/\lambda) = V_0 \exp(-x/\lambda). \quad ...(11.8)$$

Thus, λ is the electrical length constant for the continuum approximation.

This treatment relates nicely to continuum models for current spread in individual cells that have extended processes, like neurons with their dendrites or axons. These processes behave like electrical cables; current flows along the axis and leaks through the membrane. In this case of a neuronal process, we could think of it for simulation purposes as a chain of short compartments (length Δx) each with a membrane surface area and connected by axial resistances; there are no gap junctions between the fictitious adjacent compartments. If d is the diameter of the cross section, then we would have $g_m = \Delta x \pi d G_m$, where G_m is the leak conductance density (S/cm^2) and $r_c = \Delta x 4Ri(\pi d^2)$. where R_i is the cytoplasmic-specific resistivity (ohm-cm). Putting this together we obtain for the electrical length constant of a passive neuronal dendrite or axon $\lambda = \sqrt{d/(4G_m R_i)}$. As an illustration, for some neurons estimates are in the range G_m = 0.25 mS/cm^2, R_i = 100 $\Omega \cdot$cm, so that with d = 2.5 μm yields λ = 500 µm. This means that for a dendritic branch of physical length 250 µm, the corresponding electrotonic length is about 0.5. Therefore, we do not expect excessive attenuation for steady or slow voltage changes.

Transients decay more abruptly. Also, the attenuations are more severe from dendritic locations to the cell body than vice versa. In electrically coupled excitable systems with weaker gap junctions or extended electrical size (physical length larger than A) synchrony is not quickly established, and localized perturbations may lead to waves or other spatiotemporal patterning. The treatment of dendrites as cables and the compartmental method is discussed in and in more depth in Johnston and Wu (1995), Koch and Segev (1998), Koch (1999), and Keener and Sneyd (1998). We note that this discretization of neuronal cable-like processes is called the compartmental method and was first developed by Rall for treating the effect of spatiotemporally distributed synaptic inputs over a neuron's somatic-dendritic area.

Synaptic Transmission Between Neurons

In the preceding section we saw that strong gap junctions can be used for coordinating cellular electrical activity, subserving functions in which cells might work together in approximate synchrony. The bidirectionality and instantaneous nature of gap-junction coupling are well suited for achieving these goals. In the nervous system there are demands for more complex patterning. Individual cells generally are likely involved in numerous different computations, with possibly different time scales, sometimes being called into action with a brief wake-up call but in other cases only after a long barrage of inputs. In some cases the precise timing of action potentials might be important, and we would expect fast coupling mechanisms. But when firing rate, rather than spike timing, is more important for signaling, perhaps the synchronization properties of fast coupling may be less critical and slower coupling mechanisms might dominate.

Some flexibility for multiplexing of these signaling modes is attainable by exploiting the potentially large variety of intrinsic mechanisms, *i.e.,* the many different ionic channels and modulators for them. Even greater computational power can be achieved, however, by selective adjustment of the many synpses on a cell, maybe in an activity-dependent fashion. One can imagine a great many neural ensembles with possibly shared units that can be dynamically constructed, recruited, and dismissed as needed for memory storage, sensory processing and perception, and executive command functions. It is hard to imagine how instantaneous bidirectional coupling could allow in an efficient way such a rich set of alternatives.

However, the one-way signaling via chemical synapses enables the system to employ a few transmitters but yet enrich the possibilities for postsynaptic response by having many choices for the postsynaptic receptors and channels. Sites for modulatory action can be implemented on either the postsynaptic or presynaptic side. Thus while vesicle-packaged transmitter is released in punctuate fashion, like the action potential, the time scale and even the "sign" of the response are determined by the machinery on the postsynaptic cell.

Kinetics of Postsynaptic Current

We will formulate an idealized model for the current generated in a postsynaptic cell due to neurotransmitter release by a presynaptic cell in which an action potential has occurred. The sender's action potential opens voltage gated calcium channels at the axon terminal where vesicles are poised ready to fuse with the membrane and release transmitter.

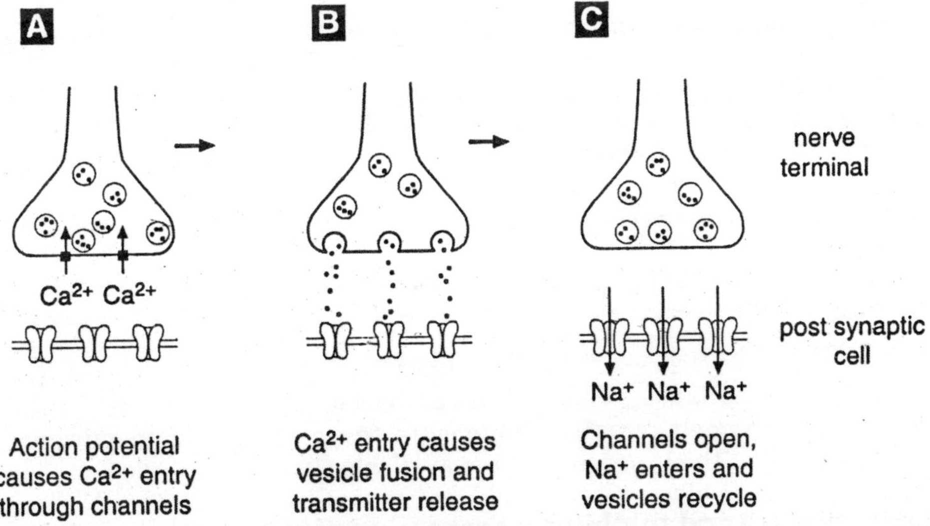

Fig. 11.5. Steps in chemical synaptic transmission. Action potential arrives at terminal of presynaptic axon; depolarization opens voltage gated calcium channels. Resulting calcium influx causes vesicle fusion and transmitter release. Transmitter diffuses, reaches postsynap-tic membrane, and binds to receptors. The activated receptors cause postsynaptic channels to open. Depending on which type of ions are allowed to pass (sodium shown here) the postsynaptic response can be depolarization or hyperpolarization.

Whether or not release occurs is a probabilistic event. The failure rate can be high at some, *e.g.*, cortical, synapses. There is also a small probability of spontaneous release. Experiments at the neuromuscular junction and central synapses use the quantal release hypothesis to analyze the statistics of postsynaptic responses for spontaneous and evoked release to estimate the number of active sites, quantal content, and release probability. We will not consider these issues here, nor the details of transmitter diffusion in and removal from the synaptic cleft, the small extracellular space between the pre- and postsynaptic sites. We suppose that transmitter is available briefly, and that during this time it can bind to receptors, actually receptor-channel complexes, on the postsynaptic membrane. For this simple model we imagine a two-state channel.

Binding of transmitter to a receptor molecule favors opening of the channel and unbinding leads to closing. A kinetic scheme for this is the following:

$$C + T \underset{\beta'}{\overset{\alpha'}{\rightleftharpoons}} O.$$

If s is the fraction of complexes in the open state O, then $1 - s$ is the fraction closed, and we have, given a transmitter concentration [T],

$$\frac{ds}{dt} = \alpha'[\mathrm{T}](1-s) - \beta' s. \qquad \text{...(11.9)}$$

This representation assumes unlimited transmitter and receptor availability. Also, it requires us to specify a time course for [T]; for example, we might assume a square pulse over some time duration (perhaps with a modest delay, on the order of a millisecond or less) when the presynaptic cell fires. We will adopt a simpler, and easy to implement, scheme, supposing that maximal transmitter is available when the presynaptic membrane potential V_{pre} is above some level, say θ_{syn}. This would mean $[\mathrm{T}] = \mathrm{T}_{max}\, \mathrm{S}_\infty$ with $= H(V_{pre} - \theta_{syn})$, where $H(x)$ is the Heaviside step function. We will smooth this out using a sigmoidal function for s_∞ Redefining the rate constants, which we view as adjustable depending on the desired time course of postsynaptic conductance, we get

$$\frac{ds}{dt} = a s_\infty (1-s) - \beta s, \qquad \text{...(11.10)}$$

with $s_\infty = 1/[1 + \exp((V - \theta_{syn})/k_{syn})]$, with k_{syn} positive. The current through the post-synaptic membrane is

$$I_{syn} = g_{syn} s (V - V_{syn}), \qquad \text{...(11.11)}$$

where the reversal potential V_{syn} depends on the concentration differences for the ion species that flow through the open receptor channels in the postsynaptic membrane; g_{syn} is the maximal conductance at the synapse if all the available channels are open.

Synapses: Excitatory and Inhibitory; Fast and Slow

This type of model has been used to describe some common types of excitatory and inhibitory synapses that have relatively simple kinetics. The different transmitters and different receptors and different ions that pass through the synaptic channels lead to a variety of synaptic types. The most commonly known transmitter that is used for excitatory synapses is glutamate. It can activate AMPA receptor channels that typically pass inward current (carried mostly by sodium ions, but also some potassium and other ions). The postsynaptic conductance is relatively fast with a rise time ($\approx \alpha^{-1}$) of order 1 millisecond and decay times ($\approx \beta^{-1}$) of a few to tens of milliseconds.

Expressions such as (11.10) have been used for these synapses. The value of V_{syn} in this case is typically about 100 mV above rest. Glutamate can also evoke slower postsynaptic responses by activating NMDA receptors (fast rise, of order milliseconds, and slow decay, of order of individual inputs. This is because the synaptic current depends on V_{post}, in particular through the driving force $V_{post} - V_{syn}$. If V_{post} is far from V_{syn}, then the dependence on V_{post} is very weak, as, say, for AMPA-mediated excitation if V_{post} is near rest. But in other cases, say for a $GABA_A$-mediated synapse, we would not get linear summation:

Doubling the number of inputs does not double the synaptic current. This is easy to see for a passive membrane system, as we considered above, with steady synaptic conductance input g_s = constant = $g_{syn}\, s$. The steady response $\overline{V}$ would be

$$\overline{V} = V_{syn}\, g_s/(g_m + g_s), \qquad ...(11.12)$$

saturating for large g_s at V_{syn}, as we expect. While increasing g_s drives V_{post} closer to V_{syn}, the increment in synaptic current diminishes for large g_s: dI_{syn}/dg_s behaves like g_s^{-2} for large g_s; this is the sublinear summation effect. When synaptic inputs are activated repeatedly the response may not remain uniform from one stimulus to the next in a train. This dynamic aspect may involve facilitation or depression of successive responses on a range of time scales.

For example, the response might be depressed by 50% or more to a steady level after less than ten stimulus repeats (say at 50 Hz), and then recovery may take a second or soe. A favoured mechanism for depression is presynaptic depletion of ready-to-release transmitter. Synaptic depression is analogous to receptor desensitization in other contexts, such as hormonal communication. Various possible functional roles for such synaptic dynamics in network activity have been suggested.

One hypothesis is that synaptic depression could mediate phasic behaviour of a feedforward network. That is, if the steady response saturates beyond some input rate (so the tonic output conveys no information about the input rate), the network could still be responsive to transient changes in input rate; just after a sudden increase in input rate the synapses would not yet be adapted to their new target level of depression, so the network's output would briefly rise and then fall back to the saturation level. In the context of recurrently connected networks synaptic depression provides a mechanism of slow negative feedback for the intercellular coupling. This can underlie rhythmic population activity even if isolated units are capable only of tonic firing for steady inputs; here the autocatalysis could be provided by recurrent excitatory connections.

When Synapses Might (or Might Not) Synchronize Active Cells

In order to illustrate some effects of mutual synaptic coupling between cells we will consider the simple case of two identical cells, each of which is autorhythmic, and examine conditions for which the cells tend to fire together or apart. We are asking about the synchronization patterns in this simple two-cell network. Classical notions are that mutual excitation tends to make cells synchronous (in phase), while mutual inhibition pushes them apart, leading to antiphase behaviour. That is, if two excitatory many tens to hundreds of ms) or a number of different types of metabotropic glutamate receptors with very slow rise and slow decay of order seconds.

The conductance associated with NMDA receptors is, curiously, also gated by the postsynaptic voltage V_{post}. If V_{post} is not adequately above V_{pest}, the NMDA-associated channels are blocked by magnesium ions from the outside; sufficient depolarization relieves this block, and the channels can open. The dependence of synaptic current on pre- and postsynaptic activity implicate the NMDA conductance in various models for associative learning and synaptic plasticity generally. The very slow metabotropic glutamate-mediated effects are believed to be primarily modulatory, acting, for example, to effectively change the firing threshold of a cell. In the examples below of excitatory synaptic coupling we will be thinking primarily of AMPA-mediated excitatory synapses, for which (11.10) is a good first approximation.

More complete descriptions, even for this simple receptor-channel complex, might involve kinetic models with additional states. Fast inhibitory synapses are activated by the transmitter GABA when they bind to $GABA_A$ receptor channels. The postsynaptic conductance has a fast rise time, one or a few milliseconds, and somewhat slower decay than the AMPA-excitatory conductance, on the order of tens of milliseconds. The current is carried largely by chloride ions. Interestingly, the value for V_{syn} can be quite variable, typically –60 mV to –80 mV but in some cases as depolarized as — 35 mV; this variability likely reflects different types or activities of chloride pumps that regulate intracellular Cl^- concentration. If a cell is sitting near its resting potential of say — 65 mV, the synaptic current that is generated by activating these channels would be outward (hyperpolarizing) if V_{syn} is below — 65 mV or inward (depolarizing) if V_{syn} is above —65 mV. It is not uncommon for $GABA_A$ - mediated currents to be depolarizing in developing neural tissue. Thus it is not strictly correct, although frequently done, to refer to GABA as an inhibitory transmitter.

One should take care in describing a $GABA_A$ - mediated synapse as inhibitory; the currents sign depends on where V_{syn} is relative to V. Note that if V is close to V_{syn}, little synaptic current will be generated, even if a large conductance $g_{syn}s$ is activated. On the other hand, the membrane potential would be effectively clamped to V_{syn} until the conductance deactivates. In this case the $GABA_A$ synapse acts as a strong shunt in the membrane; other modest-sized synaptic inputs would be ignored during this time. The transmitter GABA acting through $GABA_B$ receptors can lead to a very slow inhibition with a conductance that rises and decays on the order of 100 or more milliseconds. The current is carried primarily by K^+ ions and so the reversal potential may be –70 to –90 mV. The simple model in (11.10) cannot account for this slow current. One shortcoming is that the model predicts a peak shortly after the depolarization of V_{pre}, while $GABA_B$-mediated inhibition peaks only much later.

A minimal model for this current would involve at least two dynamic variables. Typical sizes of I_{syn} at a single synapse are a few picoamperes and may evoke responses (*e.g.,* in cortical neurons) of 0.1 to 1 mV. A number of inputs must be summed in order to bring the neuron to firing threshold. Generally, the postsynaptic response to multiple inputs, even in the subthreshold regime, is not, however, a linear summation cells are somewhat out of phase, the firing of the leading cell encourages the follower to fire sooner, thus bringing the cells more nearly in phase, while for inhibitory cells that are somewhat out of phase the leading cell's firing will delay the follower, increasing their phase difference.

Successive cycles would increase their phase difference to 180° but not beyond, since then the follower would act as the leader, pushing the phase difference back toward 180°. These expectations and some surprises will be illustrated below when we consider the effects of synaptic time scales. Here we will use the Morris-Lecar model in Type I mode so that near the threshold for repetitive firing the steady firing frequency can be made arbitrarily low. We dictate here that the gating variables kinetics are relatively slow and that the stimulating current is adjusted so that the cell model is firing slowly, about 15 Hz. First we consider the case of mutual excitation. In panels A and B of Fig. 11.6 we confirm the expectation that the cells will fire together.

The cells are substantially out of phase before we actually implement the coupling, but then afterwards they converge to a pattern of near synchrony, with one cell preceding the other by just a slight bit. When the coupling is first turned on the follower cell immediately advances, and throughout the transient phase of synchronization the cells are causing each other to fire faster. Interestingly, as synchrony is established the cells slow down to nearly their intrinsic frequencies, as if the coupling is only to synchronize them. When comparing the time courses we see that during the transient phase the leader cell's voltage is more affected than the follower's.

The synaptic current from the follower at first reduces the leader's postspike hyperpolarization. Then, as the follower catches up, this synaptic current is delivered when the leader is still strongly depolarized, and its intrinsic conductances swamp the perturbing effect of the synaptic input. For the preceding example the synaptic conductance time course has fast rise and fast decay phases; the conductance is essentially activated only during the presynaptic depolarization. If we allow for the synaptic decay to be much slower, with all other parameters and initial conditions unchanged, we find that these two cells now fire in antiphase, as shown in Fig. 11.6C and D. The frequency of each cell is nearly twice that of an isolated cell. This is understandable.

Since the synaptic current decays more slowly, it provides a longer-lived depolarizing influence, and it is strongest during a cell's trajectory as it rises toward threshold, when it is most responsive to depolarizing influences. Note that if the output of this two-cell network converges onto a common target, the effective delivery rate, because of the antiphase pattern, is twice that of each, four times that of an individual cell. For the examples in this section we have chosen parameter values in order to emphasize the importance of synaptic time scales in determining firing patterns in networks. This is an active area of research.

Deeper understanding will be achieved, and for now, we caution that it should not be taken as universal that fast (slow) excitatory synaptic coupling leads to in-phase (antiphase) firing. In fact, for the example in Fig. 11.6. A and B, different initial conditions to such coupled cells lead to antiphase locking even for these fast synapses. Next we consider the case of two cells (the same cells as above) coupled with inhibitory synapses. We start the cells with different initial voltages and watch them settle slowly into an antiphase rhythm (Fig. 11.7) during the first 800 ms of the simulation. During this portion the inhibitory synapses have fast kinetics, and we are confirming the classical expectation that inhibition leads to antiphase firing between a cell pair.

In these voltage time courses one can clearly see the hyperpolarizing effect of the synaptic inputs. As a cell's voltage rises from its minimum the partner fires and sends a brief pulse of outward current, which halts transiently the rising voltage. At t = 800 ms the decay rate of inhibition is slowed from $\beta = 1$/ms to $\beta - 0.1$/ms. Within a few cycles the cells lock into perfect synchrony. Slowly decaying mutual inhibition can lead to in-phase locking among neural oscillators. This behavior has been proposed as the mechanism for gamma rhythms that are seen in various brain regions, and are believed to have a functional role in some cognitive processes.

As a secondary note, although we do not show it here, this particular network with the fast inhibitory synapses is bistable. In addition to the antiphase behavior (for $t < 800$ ms) it also has a stable in-phase behavior for some set of initial states.

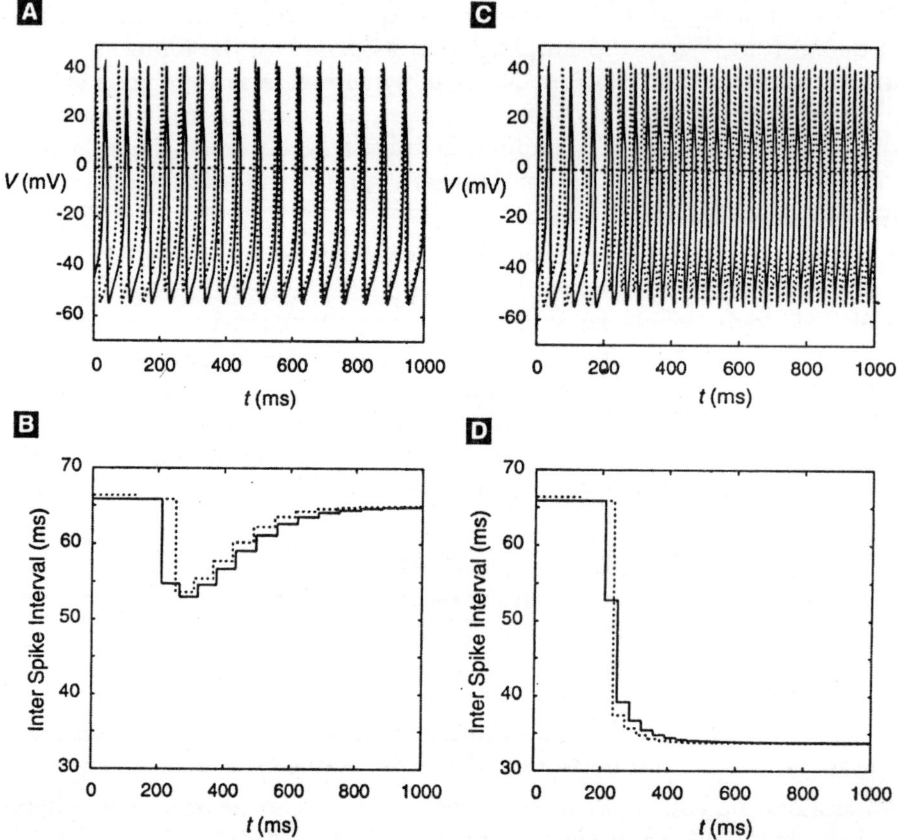

Fig. 11.6. The kinetics of synaptic excitation affect synchronization properties of cells, as illustrated with two identical coupled cells. Neuron models are uncoupled until t = 200 ms. áA and B) fast excitatory synapses lead to synchrony; time courses of V_1, V_2 in ⟨A⟩ and of instantaneous interspike interval in ⟨B⟩. ⟨C and D⟩ antiphase pattern evolves with slow excitatory synapses; time courses of V_1, V_2 in ⟨C⟩ and of instantaneous interspike interval in ⟨D⟩. Parameter values for $v3$, $v4$, g_{Ca} are as in Fig. 11.6-Fig. 11.3, except that here, $\phi = 0.1$ and $I_{app} = 15\ \mu A/cm^2$; for the synaptic kinetics, $\theta_{syn} = 20$ mV, $k_{syn} = 2$ mV with $\alpha = 3$/ms, $\beta = 1$/ms in ⟨A and B⟩ and $\alpha = 3$/ms, $\beta = 0.1$/ms in ⟨C and D⟩. Initial conditions: $V_1 = -44$ mV, $w_1 = s_1 = 0$, $V_2 =$ 0 mV, $w_2 = 0.3$, $s_2 = 0.3$.

The examples above are highly idealized and primarily directed toward rhythmicity. The parameters have been chosen in order for us to illustrate several features about temporal patterning in mutually coupled pairs with just minimal adjustments from one case to the next. Important questions arise about how these features might carry over to larger networks. In a case of mixed fast and slow synaptic coupling are the synchronizing or desynchronizing effects more important? Are the patterns robust to effects of noise and heterogeneity? What if there is a mixture of gap-junctional and synaptic coupling, in regimes where their effects counteract each other?

Neural Circuits as Computational Devices

Up to this point we have examined how various coupling mechanisms can coordinate cellular electrical activity in the context of stereotypical steady firing patterns :

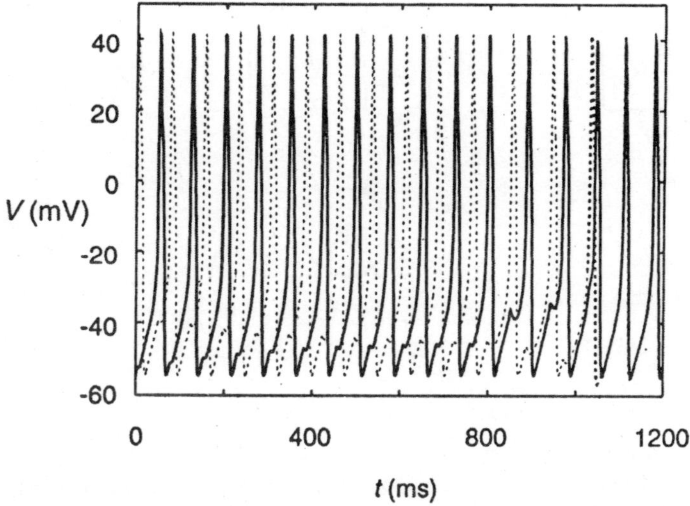

Fig. 11.7. Kinetics of synaptic inhibition can affect synchronization. Fast inhibition leads to antiphase behavior for $t < 800$ ms; switching to slowly decaying inhibition for $t > 800$ ms leads to in-phase behavior. Parameters are the same as in Fig. 11.6 except that here $\beta = 1$/ms for $t < 800$ ms and $\beta = 0.1$/ms for $t > 800$ ms. Initial conditions: $V_1 = -55$ mV, $w_1 = s_1 = 0$, $V_2 = -20$ mV, $w_2 = s_2 = 0$.

synchronous or antisynchronous oscillations. Beyond these examples one seeks to understand, in the context of neural systems, how circuits (cells and their connectivity) might be implemented to carry out various computations or computational strategies. We will develop examples to illustrate idealized mechanisms for carrying out a few computational tasks: temporally specific addition or coincidence detection, temporally specific subtraction or vetoing, and direction selectivity. The setup will be similar in each case. There will be two input units, R_1 and R_2, that we might think of as sensory neurons, receptors to the outside world. They feed into a target neuron that does the computation. (Fig. 11.8A). This is a feedforward circuit.

The information will be passed by means of chemical synapses. Depending on the combined synaptic input (triggered by the firings of R_1 and R_2) the target neuron may or may not elicit a spike. The "answer" from the computation will simply be yes or no, whether the target neuron fires or does not fire an action potential for our specified input pattern. This will depend on whether R_1 and R_2 are both activating excitatory synapses (Fig. 11.8A) or one of them is sending inhibition and the other excitation (Fig. 11.8B), and of course the relative timing will matter. For a neuron in which inputs are distributed over the dendritic tree, the spatial as well the temporal distribution of incoming signals determines the net response. For example, suppose there is simultaneously timed excitation and inhibition delivered at different locations along one dendritic branch.

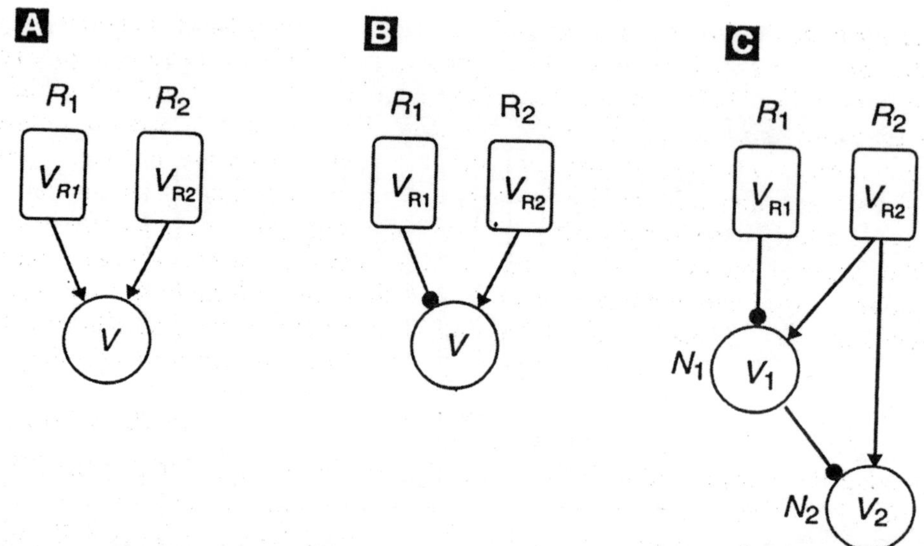

Fig. 11.8. Schematic of receptors (R_1, R_2), neurons and connections for implementing neural computations. (A) R_1 and R_2 activate excitatory inputs on target neuron. (B) R_1 sends inhibition; R_2 sends excitation. (C) First-layer neuron N_1 (same configuration as in (B) sends inhibition to second-layer neuron N_2, which also receives direct excitation from R_2.

An inhibitory synapse located closer to the soma will be more effective at reducing the response to the excitatory input than if it is located more distally on the branch. Let s_1 and s_2 be the (fractional) postsynaptic conductances that arise in the target cell from activity of R_1 and R_2, respectively. We will consider the results of only a single firing of each of the input neurons. The conductances s_1 and s_2 will be determined from the kinetics as in (11.10).

The kinetic parameters are chosen so that an excitatory conductance time course is brief and an inhibitory conductance is slower, on time scales comparable to spike dynamics. For simplicity we do not describe the action potentials of R_1 and R_2 in detail. Instead, we represent the receptor voltage responses, v_{R1} and $V_{R2,}$ as brief (2 ms), square-pulse, depolarizing events. These will yield appropriate time courses for s_1 and s_2. The parameters for our Morris-Lecar target neurons are the standard ones from Table 11.4 for Type II behavior; here, we set $I_{app} = 0$. The equations that govern the response of the target neuron then are as follows:

$$C\frac{dV}{dt} = -I_{ion} - g_1 s_1(V - V_{syn,1}) - g_2 s_2(V - V_{syn,2}) \quad ...(11.13)$$

$$\frac{dw}{dt} = \frac{\phi[w_\infty - w]}{\tau w} \quad \text{...(11.14)}$$

$$\frac{ds_1}{dt} = \alpha_1 s_{\infty,1}(V_{R1})(1 - s_1) - \beta_1 s_1, \quad \text{...(11.15)}$$

$$\frac{ds_2}{dt} = \alpha_2 s_{\infty,2}(V_{R2})(1 - s_2) - \beta_2 s_2, \quad \text{...(11.16)}$$

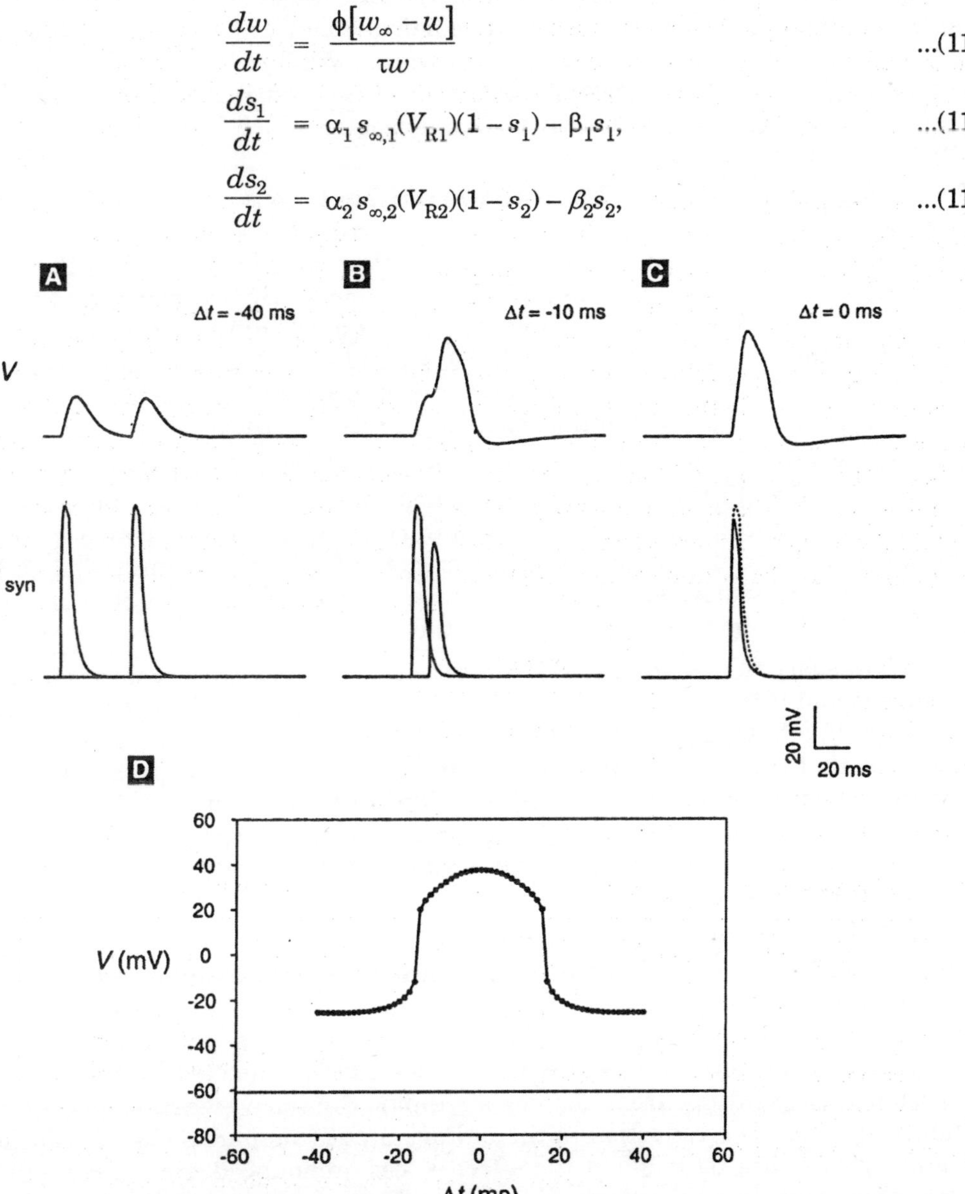

Fig. 11.9. Coincidence detection for excitatory inputs. V time courses of target neuron ⟨upper records⟩ for excitatory inputs ⟨lower records are synaptic currents⟩ from receptor cells R_1, R_2 that are offset in time by Δt. Results for three different values of Δt ⟨in ms⟩: –40 in ⟨A⟩, –10 in ⟨B⟩, 0 in ⟨C⟩, Note: The current record shown dashed in ⟨C⟩ is reproduced from ⟨A⟩ to show the effect of reduced driving force when inputs occur simultaneously. ⟨D⟩ Response plot of peak of V ⟨following second input⟩ versus Δt. Parameter values for excitatory synaptic inputs: $\alpha_1 = \alpha_2 = 1.0$/ms, $\beta_1 = \beta_2 = 0.3$/ms, $V_{sny,1} = V_{sny,2} = 100$ mV, $\theta_{syn} = 20$ mV, $k_{syn} = 2$ mV; $g_1 = g_2$ 1.0 mS/cm^2. At $t = 0$ neurons are at rest, $V = -60.9$ mV, $w = 0.0149$ and synaptic gating variables are zero.

The parameters g_1 and g_2 represent the synaptic weights, the maximum postsynaptic conductances that can be evoked by activity coming from R_1 and R_2. The values of $V_{syn,1}$ and $V_{Syn,2,}$ the synaptic reversal potentials, will determine whether the synapse is excitatory (with V_{syn} far above V_{rest}, so that the postsynaptic current is inward) or inhibitory (with V_{syn} below V_{rest}, say equal to V_k). We begin with a simple case in which R_1 and R_2 both deliver brief excitatory input to the target neuron.

The initiation times of v_{R1} and V_{R2} are respectively t_1 and t_2. In our examples, we will fix t_2 and vary t_1, where $t_1 = t_2 + \Delta t$ so that the input from R_1 precedes by $|\Delta t|$ the input from R_2 (if Δt is negative; a positive value of Δt means that R_2 fired before R_1). Suppose that the synapses are excitatory and identical, having the same synaptic weights and the same time courses except for initiation times. Assume that each input is subthreshold, but when delivered simultaneously they can fire the target cell. We see in Fig. 11.9A that when R_1 fires far earlier than R_2 then two nearly identical input current events are seen in the target cell, and neither leads to a target cell firing. No nonlinearities are evident in this case.

However, when the inputs are delivered within a restricted time window they can cause the target cell to fire (Fig. 11.9B,C). Thus the neuron's integration and threshold properties make it act like a coincidence detector for excitatory input. The regenerative spike-generating mechanism is quite nonlinear; its output does not reflect linear summation of inputs: one plus one is not necessarily equal to two. The response plot (Fig. 11.9D) summarizes the sensitivity to Δt; it is the tuning curve for the target cell as a coincidence detector.

Notice how sharp the sensitivity is. If this target neuron activates synapses that feed higher order neurons and if the synapses require pre-synaptic depolarization above 0 mV, then the next level would effectively see the yes/no answer. Emphasizing the computational analogy one might say that the target neuron is performing the logical AND operation. Notice further that the tuning curve is symmetric about $\Delta t = 0$, as expected. Since the inputs are identical, the target cell's response is the same for R_1 preceding or following R_2 by Δt. The feature of sublinear summation of synaptic inputs that are generated by conductance changes has already been demonstrated.

The basis for this is that the synaptic current depends on the membrane potential through the driving potential $V - V_{syn}$. Thus the current for a given postsynaptic conductance is reduced when V operates closer to V_{syn}. This effect is seen in Fig. 11.9B for the case of nearly (but not exactly) simultaneous transient inputs. When $|\Delta t|$ is small enough the target neuron is still depolarized when the second input arrives. Hence current injected by this second input is smaller than that resulting from the earlier input. Similarly, when the two inputs occur simultaneously there is a reduction in the unitary input (compare solid with dashed records in the lower part of Fig. 11.9C), but the effect is less pronounced, since the input current is complete before significant depolarization occurs. Next we illustrate the vetoing power that inhibition can exercise over a suprathreshold excitatory input.

For this case R_2 yields an excitatory conductance as in Fig. 11.9 except doubled in size, adequate to evoke by itself a spike in the target neuron. Also, R_1 activates an inhibitory synaptic conductance that has a slower time course than an excitatory one. If inhibition is delivered sufficiently in advance of excitation, then it has no deleterious effect on the response to excitation (Fig. 11.10A1). However, if it occurs in a narrow temporal window that starts just before excitation, it can overrule the excitation, vetoing an action potential in the target cell (Fig. 11.10A2). Even if it occurs slightly after the excitatory input, it can still prevent spiking or at least shorten the target neuron's action potential (Fig. 11.10A3). The tuning curve for this vetoing or subtraction effect is not symmetric (Fig. 11.10A4).

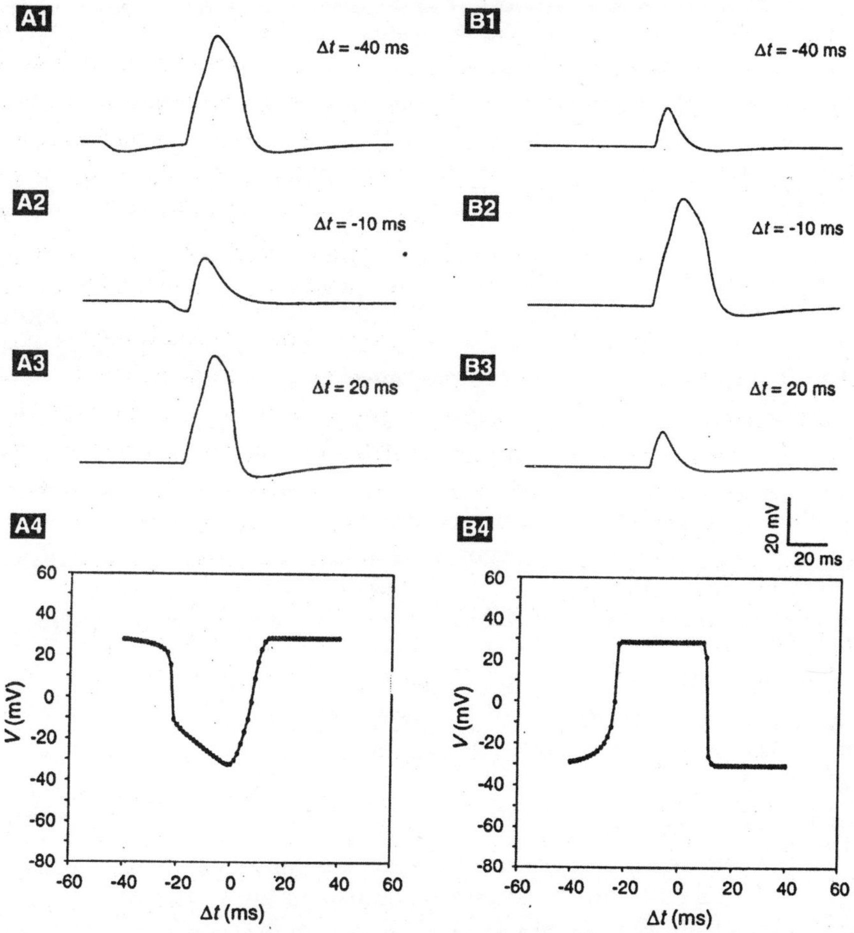

Fig. 11.10. (A1–A4) Implementing subtraction with excitation and inhibition. (A1–A3) membrane potential time courses V of target neuron for excitatory input from receptor cell R_2 and inhibitory input from R_1, offset in time by Δt (see schematic in Fig. 11.8B). Results for different values of Δt, –40, –10, 20 ms, are shown in (A1–A3), respectively. (A4) Response plot of peak of V_1 (associated with input from R_1) versus Δt. Parameter values for excitatory inputs as in Fig. 11.9; for inhibitory input kinetics: $\alpha_1 = 0.5$/ms, $\beta_1 = 0.1$/ms, $V_{syn,1} = -84$ mV, $\theta_{syn} = 20$ mV, $k_{syn} = 2$ mV, $g_1 = 2.5$ mS/cm^2, $g_2 = 2.0$ mS/cm^2. (B1–B4) Directionally selective, motion–detecting neuron. (B1–B3) Membrane potential time courses V_2 fo second layer neuron N_2 (Fig. 10.8C) that receives excitatory input from receptor cell R_2 and inhibitory input from N_1 (A1–A4). Results for different values of Δt, –40, –10, 20 ms, are shown in panels (B1–B3), respectively. (B4) Response plot of peak of V_2 versus Δt. Parameter values for N_1 and the inputs to N_1 are same as in the left panels; for N_2 and inputs, the parameters are identical as onto N_1 (with $g_{N2,N1} = 2.5$ mS/cm^2, $g_{N2,R2} = 2.0$ mS/cm^2). Input from R_2 to N_2 is delayed in each case by 10 ms. At t = 0 neurons are at rest, $V = -609$ mV, $w = 0.0149$ and synaptic gating variables are zero.

The inhibition with its slowly decaying tail is more effective if delivered with or reasonably before the excitation. But it is ineffective if delivered afterward even by a modest amount. Once the intrinsic spike-generating conductances are activated only a very strong and not too delayed inhibition can cancel the spike. Such an input-output relation might be likened to the logical AND-NOT operation: An output spike occurs only if R_2 fires and R_1 does not. If one imagines that the receptor cells R_1 and R_2 are activated in temporal sequence (with delay Δt) by a moving stimulus in the environment, our target neuron shows sensitivity to the direction of this movement. Suppose that a stimulus moving from left to right activates R_1 and then R_2. In this case our tuning curve shows that vetoing occurs far more effectively for movement from left to right (Δt negative) than from right to left (Δt positive). Now suppose that we want to go the next step by having an output neuron that acts as a directionally selective motion detector.

Can we create a unit that fires only when the stimulus moves across the receptor pair from left to right and in a confined range of Δt (this means either that the stimulus is moving fast enough or it is assumed of moderate speed and in the spatially localized receptive fields of these receptors)? We want a tuning curve that looks like Fig. 11.10B4.

A simple way to solve conceptually this design task is to imagine subtracting the tuning curve of Fig. 11.10A4 from the response curve of R_2 delivered alone to the target neuron (the latter being just flat at 32 mV, independent of Δt). One way to implement this in neural ware is by introducing a downstream neuron N_2 (with voltage V_2) whose tuning curve will have this property. Let us refer to the target neuron in earlier examples as N_1 (with voltage V_1). The two Morris-Lecar units N_1,N_2 have identical intrinsic properties. To carry out the subtraction we now feed excitatory input from R_2 into N_2, with some propagation delay, and have N_1 inhibit N_2 (Fig. 11.8C). Thus for large $|\Delta t|$ the inhibition from N_1 vetoes the excitation from R_2 with the result that N_1 does not fire (Fig. 11.10B1, B3). For modest $|\Delta t|$ the inhibition is preferentially absent for Δt negative and N_2 can fire (Fig. 11.10B2). Fig. 11.10B4 shows the tuning curve of the output neuron N_2. With this configuration we identify N_1 and N_2 as neurons in the first and second processing layers after the networks input layer of receptor units.

Other types of feedforward connectivities can lead to different feature-detecting properties; *e.g.*, in the primary visual cortex there are neurons ("simple cells") that respond to bars of light with specific orientation, and the sensory signals coming to these neurons are mapped with specific alignments from the thalamus (which shows, to first approximation, no orientation specificity). The examples above have exploited the sensitivity of the model neuron's responsiveness to timing of its inputs. There is good evidence that sound localization as carried out by the auditory system (in some frequency ranges) can involve neurons capable of coincidence discriminations at submillisecond time scales. These neurons have been shown in some species to have potassium currents that are believed to render them unresponsive to tonic inputs but highly sensitive to transient inputs.

Whether or not neural computations generally are performed based on precise spike timing raises interesting questions about the nature of the neural code and mechanisms. How could spike timing be preserved in the face of neural variability, including the possible sources of heterogeneity of neuronal and coupling properties, sparseness of coupling, and the stochastic nature of synaptic transmission? Can some type of population averaging or multi-layer processing recover timing information, or must it necessarily be degraded as activity passes through successive layers? In some contexts firing rate (short time averaged) is likely the code carrier.

Large-Scale Networks

The previous sections have dealt with simple 2 or 3-cell networks, but how does one develop and implement formulations to treat large cellular ensembles? One approach, of course, is just to extend to N units the cell-based descriptions that we have been using above. This means specifying the biophysical details of each unit, the coupling dynamics (*e.g.*, synaptic conductances or diffusion of intercellular messengers), and the coupling topology (*e.g.*, nearest-neighbour or all-to-all coupling or something intermediate). In the case of spiking units this would mean that we are attempting to keep track of each spike in the network and time course of intrinsic variables (membrane potential and gating variable for ionic conductances at all times) and coupling variables associated with the intercellular communication. Such approaches have been used to model electrical activity in the heart, the dynamics of hormonal secretory systems like the pancreas, and behaviours of neural systems like epileptic waves in brain slices or receptive field properties of visual cortex. An alternative approach to such detailed cell-based models is to use mean field descriptions. For this one assumes that the space and time scales of interest do not demand knowing what each individual unit is doing at every moment in time; that the amount of heterogeneity and fluctation in the system is enough but not too much so that averaging over cellular properties and small time and space scales is allowable to yield a description for the dynamics of the ensemble averages.

For example, instead of instantaneous membrane potential (for each cell, on the time scale of milliseconds) being the primary observable, we might use the firing rate of a cell on the time scale of tens of milliseconds and a spatial scale that includes several cells in a unit length. If you think about it, the Hodgkin-Huxley model of electrical excitability is a mean field description at the cellular level with respect to the population of ionic channels. The random openings and closings of individual channels are not described; only the probability or fraction of gating variables being in a particular state is represented.

An implicit assumption in using the mean field is that the individuals are behaving asynchronously. We would not use the HH model if we thought that the channels were all opening and closing precisely together in lock step. Similarly, using a mean field model for a cellular network means that we assume that the cells are not synchronized on a fine temporal scale, say spike for spike. Consequently, we would not use the mean field models to answer questions about precise spike timing, as we have been doing in the previous sections. So how does one obtain the mean field model for an ensemble? This is a difficult and significant question, or several questions, actually: How do you retain some information about biophysical properties of cells and synapses when you go from a "micro" to a "macro" description?

Can you be sure if you do a cell-based network simulation that your units will actually behave asynchronously? How do you decide whether your mean field model should retain explicitly some form of cell heterogeneity? How does one average over the different sources of stochasticity? These issues will keep researchers in computational cell biology quite busy over the next several years. In practice, mean field or population-averaged models in cell/neurobiology are typically not derived from first principles. They are more or less developed from a set of assumptions that involve disregarding some of the difficult issues. One may lose some connection with the biophysical details in this way. Here's one example along these lines of a derivable mean field description.

Suppose we treat a spike as an event without considering the ionic currents that generate the spike. The leaky integrate-and-fire model of neuronal excitability does this. Here one assumes that below "firing threshold" the neuron has a linear I-V relation and the membrane potential is described by an RC circuit:

$$C_m \frac{dV}{dt} = -\frac{V}{R_m} - I_{syn} + I_{app}, \qquad \text{...(11.17)}$$

where I_{syn} is the summed synaptic input from the other cells making synaptic contact with this one.

When V reaches threshold *Vth* then we say that a spike occurred, and *V* is reset to some level, typically to V_{rest}. In a case like this the model is linear except for the resetting events, and on the face of it each cell has one variable, *V*. So, what about the synaptic variables? One could assume a very simple model for a synaptic input that given a presynaptic spike the postsynaptic *V* jumps instantaneously up or down by a specified amount. This assumption throws away the time course of the postsynaptic conductance. However, on the plus side one may now derive legitimately a mean field population model, getting a (partial) differential equation for the probability density function $p(v, t)$ of membrane potential (say for $V_{rest} \leq V \leq Vth$) as a function of time. This description does indeed well describe a network of n such integrate-and-fire cells, sparsely coupled, in the limit of large n. It has been used to describe a variety of behaviours. However, as one seeks to restore some of the biophysical variables, the description can become impractical.

Each new dynamic variable, synaptic or intrinsic, that one includes becomes another independent variable in $p(v,..., t)$ leading to a high-dimensional system growing geometrically in size. In contrast, the problem size grows linearly if one adds additional variables in the case of an n-cell integrate-and-fire system. One of the insights that emerge from approaches like these is that if cells are really asynchronous, then at each instant in time they will have *V*-values spread over the interval, and one should expect that some fraction (perhaps small) of cells are near to V^{th}. Consequently, a transient excitatory input will lead to almost instantaneous firing of some cells. This means that the network can exhibit very fast response times for some inputs. A less rigorously supported but attractive mean field model was popularized by Wilson and Cowan.

This formulation jumps right to mean field quantities, which in the simplest interpretation are the probabilities of firing (per unit time) of the typical, or "mean," cell in the given populations: *E* and *I* for, say, excitatory and inhibitory subpopulations of interacting neurons. Note that this Wilson-Cowan formulation implicitly assumes random, sparse coupling without spatial structure. One assumes that the neurons have input/output relations represented by f_E and f_I so that under steady input to, say, the "*E*" cells, they would fire with steady rate f_E (and analogously for "*I*" cells). The shape of f_E is usually taken to be sigmoidal, the rate increasing to a saturation level as input_E grows large. The network is considered not able to equilibrate instantaneously to dynamic inputs but that it "integrates" with a time constant τ_E.

For now, consider an isolated population of mutually excitatory cells. We would have

$$\tau_E \frac{dE}{dt} = -E + f_E, \qquad \text{...(11.18)}$$

where f_E is a function (such as that explored in Fig. 11.12) of input to the E population from both extrinsic sources P_E and from the other cells in the network. In the mean field approach the other cells are also firing at rate E, so the synaptic field is proportional to E. Thus,

$$input_E = a \cdot E + P_E. \qquad \text{...(11.19)}$$

Note that the parameter a incorporates the transformation from presynaptic firing rate to postsynaptic conductance and to postsynaptic current, etc. One should also appreciate that in using the sigmoidal-like input/output relation there is not a strict threshold for cell firing. The firing probability decreases toward zero for decreasing input strength, but it is never really zero even for negative inputs.

An implicit assumption here is that noise in the system smoothes the abrupt thresholds as found in frequency-current relations of, say, cellular models like Morris-Lecar without noise. Also, in any real network cells are not prefectly identical, so this heterogeneity also contributes to such smoothing in the mean field models. Now let us warm up by considering just the isolated E network and ask what steady states of activity are possible. At steady state we must have

$$\frac{dE}{dt} = 0, \qquad \text{...(11.20)}$$

so that from (11.18),

$$E = f_E(a \cdot E + P_E). \qquad \text{...(11.21)}$$

The solutions to this nonlinear transcendental equation are most easily analyzed graphically. That is, plot the left-hand side and the right-hand side on the same axes as functions of E. Their intersections correspond to the possible solutions. The parameter a acts primarily on the steepness of the sigmoidal curve, while P_E is more related to left/right translation. In Fig. 11.11 we can see that for each value of a there is an intersection near E = 0. That is, the network has a stable rest state for these parameter values. On the other hand, we see that if a is large enough, then there are multiple steady states.

For this one-dimensional dynamical system the middle state of moderate activity is unstable, while the upper state of high activity is stable. In this parameter regime the network is bistable. If the cells are at rest (very low firing rate) and no transient input is given, then they will remain at rest. However, if a sufficiently strong brief input is delivered, then the cells will become active and because of the recurrent excitation in the network the cells will continue to drive each other, and the network reaches a steady state of high activity. This recurrent excitation is the autocatalytic mechanism that underlies the bistability. We may view a as the relative weight of the autocatalytic feedback, the strength of excitatory interactions. Note that because the input/output relation is saturating the activity level does not grow without bound.

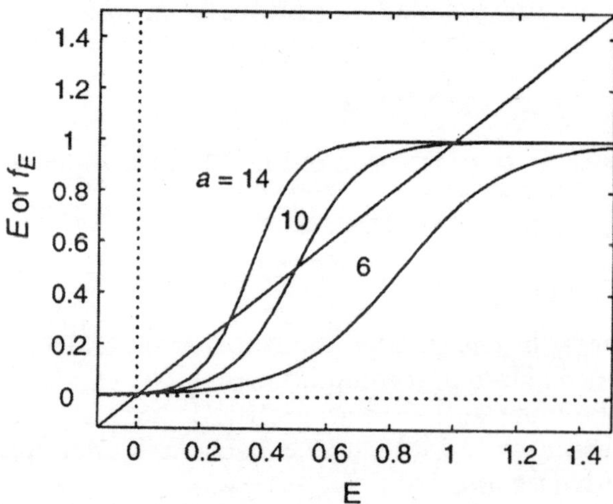

Fig. 11.11. Graphical depiction for determi-nation of steady states to single population of recurrently connected excitatory neurons. Left-and right-hand sides of (11.21), plotted as functions of E, firing rate. Definitions: f_E = 1/[1 + exp(-input$_E$)] and P_E= –5; right-hand side plotted for three different values a = 6, 10, 14.

Now we are ready to consider the two populations of excitatory and inhibitory cells. Extending the development above, we write down the model equations as

$$\tau_E \frac{dE}{dt} = -E + f_E, \quad \text{...(11.22)}$$

$$\tau_I \frac{dI}{dt} = -I + f_I, \quad \text{...(11.23)}$$

where f_E is a function of $input_E = a \cdot E - b \cdot I + p_E$ and f_I is a function of $input_I = c \cdot E - d \cdot I_f,$ p_I. This pair of ordinary differential equations governs the behavior, in the mean field sense, of the interacting E and I populations. Note the signs of the interaction terms inside of $input_E$ and $input_I$. Positive a means recurrent excitation, autocatalysis; positive c means that I is driven by E; positive c • b (with the minus sign on b) means that negative feedback through / onto E; and – d • I means I cells inhibit each other. The model equations can be analyzed with phase plane methods. For the case shown in Fig. 11.12 the network has a stable rest state (the only intersection of the E- and I-nullclines) with both populations at very low firing rate. This state is globally attracting. For any initial conditions the network has a transient response and then returns to rest.

The three different trajectories correspond to responses of E and I for brief stimuli to the E population. For a small stimulus only a few E cells fire, not enough to amplify the excitatory activity. However, with a larger initial level of E we see some regenerative effects. First, E grows beyond its starting value: Some nonlinear amplification occurs. Then, E peaks as the I population is recruited, driven by the firing of the excitatory cells. Eventually, all cells return to the rest state. This illustrates the excitability of the network. The phase plane portrait of the network is qualitatively similar to that of the basic Morris-Lecar model of cellular electrical excitability. The activator variable (in this case E) has a cubic-shaped nullcline; the left, middle, and right branches correspond to the three roots of (11.21) for a fixed level of $P_E - b.\ I$. The variable I here corresponds to a recovery process like w in the Morris-Lecar model. Its nullcline is monotonic.

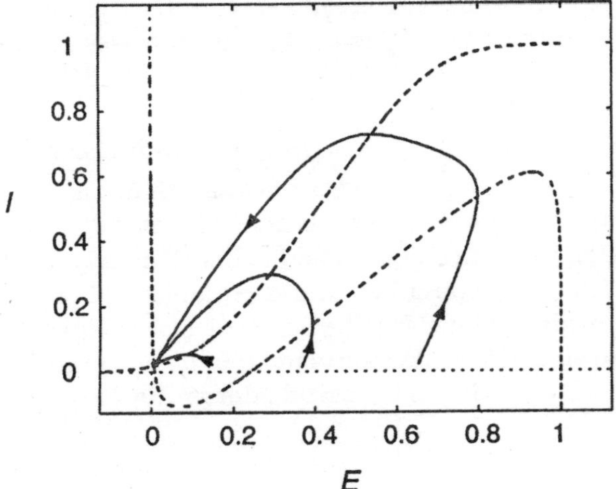

Fig. 11.12. Network excitability. Phase plane portrait for firing-rate model of neuronal net-work with interacting excitatory and inhibitory subpopulations, based on ⟨11.23⟩. The cellu-lar input/ output relations ⟨rate vs. input⟩ are $f_E = 1/[1 + \exp\langle input_E\rangle]$ as in Fig. 11.11 and $f_I = 1/[1 + \exp\langle -input_I\rangle]$. The E-nullcline is the cubic-shaped dashed line; I-nullcline is the sigmoidal dashed line. Trajectories (solid), for three different initial conditions that repre-sent perturbations from rest to the E cells, all tend back to the rest state for large time. Parameter values: a = 16, b = 12, c = 16, d = 5, P_E = –5, P_I = –4, $\tau_E = \tau_I$ = 1. Note that the model equations have been expressed in dimensionless form.

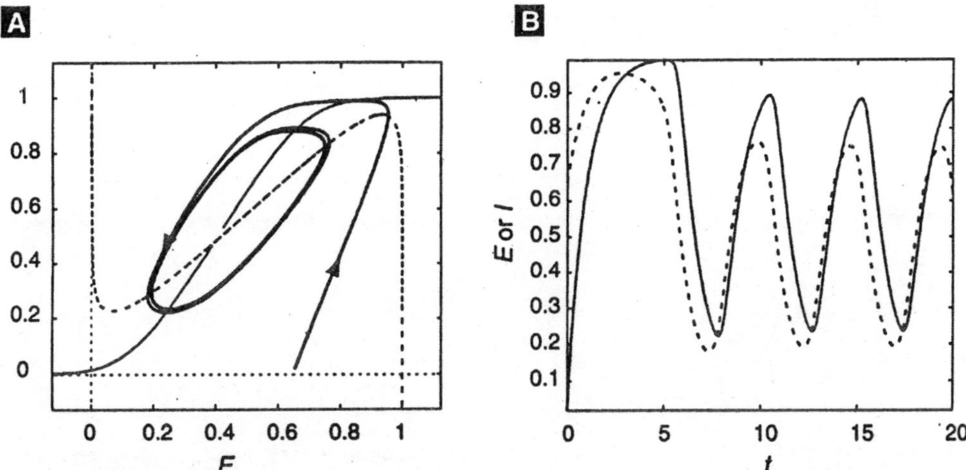

Fig. 11.13. Network oscillations. ⟨A⟩ Phase plane and ⟨B⟩ time courses for the firing rate model, with increased input to the excitatory population: P_E = –1. Other parameters are as in Fig. 11.12 Dashed line is for E-nullcline in ⟨A⟩ and E vs. t in ⟨B⟩ solid line is for I-nullcline in ⟨A⟩ and for I vs. t in ⟨B⟩. Darker solid curve in ⟨A⟩ is the phase plolt.

If we vary the parameter P_E, the E-nullcline changes, while the I-nullcline remains fixed. For example, increasing a bit the extrinsic input to the E population, say imagining more feedforward excitation onto the E cells from some other network, will cause them to fire faster, and thereby recruit more inhibition. The steady state will migrate to the right and upward along the I-nullcline. If one increases P_E enough, then the steady state will move onto the middle branch and perhaps destabilize. This is seen in Fig. 11.13A, where the network is now in an oscillatory mode, having gone through a Hopf bifurcation.

The negative-feedback inhibition lags somewhat behind the excitatory activity, seen in this phase plane portrait as well as in the time courses shown in Fig. 11.13B. If the E-population is driven hard enough, the system stops oscillating. The steady state, for large enough P_E, restabilizes and is on the right branch of the E-nullcline. This corresponds to a balanced state with strong steady firing in each population. These examples are only illustrative of this class of mean field firing-rate models. Some specific neural systems where such formulations have been applied include Pinto et al. (1996), Ermentrout (1998), and Hansel and Sompolinsky (1998); in the paper by Pinto et al. you will find a pseudo-derivation of the model based on cellular properties.

WHOLE CELL MODELS

It is still an open question whether or not computer simulation of a whole cell by modeling all its metabolic pathways and other machineries is feasible. Yet the time has come that we can begin working toward this ultimate goal. The E-CELL project was launched in 1996 at Keio University in order to model and simulate various cellular processes, with the ultimate goal of simulating the cell as a whole. The E-CELL simulation system is a generic software package for cell modeling, and its first version was completed in 1997. E-CELL simulates the behaviour of a model cell by computing interactions among molecular species it contains. E-CELL allows the user to define chemically discrete compartments such as membranes, chromosomes, and cytoplasm.

The total amounts of substances in all of the cellular compartments comprise the internal representation of the cell state, which can be monitored and/or manipulated by means of various graphical user interfaces. The E-CELL system enables us to model not only metabolic path-ways but also other higher-order cellular processes, such as protein synthesis and membrane transport, within the same framework. E-CELL also attempts to provide a framework for higher-order cellular phenomena such as the gene regulation network, DNA replication, and other events in the cell cycle. These various processes can then be integrated into a single simulation model.

A major challenge of the project is to develop a method of modeling that is sufficiently robust to accommodate a realistically large scale model consisting of hundreds and thousands of processes with drastic differences in behavior and time scale among them. Using the E-CELL system, we have successfully constructed a virtual cell with 127 genes sufficient for "self-support". The gene set was selected from the genome of *Mycoplasma genitalium,* the organism having the smallest known genome. The set includes genes for transcription, translation, the glycolysis pathway for energy production, membrane transport, and the phospholipids biosynthesis pathway for membrane structure. The rest of this section will review previous work related to computer simulation of cellular processes. Following section describe the E-CELL simulation system and present the first "virtual cell" with 127 genes.

The following cell models are currently being constructed using the E-CELL system:

- Kinetic model of a human erythrocyte
- Signal transduction for bacterial chemotaxis
- Gene regulation network for *E. coli's lac* and *ara* operons and the lytic-lysogenic switch network of bacteriophage lambda

- Energy metabolism in mitochondria.

Many attempts have been made to simulate molecular processes in both cellular and viral systems. Perhaps the most active area of cellular simulation is the kinetics of metabolic pathways. Software packages for quantitative simulation of cellular processes, based on numerical integration of rate equations, have been developed. KINSIM MetaModel SCAMP and MIST deal with steady states and their metabolic control analysis (MCA) coefficients. GEPASI calculates steady states as well as reaction time behavior and characterizes the steady state with MCA. SIMFIT and KINSIM integrate the simulation of dynamic models and the parameter fitting program. V-Cell is a solver for nonlinear PDE/ ODE /algebraic systems, and can represent the cellular geometry. Dbsolve combines continuation and bifurcation analysis. Several groups have proposed and analyzed gene regulation and expression models by simulation and signal transduction mechanisms have also been areas of research for biological modeling and simulation.

The E-Cell Software Architecture

The E-CELL software employs a *structured substance-reactor model* to construct and simulate the cell model. The structured substance-reactor model consists of three classes of objects—*Substance, Reactor,* and *System*—representing molecular species, reactions, and functional/physical compartments, respectively. To model chromosomes and other genetic materials, the system also has sophisticated data structures such *as Genome, Genomic Element,* and *Gene.* Assuming rapid equilibrium in each compartment, the state of the cell at a point in time and its dynamics are represented as a set of concentration vectors and reaction kinetics. Therefore, simulation of the cell can be viewed as integration of ordinary differential equations (ODEs).

The numerical integration is performed in distributed form with *Stepper* and *Integrator* class objects in the system. With this design, the system can be extended to handle multicompartment, multischeme, and multiphase numerical integration. In other words, time-step sizes and integration algorithms can vary among different compartments, and therefore parallel computation can be applied to cope with the *stiffness* problem. The problem of stiffness occurs in a system like a cell where there are very different scales of the time variable on which the dependent variables are changing. The stiffness often results in various kinds of numerical errors, some of which are fatal for the simulation. The E-CELL system is written in C++ and designed with the object-oriented MVC (model, view, and control) model, which facilitates independent development of the reaction rules (Model), user interfaces (View), and simulation engine (Control).

The kinetics of each reaction is encapsulated into the Reactor class objects. In addition to general reaction schemes provided by the system, task-specific reaction schemes can be defined by a user without detailed knowledge of implementation of the other parts of the system. The E-CELL simulation environment consists of several software components: the core simulation system (E-CELL system), the E-CELL manager, the rule file compiler, and other data converters/processors. Besides the system-specific E-CELL rule format, the user can write simulation rules using any commercially/publicly available spread-sheet software. The rules are compiled into .eri (E-CELL Rule Intermediate) format, which can be interpreted by the simulation system. The core system reads a compiled rule, starts simulation, and outputs logged data. The simulation process can be automated using E-CELL Script language (.ecs). The process of rule writing, simulation, and analysis is managed by the E-CELL manager.

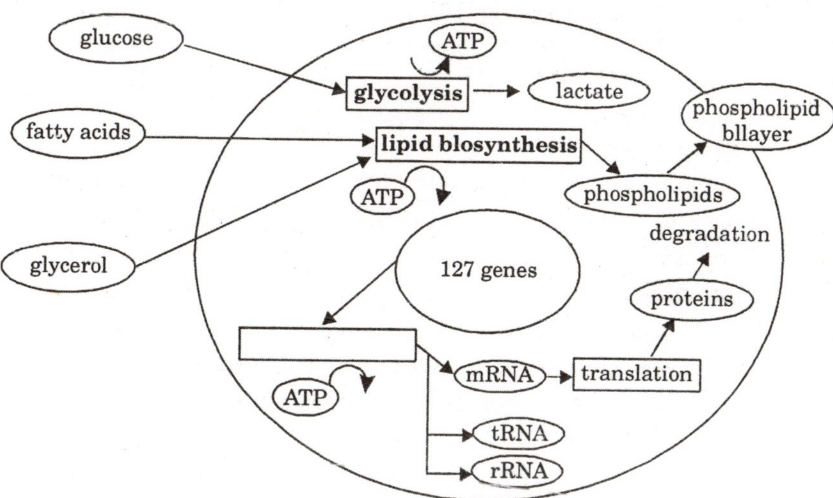

Fig. 12.1. The self-surviving cell model with 127 genes. It uptakes glucose and excretes lactate, generating ATP to be used to synthesize proteins. Since proteins degrade over time, the cell has to keep synthesizing proteins in order to sustain its life.

THE SELF-SUPPORTING CELL MODEL

Using the E-CELL system, we have successfully constructed a virtual cell with 127 genes sufficient for "self-support". The gene set was selected from the genome of *Mycoplasma genitalium,* the organism having the smallest known genome. The set includes genes for transcription, translation, the glycolysis pathway for energy production, membrane transport, and the phospholipid biosynthesis pathway for membrane structure. The hypothetical cell we have modeled uptakes glucose from the culture medium, using a phosphotransferase system; generates ATP by catabolizing glucose to lactate by glycolysis and fermentation pathways; and exports lactate out of the cell. Since enzymes and other proteins are modeled to degrade spontaneously over time, they must be constantly synthesized in order for the cell to sustain life.

The protein synthesis is implemented by modeling the molecules necessary for transcription and translation: RNA polymerase, ribosomal subunits, rRNAs, tRNAs, and tRNA ligases. The cell also uptakes glycerol and fatty acid, and produces phosphatidyl glycerol for membrane structure, using a phospholipid biosynthesis pathway. The E-CELL interfaces provide a means of conducting experiments *in silico.* For example, we can starve the cell by draining glucose from the culture medium. The cell will eventually die because it runs out of ATP. If glucose is added back, the cell may or may not recover, depending on the length of starvation. We can also kill the cell by knocking out an essential gene—for example, that for protein synthesis. The cell will become unable to synthesize proteins, and all enzymes will eventually disappear due to spontaneous degradation.

THE HUMAN ERYTHROCYTE MODEL

The human erythrocyte has been well studied since the 1970s, and extensive biochemical data on it's enzymes and metabolites have been accumulated. The cell uptakes glucose from the environment and processes it through the glycolysis pathway, generating ATP molecules for other cellular metabolism. The ATP molecules are consumed mostly for cation transport in order to keep the electroneutrality and osmotic balance. The cell also has several other pathways, such as nucleotide metabolism and the pentose phosphate pathway. The first prototype of our cell model (basic model), which was completed in March 1999, consists of the glycolysis pathway, the pentose phosphate pathway, and nucleotide metabolism.

Parameters of their kinetic equations are based on experimental data found in the literature. We obtained the steady state with this model and now set out to analyze the consequences of enzyme deficiencies.

The model has been extended in various ways since then, and the latest version is capable of simulating the pH dependence of enzymes, osmotic balance, electroneutrality, and oxygen and carbon dioxide transportation by hemoglobin. The current model can also represent the hemolysis and the destructive processing of abnormal erythrocytes in the spleen.

Since hemoglobin and carbon dioxide have a buffering effect for pH change, and osmotic balance can absorb quantitative changes of metabolic intermediates, these function will probably enhance the robustness and tolerance of the model. We are currently analyzing the effect of this expansion on robustness and tolerance. Since the discovery of pyruvate kinase deficiency (Tanaka and Paglia, 1995) erythroenzymopathies associated with hereditary hemolytic anemia have been extensively investigated. We are currently trying to reconstruct the erythrocyte model with these deficiencies. Simulation experiments using the basic model of the human erythrocyte were carried out. Pyruvate kinase activity was set to zero at 2000 seconds. The ATP production rate became lower, and ATP was eventually exhausted. Increases of 2-phosphoglycerate, 3-phosphoglycerate, and phosphoenplpyruvate were observed (Fig. 12.4). These phenomena were reported in human erythrocytes from patients with PK (pyruvate kinase) deficiency. We plan to use the kinetic parameters of the PK mutant to obtain the steady state of the defective erythrocyte.

SIGNAL TRANSDUCTION FOR BACTERIAL CHEMOTAXIS

Chemotaxis is the orientation of an organism in relation to the presence of a particular chemical. The chemotactic response of *E. coli* depends on the ability to modulate the flagellar motor in response to external stimuli. The flagellar motor has a switch protein that interacts with proteins called CheY. The CheY protein in a phosphorylated form is known to bind with the motor up to 20 times more frequently than that in an unphosphorylated form. In this way, the phosphorylation controls bacterial response to external stimuli by regulating the binding affinity of the CheY protein and the motor switch. The external stimulus bound to a bacterial receptor works as a initial control element of the cytoplasmic phosphorylation cascade. If the stimulus is an attractant, the phosphorylation flow is suppressed and the motor rotates counterclockwise, resulting in smooth swim. We are currently modeling the phosphorylation cascade for bacterial (*E. coli*) chemotaxis in which the attractant is aspartate.

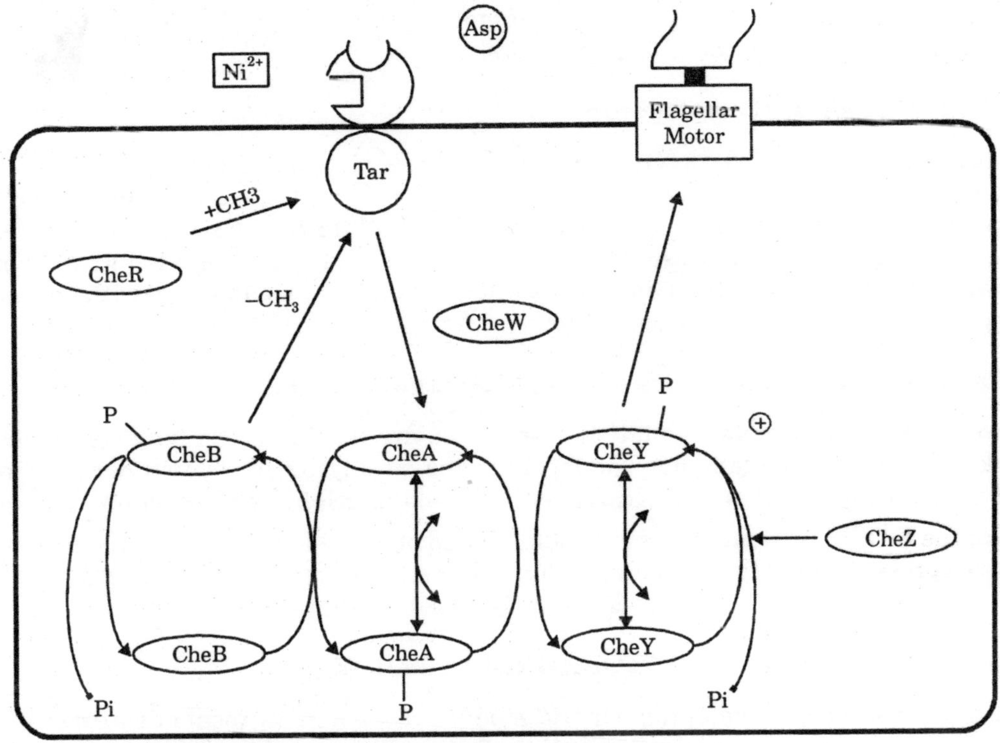

Fig. 12.2. Signal transduction for bacterial chemotaxis.

The E-CELL system allows us to perform virtual experiments on the bacterial chemotaxis model. For example, mutants of the model can be created by making a particular protein always zero. The simulated behavior of each mutant, as well as the wild type, can then be compared with the results of laboratory experiments, using indexes such as swimming direction, response time, and phosphate flux. We also plan to compare our results with other simulation systems.

GENE REGULATION NETWORK

This section presents a general framework for modeling gene expression using the E-CELL system. Using the framework, we modeled and simulated the following three gene regulation systems: the *lac* operon of *E. coli*, the *ara* operon of *E. coli,* and the lytic-lysogenic switch network in bacteriophage lambda. We had previously constructed a detailed model of the gene expression system as a part of the "self-supporting cell model." The previous model faithfully reflected the gene expression system of M. *genitalium,* involving more than 100 objects including subunits, factors, amino acids, nucleotides, and tRNAs, and their ligases.

However, in order to simulate complex gene regulation systems with a large number of genes, a more abstract and simpler model is desired for the sake of efficiency. The gene expression system used in this work consists of four basic elements: (1) production of RNAs and proteins, (2) regulation of expression by various factors, (3) time delay between regulation and production, and (4) increase of the sigmoidal curve of product over time. All gene expressions are reduced to these four elements.

E. coli's lac Operon and *ara* Operon

Based on the framework described above, we have modeled gene regulation systems of the *lac* operon and the *ara* operon in *E. coli.* We are currently integrating the *lac* operon model with the glycolysis pathway, so that we can simulate *E. coli's* sugar metabolism and regulation with lactose and glucose. The *ara* operon model is being integrated with the pentose phosphate cycle. Results of these simulations will be compared and analyzed with experimental data reported in the literature (Kepes, 1969).

Lytic-Lysogenic Switch Network of Bacteriophage Lambda

As a more complex gene regulation system, we are modeling the lytic-lysogenic switch network of bacteriophage lambda. The model of the lytic-lysogenic switch network includes the effects of glucose deprivation, ultraviolet light, and multiplicity of infection.

Since the E-CELL system allows easy substitution of individual reactions, several variations of gene expression models of bacteriophage lambda reported in the literature will be constructed and compared using the E-CELL system.

MITOCHONDRIA MODEL

We have also been developing kinetic models of various mitochondrial metabolisms, including gene expression, energy metabolism, fatty acid metabolism (beta oxidation), inner-membrane metabolite carriers, and protein carriers. In this section, we report the model of pathways for energy metabolism that was recently completed using the E-CELL system.

A mitochondrion has three major energy metabolic pathways: electron transport, TCA cycle, and beta oxidation. The basic steps of energy metabolism are listed below.

- Inner-membrane metabolite carriers transport pyruvate, fatty acid, ADP, and Pi.
- Acetyl-CoA is produced from pyruvate by pyruvate dehydrogenase.
- Acetyl-CoA is made from fatty acid by beta oxidation.
- The TCA cycle oxidizes acetyl-CoA to produce NADH, the substrate of electron transport.
- The four enzymes of electron transport oxidize NADH.
- At the end of electron transport, ATP synthase generates ATP.

There are five enzymatic reactions for electron transport, eleven enzymatic reactions, and nine other reactions for the TCA cycle, five enzymatic reactions for beta oxidation, and eight enzymatic reactions for the metabolite carrier system. All the enzymatic reactions are modeled on the basis of kinetic parameters found in the literature.

We will continue to enhance this mitochondrial model, adding other major part of metabolism, including gene expression, fatty acid metabolism, and DNA replication. Our eventual goal is to apply this model to pathological analyses of mitochondrial diseases such as Leigh syn-drome, mitochondrial myopathy and encephalopathy, lactic acidosis, and strokelike episodes.

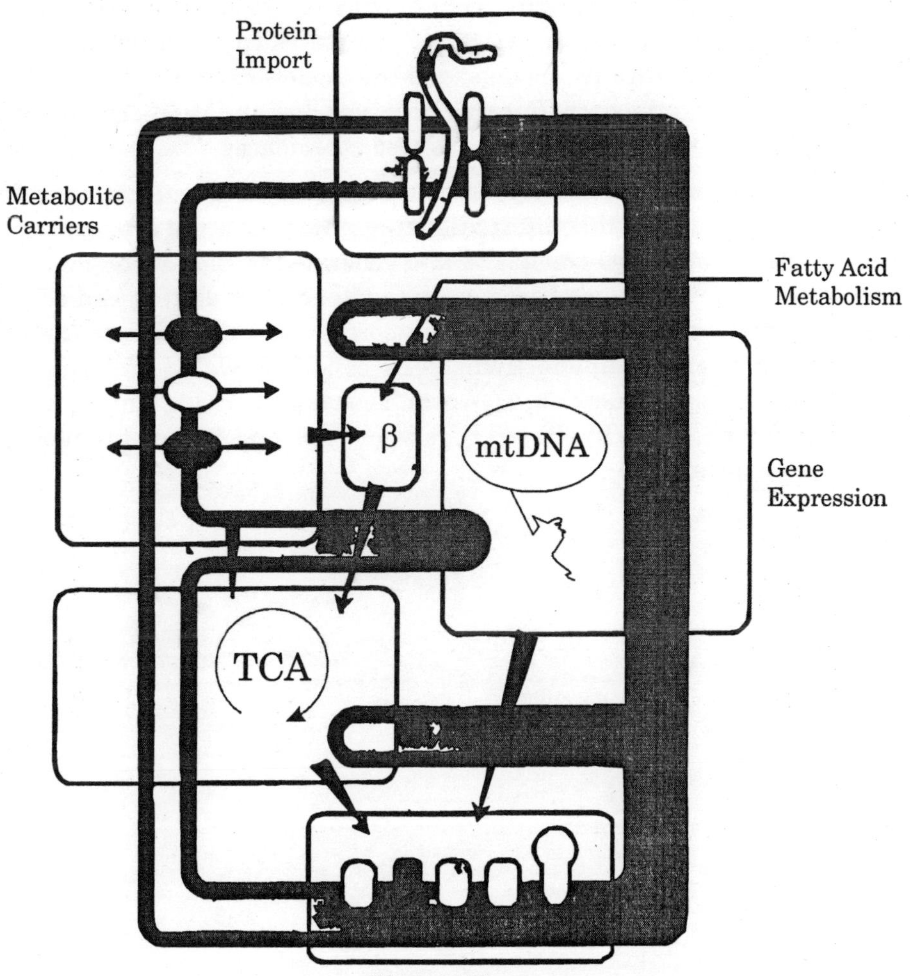

Fig. 12.3. The mitochondria model consists of gene expression system, protein transport, metabolite carriers, TCA cycle, electron transport, and fatty acid metabolism. Its genome has 37 genes, and the total of 30 enzymatic reactions are modeled.

CONCLUDING REMARKS

The E-CELL system has been available since July 1999 from our Website. Some of the models described in this chapter can be downloaded from this Website. Our simulation work with E-CELL has shown that large-scale modeling of various cellular metabolisms using the same framework appears feasible. One of the major problems in constructing large-scale cell models is lack of quantitative data. Most of the biological knowledge available is qualitative (*e.g.,* functions of genes, pathway maps, which proteins interact with what, etc.).

For simulation, however, quantitative data such as concentrations of metabolites and enzymes, flux rates, kinetic parameters, and dissociation constants are essential. A major challenge is to develop high-throughput technologies for measurement of inner-cellular metabolites. A great deal of data for a variety of cell states can then be collected with the technologies to construct quantitative models, and the models can be refined iteratively until the simulation results match the data. For this new type of simulation-oriented biology, Keio University has established the Institute for Advanced Biosciences.

The institute consists of centers for metabolome research, bioinformatics, and genome engineering. The ultimate goal of this international research institute is to construct a whole-cell model *in silica,* based on a large amount of data generated by high-throughput metabolome analyses, then design a novel genome based on the computer simulation and create real cells with the novel genome by means of genome engineering. As mentioned at the beginning of this chapter, whether or not complete living cells can be modeled and simulated at the molecular level is still an open question. However, the rapidly accumulating information on cellular metabolism increases the likelihood that whole-cell simulation of real organisms will be feasible in the near future.

Index

E

F

G

□□□